# Advances in Harmonic Analysis and Partial Differential Equations

# CONTEMPORARY MATHEMATICS

748

# Advances in Harmonic Analysis and Partial Differential Equations

AMS Special Session
Harmonic Analysis and Partial Differential Equations
April 21–22, 2018
Northeastern University, Boston, MA

Donatella Danielli
Irina Mitrea
Editors

AMERICAN MATHEMATICAL SOCIETY

2010 *Mathematics Subject Classification.* Primary 31A10, 33C10, 35G20, 35P20, 35S05, 39B72, 42B35, 46E30, 76D03, 78A05.

**Library of Congress Cataloging-in-Publication Data**

Names: AMS Special Session on Harmonic Analysis and Partial Differential Equations (2018 : Northeastern University). | Danielli, Donatella, 1966- editor. | Mitrea, Irina, editor.

Title: Advances in harmonic analysis and partial differential equations : AMS special session on Harmonic Analysis and Partial Differential Equations, April 21-22, 2018, Northeastern University, Boston, MA / Donatella Danielli, Irina Mitrea, editors.

Description: Providence, Rhode Island : American Mathematical Society, [2020] | Series: Contemporary mathematics, 0271-4132 ; volume 748 | Includes bibliographical references.

Identifiers: LCCN 2019040080 | ISBN 9781470448967 (paperback) | ISBN 9781470455163 (ebook)

Subjects: LCSH: Harmonic analysis–Congresses. | Differential equations, Partial–Congresses. | AMS: Potential theory – Two-dimensional theory – Integral representations, integral operators, integral equations methods. | Special functions. | Partial differential equations – General higher-order equations and systems – Nonlinear higher-order equations. | Partial differential equations – Spectral theory and eigenvalue problems – Asymptotic distribution of eigenvalues and eigenfunctions. | Partial differential equations – Pseudodifferential operators and other generalizations of partial differential operators – Pseudodifferential operators. | Difference and functional equations – Functional equations and inequalities – Systems of functional equations and inequalities. | Harmonic analysis on Euclidean spaces – Harmonic analysis in several variables – Function spaces arising in harmonic analysis. | Functional analysis – Linear function spaces and their duals. | Fluid mechanics – Incompressible viscous fluids – Existence, uniqueness, and regularity theory. | Optics, electromagnetic theory – General – Geometric optics.

Classification: LCC QA403 .R425 2018 | DDC 515/.2433–dc23

LC record available at https://lccn.loc.gov/2019040080

DOI: https://doi.org/10.1090/conm/748/15051



∞ The paper used in this book is acid-free and falls within the guidelines
established to ensure permanence and durability.
Visit the AMS home page at `https://www.ams.org/`

10 9 8 7 6 5 4 3 2 1 25 24 23 22 21 20

# Contents

# Preface

Problems arising in Partial Differential Equations are often models of a reality that is highly intricate from both analytic and geometric points of view. The tools required to treat these models need to be very sophisticated themselves and situated at the crossroads of several major fields of mathematics, including Harmonic Analysis (singular integral operators, Calderón-Zygmund theory), Operator Theory (spectral theory, functional calculus), Complex Analysis/Several Complex Variables (CR geometry, Cauchy-type integral operators, conformal and quasiconformal mappings), Numerical Analysis (fast Fourier transform, wavelets), Scientific Computing (interval analysis, validated numerics), and Geometric Measure Theory (classes of sets of locally finite perimeter, quantitative versions of rectifiability, etc.). Combining techniques originating in these fields has proved to be extremely potent when dealing with a host of difficult and important problems in analysis. Indeed, there are many notable achievements in this direction whose degree of technical sophistication is truly breathtaking.

The current volume focuses on new developments at the interface between Real and Complex Analysis, Harmonic Analysis, and Partial Differential Equations. It contains papers contributed by speakers in the Special Session on Harmonic Analysis and Partial Differential Equations at the American Mathematical Society Sectional Meeting at Northeastern University, Boston, MA, April 21–22, 2018. The editors believe that it is imperative to raise the level of awareness of junior mathematicians, including graduate students and post-doctoral fellows, about the necessity of having a solid background in all of these disciplines, and the current volume offers a glimpse into a variety of current topics and research problems in these areas. Specifically,

- The paper *BMO on shapes and sharp constants*, by G. Dafni and R. Gibara, introduces and establishes fundamental properties and classical inequalities (including the John-Nirenberg inequality) of a new type of BMO space depending on an integrability parameter and a basis of shapes in the Euclidean setting.
- The paper *Applications of harmonic analysis techniques to regularity problems of dissipative equations*, by M. Dai and H. Liu, is a survey on the state of the art of Partial Differential Equations in fluid mechanics, with emphasis on establishing regularity results through conditions assumed on low frequency components of the solutions.
- The paper *Two classical properties of the Bessel quotient and their implications in pde's*, by N. Garofalo, lies at the intersection between several different topics from Partial Differential Equations, Probability, Differential Geometry, and Special Functions. Here the author uses classical

properties of the Bessel quotient to establish new and sharp results for a class of degenerate partial differential equations of parabolic type in the upper half-space, which arise in connection with the analysis of fractional heat operators.

- The paper *On existence of dichromatic single element lenses*, by C. E. Gutiérrez and A. Sabra, deals with the issue of existence of solutions for a modeling problem in geometric optics. In practical terms, this amounts to finding a refracting lens that is capable of reshaping a beam of dichromatic light from a source in a prescribed manner.
- The paper *Free boundary regularity near the fixed boundary for the fully nonlinear obstacle problem*, by E. Indrei, is concerned with regularity properties of the free boundary for obstacle type problems, a line of investigation that has generated new developments in recent years. This is done through a blow-up classification procedure, which is of interest in its own.
- The paper *The Poisson integral formula for variable-coefficient elliptic systems in rough domains*, by D. Mitrea, I. Mitrea, and M. Mitrea, establishes Poisson integral formulas for solutions of second-order, homogeneous, divergence-form elliptic systems with complex-valued Lipschitz coefficients for a very general class of non-smooth domains, which is sharp from the geometric measure theoretic point of view.
- The paper *Variations on quantum ergodic theorems, II*, by M. Taylor, presents new quantum ergodic results for first-order positive, self-adjoint, elliptic pseudodifferential operators on compact Riemannian manifolds. Specifically, the work deals with scenarios in which the Hamiltonian flow generated by the symbol of the pseudodifferential operator in question is not necessarily ergodic.

We hope you will enjoy this volume!

Donatella Danielli
Irina Mitrea

Contemporary Mathematics
Volume **748**, 2020
https://doi.org/10.1090/conm/748/15054

# BMO on shapes and sharp constants

Galia Dafni and Ryan Gibara

Abstract. We consider a very general definition of BMO on a domain in $\mathbb{R}^n$, where the mean oscillation is taken with respect to a basis of shapes, i.e. a collection of open sets covering the domain. We examine the basic properties and various inequalities that can be proved for such functions, with special emphasis on sharp constants. For the standard bases of shapes consisting of balls or cubes (classic BMO), or rectangles (strong BMO), we review known results, such as the boundedness of rearrangements and its consequences. Finally, we prove a product decomposition for BMO when the shapes exhibit some product structure, as in the case of strong BMO.

## 1. Introduction

First defined by John and Nirenberg in [**37**], the space BMO of functions of bounded mean oscillation has served as the replacement for $L^\infty$ in situations where considering bounded functions is too restrictive. BMO has proven to be important in areas such as harmonic analysis, partly due to the duality with the Hardy space established by Fefferman in [**23**], and partial differential equations, where its connection to elasticity motivated John to first consider the mean oscillation of functions in [**36**]. Additionally, one may regard BMO as a function space that is interesting to study in its own right. As such, there exist many complete references to the classical theory and its connection to various areas; for instance, see [**27, 31, 39, 58**].

The mean oscillation of a function $f \in L^1_{\mathrm{loc}}(\mathbb{R}^n)$ was initially defined over a cube $Q$ with sides parallel to the axes as

$$\fint_Q |f - f_Q|, \tag{1.1}$$

where $f_Q = \fint_Q f$ and $\fint_Q = \frac{1}{|Q|}\int_Q$. A function $f$ was then said to be in BMO if the quantity (1.1) is bounded independently of $Q$. Equivalently, as will be shown, the same space can be obtained by considering the mean oscillation with respect to balls; that is, replacing the cube $Q$ by a ball $B$ in (1.1). Using either characterization, BMO has since been defined in more general settings such as on domains, manifolds, and metric measure spaces ([**5, 13, 38**]).

2010 *Mathematics Subject Classification.* 46E30, 42B35, 26D15.

The authors were partially supported by the Natural Sciences and Engineering Research Council (NSERC) of Canada, the Centre de recherches mathématiques (CRM), and the Fonds de recherche du Québec – Nature et technologies (FRQNT).

There has also been some attention given to the space defined by a mean oscillation condition over rectangles with sides parallel to the axes, either in $\mathbb{R}^n$ or on a domain in $\mathbb{R}^n$, appearing in the literature under various names. For instance, in [**43**], the space is called "anisotropic BMO" to highlight the contrast with cubes, while in papers such as [**16, 21, 22, 25**], it goes by "little BMO" and is denoted by bmo. The notation bmo, however, had already been used for the "local BMO" space of Goldberg ([**30**]), a space that has been established as an independent topic of study (see, for instance, [**8, 20, 63**]). Yet another name for the space defined by mean oscillations on rectangles - the one we prefer - is the name "strong BMO". This name has been used in at least one paper ([**47**]), and it is analogous to the terminology of strong differentiation of the integral and the strong maximal function ([**14, 32, 34, 35, 56**]), as well as strong Muckenhoupt weights ([**4, 51**]).

In this paper we consider BMO on domains of $\mathbb{R}^n$ with respect to a geometry (what will be called a basis of shapes) more general than cubes, balls, or rectangles. The purpose of this is to provide a framework for examining the strongest results that can be obtained about functions in BMO by assuming only the weakest assumptions. To illustrate this, we provide the proofs of many basic properties of BMO functions that are known in the literature for the specialised bases of cubes, balls, or rectangles but that hold with more general bases of shapes. In some cases, the known proofs are elementary themselves and so our generalisation serves to emphasize the extent to which they are elementary and to which these properties are intrinsic to the definition of BMO. In other cases, the known results follow from deeper theory and we are able to provide elementary proofs. We also prove many properties of BMO functions that may be well known, and may even be referred to in the literature, but for which we could not find a proof written down. An example of such a result is the completeness of BMO, which is often deduced as a consequence of duality, or proven only for cubes in $\mathbb{R}^n$. We prove this result (Theorem 3.9) for a general basis of shapes on a domain.

The paper has two primary focuses, the first being constants in inequalities related to BMO. Considerable attention will be given to their dependence on an integrability parameter $p$, the basis of shapes used to define BMO, and the dimension of the ambient Euclidean space. References to known results concerning sharp constants are given and connections between the sharp constants of various inequalities are established. We distinguish between shapewise inequalities, that is, inequalities that hold on any given shape, and norm inequalities. We provide some elementary proofs of several shapewise inequalities and obtain sharp constants in the distinguished cases $p = 1$ and $p = 2$. An example of such a result is the bound on truncations of a BMO function (Proposition 6.3). Although sharp shapewise inequalities are available for estimating the mean oscillation of the absolute value of a function in BMO, the constant 2 in the implied norm inequality - a statement of the boundedness of the map $f \mapsto |f|$ - is not sharp. Rearrangements are a valuable tool that compensate for this, and we survey some known deep results giving norm bounds for decreasing rearrangements.

A second focus of this paper is on the product nature that BMO spaces may inherit from the shapes that define them. In the case where the shapes defining BMO have a certain product structure, namely that the collection of shapes coincides with the collection of Cartesian products of lower-dimensional shapes, a product structure is shown to be inherited by BMO under a mild hypothesis related to the theory of differentiation (Theorem 8.3). This is particularly applicable to the case of strong BMO. It is important to note that the product nature studied here is different from that considered in the study of the space known as product BMO (see [**9**, **10**]).

Following the preliminaries, Section 3 presents the basic theory of BMO on shapes. Section 4 concerns shapewise inequalities and the corresponding sharp constants. In Section 5, two rearrangement operators are defined and their boundedness on various function spaces is examined, with emphasis on BMO. Section 6 discusses truncations of BMO functions and the cases where sharp inequalities can be obtained without the need to appeal to rearrangements. Section 7 gives a short survey of the John-Nirenberg inequality. Finally, in Section 8 we state and prove the product decomposition of certain BMO spaces.

This introduction is not meant as a review of the literature since that is part of the content of the paper, and references are given throughout the different sections. The bibliography is by no means exhaustive, containing only a selection of the available literature, but it is collected with the hope of providing the reader with some standard or important references to the different topics touched upon here.

## 2. Preliminaries

Consider $\mathbb{R}^n$ with the Euclidean topology and Lebesgue measure, denoted by $|\cdot|$. By a domain we mean an open and connected set.

Definition 2.1. We call a shape in $\mathbb{R}^n$ any open set $S$ such that $0 < |S| < \infty$. For a given domain $\Omega \subset \mathbb{R}^n$, we call a basis of shapes in $\Omega$ a collection $\mathscr{S}$ of shapes $S$ such that $S \subset \Omega$ for all $S \in \mathscr{S}$ and $\mathscr{S}$ forms a cover of $\Omega$.

Common examples of bases are the collections of all Euclidean balls, $\mathcal{B}$, all cubes with sides parallel to the axes, $\mathcal{Q}$, and all rectangles with sides parallel to the axes, $\mathcal{R}$. In one dimension, these three choices degenerate to the collection of all (finite) open intervals, $\mathcal{I}$. A variant of $\mathcal{B}$ is $\mathcal{C}$, the basis of all balls centered around some central point (usually the origin). Another commonly used collection is $\mathcal{Q}_d$, the collection of all dyadic cubes, but the open dyadic cubes cannot cover $\Omega$ unless $\Omega$ itself is a dyadic cube, so the proofs of some of the results below which rely on $\mathscr{S}$ being an open cover (e.g. Proposition 3.8 and Theorem 3.9) may not apply.

One may speak about shapes that are balls with respect to a (quasi-)norm on $\mathbb{R}^n$, such as the $p$-"norms" $\|\cdot\|_p$ for $0 < p \leq \infty$ when $n \geq 2$. The case $p = 2$ coincides with the basis $\mathcal{B}$ and the case $p = \infty$ coincides with the basis $\mathcal{Q}$, but other values of $p$ yield other interesting shapes. On the other hand, $\mathcal{R}$ is not generated from a $p$-norm.

Further examples of interesting bases have been studied in relation to the theory of differentiation of the integral, such as the collection of all rectangles with $j$ of the sidelengths being equal and the other $n - j$ being arbitrary ([**64**]), as well as the basis of all rectangles with sides parallel to the axes and sidelengths of the form $\big(\ell_1, \ell_2, \ldots, \phi(\ell_1, \ell_2, \ldots, \ell_{n-1})\big)$, where $\phi$ is a positive function that is monotone increasing in each variable separately ([**15**]).

Definition 2.2. Given two bases of shapes, $\mathscr{S}$ and $\tilde{\mathscr{S}}$, we say that $\mathscr{S}$ is comparable to $\tilde{\mathscr{S}}$, written $\mathscr{S} \trianglelefteq \tilde{\mathscr{S}}$, if there exist lower and upper comparability constants $c > 0$ and $C > 0$, depending only on $n$, such that for all $S \in \mathscr{S}$ there exist $S_1, S_2 \in \tilde{\mathscr{S}}$ for which $S_1 \subset S \subset S_2$ and $c|S_2| \le |S| \le C|S_1|$. If $\mathscr{S} \trianglelefteq \tilde{\mathscr{S}}$ and $\tilde{\mathscr{S}} \trianglelefteq \mathscr{S}$, then we say that $\mathscr{S}$ and $\tilde{\mathscr{S}}$ are equivalent, and write $\mathscr{S} \approx \tilde{\mathscr{S}}$.

An example of equivalent bases are $\mathcal{B}$ and $\mathcal{Q}$: one finds that $\mathcal{B} \trianglelefteq \mathcal{Q}$ with $c = \frac{\omega_n}{2^n}$ and $C = \omega_n \left(\frac{\sqrt{n}}{2}\right)^n$, and $\mathcal{Q} \trianglelefteq \mathcal{B}$ with $c = \frac{1}{\omega_n}\left(\frac{2}{\sqrt{n}}\right)^n$ and $C = \frac{2^n}{\omega_n}$, where $\omega_n$ is the volume of the unit ball in $\mathbb{R}^n$, and so $\mathcal{B} \approx \mathcal{Q}$. The bases of shapes given by the balls in the other $p$-norms $\|\cdot\|_p$ for $1 \le p \le \infty$ are also equivalent to these.

If $\mathscr{S} \subset \tilde{\mathscr{S}}$ then $\mathscr{S} \trianglelefteq \tilde{\mathscr{S}}$ with $c = C = 1$. In particular, $\mathcal{Q} \subset \mathcal{R}$ and so $\mathcal{Q} \trianglelefteq \mathcal{R}$, but $\mathcal{R} \ntrianglelefteq \mathcal{Q}$ and so $\mathcal{Q} \not\approx \mathcal{R}$.

Unless otherwise specified, we maintain the convention that $1 \le p < \infty$. Moreover, many of the results implicitly assume that the functions are real-valued, but others may hold also for complex-valued functions. This should be understood from the context.

## 3. BMO spaces with respect to shapes

Consider a basis of shapes $\mathscr{S}$. Given a shape $S \in \mathscr{S}$, for a function $f \in L^1(S)$, denote by $f_S$ its mean over $S$.

DEFINITION 3.1. We say that a function satisfying $f \in L^1(S)$ for all shapes $S \in \mathscr{S}$ is in the space $\mathrm{BMO}^p_{\mathscr{S}}(\Omega)$ if there exists a constant $K \ge 0$ such that

$$\left(\fint_S |f - f_S|^p\right)^{1/p} \le K, \tag{3.1}$$

holds for all $S \in \mathscr{S}$.

The quantity on the left-hand side of (3.1) is called the $p$-mean oscillation of $f$ on $S$. For $f \in \mathrm{BMO}^p_{\mathscr{S}}(\Omega)$, we define $\|f\|_{\mathrm{BMO}^p_{\mathscr{S}}}$ as the infimum of all $K$ for which (3.1) holds for all $S \in \mathscr{S}$. Note that the $p$-mean oscillation does not change if a constant is added to $f$; as such, it is sometimes useful to assume that a function has mean zero on a given shape.

In the case where $p = 1$, we will write $\mathrm{BMO}^1_{\mathscr{S}}(\Omega) = \mathrm{BMO}_{\mathscr{S}}(\Omega)$. For the classical BMO spaces we reserve the notation $\mathrm{BMO}^p(\Omega)$ without explicit reference to the underlying basis of shapes ($\mathcal{Q}$ or $\mathcal{B}$).

We mention a partial answer to how $\mathrm{BMO}^p_{\mathscr{S}}(\Omega)$ relate for different values of $p$. This question will be taken up again in a later section when some more machinery has been developed.

PROPOSITION 3.2. *For any basis of shapes,* $\mathrm{BMO}^{p_2}_{\mathscr{S}}(\Omega) \subset \mathrm{BMO}^{p_1}_{\mathscr{S}}(\Omega)$ *with* $\|f\|_{\mathrm{BMO}^{p_1}_{\mathscr{S}}} \le \|f\|_{\mathrm{BMO}^{p_2}_{\mathscr{S}}}$ *for* $1 \le p_1 \le p_2 < \infty$. *In particular, this implies that* $\mathrm{BMO}^p_{\mathscr{S}}(\Omega) \subset \mathrm{BMO}_{\mathscr{S}}(\Omega)$ *for all* $1 \le p < \infty$.

PROOF. This follows from Jensen's inequality with $p = \frac{p_2}{p_1} \ge 1$. □

Next we show a lemma that implies, in particular, the local integrability of functions in $\mathrm{BMO}^p_{\mathscr{S}}(\Omega)$.

LEMMA 3.3. *For any basis of shapes,* $\mathrm{BMO}^p_{\mathscr{S}}(\Omega) \subset L^p_{loc}(\Omega)$.

PROOF. Fix a shape $S \in \mathscr{S}$ and a function $f \in \mathrm{BMO}^p_{\mathscr{S}}(\Omega)$. By Minkowski's inequality on $L^p(S, \frac{dx}{|S|})$,

$$\left(\fint_S |f|^p\right)^{1/p} \le \left(\fint_S |f - f_S|^p\right)^{1/p} + |f_S|$$

and so

$$\left(\int_S |f|^p\right)^{1/p} \leq |S|^{1/p}\left(\|f\|_{\mathrm{BMO}^p_{\mathscr{S}}} + |f_S|\right).$$

As $\mathscr{S}$ covers all of $\Omega$, for any compact set $K \subset \Omega$ there exists a collection $\{S_i\}_{i=1}^N \subset \mathscr{S}$ for some finite $N$ such that $K \subset \bigcup_{i=1}^N S_i$. Hence, using the previous calculation,

$$\left(\int_K |f|^p\right)^{1/p} \leq \sum_{i=1}^N \left(\int_{S_i} |f|^p\right)^{1/p} \leq \sum_{i=1}^N |S_i|^{1/p}\left(\|f\|_{\mathrm{BMO}^p_{\mathscr{S}}} + |f_{S_i}|\right) < \infty.$$

□

In spite of this, a function in $\mathrm{BMO}^p_{\mathscr{S}}(\Omega)$ need not be locally bounded. If $\Omega$ contains the origin or is unbounded, $f(x) = \log|x|$ is the standard example of a function in $\mathrm{BMO}(\Omega) \setminus L^\infty(\Omega)$. The reverse inclusion, however, does hold:

PROPOSITION 3.4. *For any basis of shapes, $L^\infty(\Omega) \subset \mathrm{BMO}^p_{\mathscr{S}}(\Omega)$ with*

$$\|f\|_{\mathrm{BMO}^p_{\mathscr{S}}} \leq \begin{cases} \|f\|_{L^\infty}, & 1 \leq p \leq 2; \\ 2\|f\|_{L^\infty}, & 2 < p < \infty. \end{cases}$$

PROOF. Fix $f \in L^\infty(\Omega)$ and a shape $S \in \mathscr{S}$. For any $1 \leq p < \infty$, Minkowski's and Jensen's inequalities give

$$\left(\fint_S |f - f_S|^p\right)^{1/p} \leq 2\left(\fint_S |f|^p\right)^{1/p} \leq 2\|f\|_\infty.$$

Restricting to $1 \leq p \leq 2$, one may use Proposition 3.2 with $p_2 = 2$ to arrive at

$$\left(\fint_S |f - f_S|^p\right)^{1/p} \leq \left(\fint_S |f - f_S|^2\right)^{1/2}.$$

Making use of the Hilbert space structure on $L^2(S, \frac{dx}{|S|})$, observe that $f - f_S$ is orthogonal to constants and so it follows that

$$\fint_S |f - f_S|^2 = \fint_S |f|^2 - |f_S|^2 \leq \fint_S |f|^2 \leq \|f\|_\infty^2.$$

□

A simple example shows that the constant 1 obtained for $1 \leq p \leq 2$ is, in fact, sharp:

EXAMPLE 3.5. Let $S$ be a shape on $\Omega$ and consider a function $f = \chi_E - \chi_{E^c}$, where $E$ is a measurable subset of $S$ such that $|E| = \frac{1}{2}|S|$ and $E^c = S \setminus E$. Then $f_S = 0$, $|f - f_S| = |f| \equiv 1$ on $S$ and so

$$\fint_S |f - f_S|^p = 1.$$

Thus, $\|f\|_{\mathrm{BMO}^p_{\mathscr{S}}} \geq 1 = \|f\|_{L^\infty}$.

There is no reason to believe that the constant 2 for $2 < p < \infty$ is sharp, however, and so we pose the following question:

PROBLEM 3.6. What is the smallest constant $c_\infty(p, \mathscr{S})$ such that the inequality $\|f\|_{\mathrm{BMO}^p_{\mathscr{S}}} \leq c_\infty(p, \mathscr{S})\|f\|_{L^\infty}$ holds for all $f \in \mathrm{BMO}^p_{\mathscr{S}}(\Omega)$?

The solution to this problem was obtained by Leonchik in the case when $\Omega \subset \mathbb{R}$ and $\mathscr{S} = \mathcal{I}$.

THEOREM 3.7 ([**43**, **46**]).

$$c_\infty(p, \mathcal{I}) = 2 \sup_{0<h<1} \{h(1-h)^p + h^p(1-h)\}^{1/p}.$$

An analysis of this expression ([**43**]) shows that $c_\infty(p, \mathcal{I}) = 1$ for $1 \leq p \leq 3$, improving on Proposition 3.4 for $2 < p \leq 3$. Moreover, $c_\infty(p, \mathcal{I})$ is monotone in $p$ with $1 < c_\infty(p, \mathcal{I}) < 2$ for $p > 3$ and $c_\infty(p, \mathcal{I}) \to 2$ as $p \to \infty$.

It is easy to see that $\|\cdot\|_{\mathrm{BMO}^p_{\mathscr{S}}}$ defines a seminorm. It cannot be a norm, however, as a function $f$ that is almost everywhere equal to a constant will satisfy $\|f\|_{\mathrm{BMO}^p_{\mathscr{S}}} = 0$. What we can show is that the quantity $\|\cdot\|_{\mathrm{BMO}^p_{\mathscr{S}}}$ defines a norm modulo constants.

In the classical case of $\mathrm{BMO}(\mathbb{R}^n)$, the proof (see [**31**]) relies on the fact that $\mathcal{B}$ contains $\mathcal{C}$ and so $\mathbb{R}^n$ may be written as the union of countably-many concentric shapes. When on a domain that is also a shape, the proof is immediate. In our general setting, however, we may not be in a situation where $\Omega$ is a shape or $\mathscr{S}$ contains a distinguished subcollection of nested shapes that exhausts all of $\Omega$; for an example, consider the case where $\Omega$ is a rectangle that is not a cube with $\mathscr{S} = \mathcal{Q}$. As such, the proof must be adapted. We do so in a way that relies on shapes being open sets that cover $\Omega$, and on $\Omega$ being connected and Lindelöf.

PROPOSITION 3.8. *For any basis of shapes, $\|f\|_{\mathrm{BMO}^p_{\mathscr{S}}} = 0$ if and only if $f$ is almost everywhere equal to some constant.*

PROOF. One direction is immediate. For the other direction, fix $f \in \mathrm{BMO}^p_{\mathscr{S}}(\Omega)$ such that $\|f\|_{\mathrm{BMO}^p_{\mathscr{S}}} = 0$. It follows that $f = f_S$ almost everywhere on $S$ for each $S \in \mathscr{S}$.

Fix $S_0 \in \mathscr{S}$ and let $C_0 = f_{S_0}$. Set

$$U = \bigcup \{S \in \mathscr{S} : f_S = C_0\}.$$

Using the Lindelöf property of $\Omega$, we may assume that $U$ is defined by a countable union. It follows that $f = C_0$ almost everywhere on $U$ since $f = f_S = C_0$ almost everywhere for each $S$ comprising $U$. The goal now is to show that $\Omega = U$.

Now, let

$$V = \bigcup \{S \in \mathscr{S} : f_S \neq C_0\}.$$

Since $\mathscr{S}$ covers $\Omega$, we have $\Omega = U \cup V$. Thus, in order to show that $\Omega = U$, we need to show that $V$ is empty. To do this, we note that both $U$ and $V$ are open sets and so, since $\Omega$ is connected, it suffices to show that $U$ and $V$ are disjoint.

Suppose that $x \in U \cap V$. Then there is an $S_1$ containing $x$ which is in $U$ and an $S_2$ containing $x$ which is in $V$. In particular, $f = C_0$ almost everywhere on $S_1$ and $f \neq C_0$ almost everywhere on $S_2$. However, this is impossible as $S_1$ and $S_2$ are open sets of positive measure with non-empty intersection and so $S_1 \cap S_2$ must have positive measure. Therefore, $U \cap V = \emptyset$ and the result follows. □

As Proposition 3.8 implies that $\mathrm{BMO}^p_{\mathscr{S}}(\Omega)/\mathbb{C}$ is a normed linear space, a natural question is whether it is complete. For $\mathrm{BMO}(\mathbb{R}^n)/\mathbb{C}$ this is a corollary of Fefferman's theorem that identifies it as the dual of the real Hardy space ([**23**, **24**]).

A proof that $\mathrm{BMO}(\mathbb{R}^n)/\mathbb{C}$ is a Banach space that does not pass through duality came a few years later and is due to Neri ([**53**]). This is another example of a proof that relies on the fact that $\mathcal{B}$ contains $\mathcal{C}$ (or, equivalently, that $\mathcal{Q}$ contains an analogue of $\mathcal{C}$ but for cubes). The core idea, however, may be adapted to our more general setting. The proof below makes use of the fact that shapes are open and that $\Omega$ is path connected since it is both open and connected.

THEOREM 3.9. *For any basis of shapes,* $\mathrm{BMO}^p_{\mathscr{S}}(\Omega)$ *is complete.*

PROOF. Let $\{f_i\}$ be Cauchy in $\mathrm{BMO}^p_{\mathscr{S}}(\Omega)$. Then, for any shape $S \in \mathscr{S}$, the sequence $\{f_i - (f_i)_S\}$ is Cauchy in $L^p(S)$. Since $L^p(S)$ is complete, there exists a function $f^S \in L^p(S)$ such that $f_i - (f_i)_S \to f^S$ in $L^p(S)$. The function $f^S$ can be seen to have mean zero on $S$: since $f_i - (f_i)_S$ converges to $f^S$ in $L^1(S)$, it follows that

$$\fint_S f^S = \lim_{i\to\infty} \fint_S f_i - (f_i)_S = 0. \tag{3.2}$$

If we have two shapes $S_1, S_2 \in \mathscr{S}$ such that $S_1 \cap S_2 \neq \emptyset$, by the above there is a function $f^{S_1} \in L^p(S_1)$ such that $f_i - (f_i)_{S_1} \to f^{S_1}$ in $L^p(S_1)$ and a function $f^{S_2} \in L^p(S_2)$ such that $f_i - (f_i)_{S_2} \to f^{S_2}$ in $L^p(S_2)$. Since both of these hold in $L^p(S_1 \cap S_2)$, we have

$$(f_i)_{S_2} - (f_i)_{S_1} = [f_i - (f_i)_{S_1}] - [f_i - (f_i)_{S_2}] \to f^{S_1} - f^{S_2} \quad \text{in } L^p(S_1 \cap S_2).$$

This implies that the sequence $C_i(S_1, S_2) = (f_i)_{S_2} - (f_i)_{S_1}$ converges as constants to a limit that we denote by $C(S_1, S_2)$, with

$$f^{S_1} - f^{S_2} \equiv C(S_1, S_2) \quad \text{on } S_1 \cap S_2.$$

From their definition, these constants are antisymmetric:

$$C(S_1, S_2) = -C(S_2, S_1).$$

Moreover, they possess an additive property that will be useful in later computations. By a finite chain of shapes we mean a finite sequence $\{S_j\}_{j=1}^k \subset \mathscr{S}$ such that $S_j \cap S_{j+1} \neq \emptyset$ for all $1 \leq j \leq k-1$. Furthermore, by a loop of shapes we mean a finite chain $\{S_j\}_{j=1}^k$ such that $S_1 \cap S_k \neq \emptyset$. If $\{S_j\}_{j=1}^k$ is a loop of shapes, then

$$C(S_1, S_k) = \sum_{j=1}^{k-1} C(S_j, S_{j+1}). \tag{3.3}$$

To see this, consider the telescoping sum

$$(f_i)_{S_k} - (f_i)_{S_1} = \sum_{j=1}^{k-1} (f_i)_{S_{j+1}} - (f_i)_{S_j},$$

for a fixed $i$. The formula (3.3) follows from this as each $(f_i)_{S_{j+1}} - (f_i)_{S_j}$ converges to $C(S_j, S_{j+1})$ since $S_j \cap S_{j+1} \neq \emptyset$ and $(f_i)_{S_k} - (f_i)_{S_1}$ converges to $C(S_1, S_k)$ since $S_1 \cap S_k \neq \emptyset$.

Let us now fix a shape $S_0 \in \mathscr{S}$ and consider another shape $S \in \mathscr{S}$ such that $S_0 \cap S = \emptyset$. Since $\Omega$ is a path-connected set, for any pair of points $(x, y) \in S_0 \times S$ there exists a path $\gamma_{x,y} : [0,1] \to \Omega$ such that $\gamma_{x,y}(0) = x$ and $\gamma_{x,y}(1) = y$. Since $\mathscr{S}$ covers $\Omega$ and the image of $\gamma_{x,y}$ is a compact set, we may cover $\gamma_{x,y}$ by a finite number of shapes. From this we may extract a finite chain connecting $S$ to $S_0$.

We now come to building the limit function $f$. If $x \in S_0$, then set $f(x) = f^{S_0}(x)$, where $f^{S_0}$ is as defined earlier. If $x \notin S_0$, then there is some shape $S$

containing $x$ and, by the preceding argument, a finite chain of shapes $\{S_j\}_{j=1}^k$ where $S_k = S$. In this case, set

$$f(x) = f^{S_k}(x) + \sum_{j=0}^{k-1} C(S_j, S_{j+1}). \tag{3.4}$$

The first goal is to show that this is well defined. Let $\{\tilde{S}_j\}_{j=1}^{\ell}$ be another finite chain connecting some $\tilde{S}_\ell$ with $x \in \tilde{S}_\ell$ to $S_0 = \tilde{S}_0$. Then we need to show that

$$f^{S_k}(x) + \sum_{j=0}^{k-1} C(S_j, S_{j+1}) = f^{\tilde{S}_\ell}(x) + \sum_{j=0}^{\ell-1} C(\tilde{S}_j, \tilde{S}_{j+1}). \tag{3.5}$$

First, we use the fact that $x \in S_k \cap \tilde{S}_\ell$ to write $f^{S_k}(x) - f^{\tilde{S}_\ell}(x) = C(S_k, \tilde{S}_\ell)$. Then, from the antisymmetry property of the constants, (3.5) is equivalent to

$$C(S_k, \tilde{S}_\ell) = C(S_k, S_{k-1}) + \cdots + C(S_1, S_0) + C(S_0, \tilde{S}_1) + \cdots + C(\tilde{S}_{\ell-1}, \tilde{S}_\ell),$$

which follows from (3.3) as $\{S_k, S_{k-1}, \ldots, S_1, S_0, \tilde{S}_1, \tilde{S}_2, \ldots, \tilde{S}_\ell\}$ forms a loop.

Finally, we show that $f_i \to f$ in $\mathrm{BMO}^p_{\mathscr{S}}(\Omega)$. Fixing a shape $S \in \mathscr{S}$, choose a finite chain $\{S_j\}_{j=1}^k$ such that $S_k = S$. By (3.4), on $S$ we have that $f = f^S$ modulo constants, and so, using (3.2) and the definition of $f^S$,

$$\begin{aligned} \fint_S |(f_i(x) - f(x)) - (f_i - f)_S|^p \, dx &= \fint_S |(f_i(x) - f^S(x)) - (f_i - f^S)_S|^p \, dx \\ &= \fint_S |f_i(x) - (f_i)_S - f^S(x)|^p \, dx \\ &\to 0 \qquad \text{as } i \to \infty. \end{aligned}$$

□

## 4. Shapewise inequalities on BMO

A "shapewise" inequality is an inequality that holds for each shape $S$ in a given basis. In this section, considerable attention will be given to highlighting those situations where the constants in these inequalities are known to be sharp. A recurring theme is the following: sharp results are mainly known for $p = 1$ and for $p = 2$ and, in fact, the situation for $p = 2$ is usually simple. The examples given demonstrating sharpness are straightforward generalisations of some of those found in [**43**].

For this section, we assume that $\mathscr{S}$ is an arbitrary basis of shapes and that $f \in L^1(S)$ for every $S \in \mathscr{S}$.

We begin by considering inequalities that provide equivalent characterizations of $\mathrm{BMO}^p_{\mathscr{S}}(\Omega)$. As with the classical BMO space, one can estimate the mean oscillation of a function on a shape by a double integral that is often easier to use for calculations but that comes at the loss of a constant.

PROPOSITION 4.1. *For any shape $S \in \mathscr{S}$,*

$$\frac{1}{2}\left(\fint_S \fint_S |f(x) - f(y)|^p \, dy \, dx\right)^{\frac{1}{p}} \le \left(\fint_S |f - f_S|^p\right)^{\frac{1}{p}} \le \left(\fint_S \fint_S |f(x) - f(y)|^p \, dy \, dx\right)^{\frac{1}{p}}.$$

PROOF. Fix a shape $S \in \mathscr{S}$. By Jensen's inequality we have that

$$\fint_S |f(x) - f_S|^p\,dx = \fint_S \left|\fint_S (f(x) - f(y))\,dy\right|^p dx \le \fint_S \fint_S |f(x) - f(y)|^p\,dy\,dx$$

and by Minkowski's inequality on $L^p(S \times S, \frac{dxdy}{|S|^2})$ we have that

$$\begin{aligned}\left(\fint_S \fint_S |f(x) - f(y)|^p\,dy\,dx\right)^{1/p} &= \left(\fint_S \fint_S \big|(f(x) - f_S) - (f(y) - f_S)\big|^p\,dy\,dx\right)^{1/p} \\ &\le 2\left(\fint_S |f - f_S|^p\right)^{1/p}.\end{aligned}$$

□

When $p = 1$, the following examples show that the constants in this inequality are sharp.

EXAMPLE 4.2. Let $S$ be a shape in $\Omega$ and consider a function $f = \chi_E$, where $E$ is a measurable subset of $S$ such that $|E| = \frac{1}{2}|S|$. Then $f_S = \frac{1}{2}$ and so $|f - f_S| = \frac{1}{2}\chi_S$, yielding

$$\fint_S |f - f_S| = \frac{1}{2}.$$

Writing $E^c = S \setminus E$, we have that $|f(x) - f(y)| = 0$ for $(x, y) \in E \times E$ or $(x, y) \in E^c \times E^c$, and that $|f(x) - f(y)| = 1$ for $(x, y) \in E^c \times E$; hence,

$$\fint_S \fint_S |f(x) - f(y)|\,dy\,dx = \frac{2}{|S|^2}\int_{E^c}\int_E |f(x) - f(y)|\,dy\,dx = \frac{2|E^c||E|}{|S|^2} = \frac{1}{2}.$$

Therefore, the right-hand side constant 1 is sharp.

EXAMPLE 4.3. Now consider a function $f = \chi_{E_1} - \chi_{E_3}$, where $E_1, E_2, E_3$ are measurable subsets of $S$ such that $S = E_1 \cup E_2 \cup E_3$ is a disjoint union (up to a set of measure zero) and $|E_1| = |E_3| = \beta|S|$ for some $0 < \beta < \frac{1}{2}$. Then, $f_S = 0$ and so

$$\fint_S |f - f_S| = \fint_S |f| = 2\beta.$$

Since $|f(x) - f(y)| = 0$ for $(x, y) \in E_j \times E_j$, $j = 1, 2, 3$, $|f(x) - f(y)| = 1$ for $(x, y) \in E_i \times E_j$ when $|i - j| = 1$, and $|f(x) - f(y)| = 2$ for $(x, y) \in E_i \times E_j$ when $|i - j| = 2$, we have that

$$\fint_S \fint_S |f(x) - f(y)|\,dy\,dx = \frac{4\beta|S|(|S| - 2\beta|S|) + 4\beta^2|S|^2}{|S|^2} = 4\beta(1 - 2\beta) + 4\beta^2.$$

As

$$\frac{2\beta}{4\beta(1 - 2\beta) + 4\beta^2} = \frac{1}{2 - 2\beta} \to \frac{1}{2}$$

as $\beta \to 0^+$, the left-hand side constant $\frac{1}{2}$ is sharp.

For other values of $p$, however, these examples tell us nothing about the sharpness of the constants in Proposition 4.1. In fact, the constants are not sharp for $p = 2$. In the following proposition, as in many to follow, the additional Hilbert space structure afforded to us yields a sharp statement (in this case, an equality) for little work.

PROPOSITION 4.4. *For any shape $S \in \mathscr{S}$,*

$$\left(\fint_S \fint_S |f(x) - f(y)|^2 \, dy \, dx\right)^{1/2} = \sqrt{2} \left(\fint_S |f - f_S|^2\right)^{1/2}.$$

PROOF. Observe that as elements of $L^2(S \times S, \frac{dx\,dy}{|S|^2})$, $f(x) - f_S$ is orthogonal to $f(y) - f_S$. Thus,

$$\fint_S \fint_S |f(x) - f(y)|^2 \, dy \, dx = \fint_S \fint_S |f(x) - f_S|^2 \, dy \, dx + \fint_S \fint_S |f(y) - f_S|^2 \, dy \, dx$$

and so

$$\left(\fint_S \fint_S |f(x) - f(y)|^2 \, dy \, dx\right)^{1/2} = \sqrt{2} \left(\fint_S |f - f_S|^2\right)^{1/2}.$$

□

In a different direction, it is sometimes easier to consider not the oscillation of a function from its mean, but its oscillation from a different constant. Again, this can be done at the loss of a constant.

PROPOSITION 4.5. *For any shape $S \in \mathscr{S}$,*

$$\inf_c \left(\fint_S |f - c|^p\right)^{1/p} \leq \left(\fint_S |f - f_S|^p\right)^{1/p} \leq 2 \inf_c \left(\fint_S |f - c|^p\right)^{1/p},$$

*where the infimum is taken over all constants $c$.*

PROOF. The first inequality is trivial. To show the second inequality, fix a shape $S \in \mathscr{S}$. By Minkowski's inequality on $L^p(S, \frac{dx}{|S|})$ and Jensen's inequality,

$$\left(\fint_S |f - f_S|^p\right)^{1/p} \leq \left(\fint_S |f - c|^p\right)^{1/p} + (|f_S - c|^p)^{1/p} \leq 2 \left(\fint_S |f - c|^p\right)^{1/p}.$$

□

As with Proposition 4.1, simple examples show that the constants are sharp for $p = 1$.

EXAMPLE 4.6. Let $S$ be a shape on $\Omega$ and consider the function $f = \chi_E - \chi_{E^c}$ as in Example 3.5, with $|E| = |S|/2$, $E^c = S \setminus E$. Then

$$\fint_S |f - f_S| = 1.$$

and for any constant $c$ we have that

$$\fint_S |f - c| = \frac{|1 - c||E| + |1 + c||E^c|}{|S|} = \frac{|1 - c| + |1 + c|}{2} \geq 1$$

with equality if $c \in [-1, 1]$. Thus

$$\inf_c \fint_S |f - c| = \fint_S |f - f_S|,$$

showing the left-hand side constant 1 in Proposition 4.5 is sharp when $p = 1$.

EXAMPLE 4.7. Consider, now, the function $f = \chi_E$ where $E$ is a measurable subset of $S$ such that $|E| = \alpha|S|$ for some $0 < \alpha < \frac{1}{2}$. Then, $f_S = \alpha$ and

$$\fint_S |f - f_S| = \frac{(1-\alpha)\alpha|S|}{|S|} + \frac{(|S| - \alpha|S|)\alpha}{|S|} = 2\alpha(1-\alpha).$$

For any constant $c$, we have that

$$\fint_S |f - c| = \frac{|1 - |c||E|}{|S|} + \frac{|c|(|S| - |E|)}{|S|} = \alpha|1 - |c|| + |c|(1-\alpha).$$

The right-hand side is at least $\alpha(1 - |c|) + |c|(1-\alpha) = \alpha + |c|(1-2\alpha)$, which is at least $\alpha$ with equality for $c = 0$, and so

$$\inf_c \fint_S |f - c| = \alpha.$$

As $\alpha \to 0^+$,

$$\frac{2\alpha(1-\alpha)}{\alpha} \to 2,$$

showing the right-hand side constant 2 in Proposition 4.5 is sharp when $p = 1$.

When $p = 1$, it turns out that we know for which constant the infimum in Proposition 4.5 is achieved.

PROPOSITION 4.8. *Let $f$ be real-valued. For any shape $S \in \mathscr{S}$,*

$$\inf_c \fint_S |f - c| = \fint_S |f - m|,$$

*where $m$ is a median of $f$ on $S$: that is, a (possibly non-unique) number such that $|\{x \in S : f(x) > m\}| \le \frac{1}{2}|S|$ and $|\{x \in S : f(x) < m\}| \le \frac{1}{2}|S|$.*

Note that the definition of a median makes sense for real-valued measurable functions. A proof of this proposition can be found in the appendix of [**17**], along with the fact that such functions always have a median on a measurable set of positive and finite measure (in particular, on a shape). Also, note that from the definition of a median, it follows that

$$|\{x \in S : f(x) \ge m\}| = |S| - |\{x \in S : f(x) > m\}| \ge |S| - \frac{1}{2}|S| = \frac{1}{2}|S|$$

and, likewise,

$$|\{x \in S : f(x) \le m\}| \ge \frac{1}{2}|S|.$$

PROOF. Fix a shape $S \in \mathscr{S}$ and a median $m$ of $f$ on $S$. For any constant $c$,

$$\begin{aligned}\int_S |f(x) - m|\,dx &= \int_{\{x \in S: f(x) \ge m\}} \big(f(x) - m\big)\,dx + \int_{\{x \in S: f(x) < m\}} \big(m - f(x)\big)\,dx \\ &= \int_{\{x \in S: f(x) \ge m\}} \big(f(x) - c\big)\,dx + (c - m)|\{x \in S : f(x) \ge m\}| \\ &\quad + \int_{\{x \in S: f(x) < m\}} \big(c - f(x)\big)\,dx + (m - c)|\{x \in S : f(x) < m\}|.\end{aligned}$$

Assuming that $m > c$, we have that

$$\int_{\{x \in S: f(x) \ge m\}} \big(f(x) - c\big)\,dx \le \int_{\{x \in S: f(x) \ge c\}} \big(f(x) - c\big)\,dx$$

and

$$\int_{\{x\in S:c\le f(x)<m\}}(c-f(x))\,dx\le 0,$$

and so we can write

$$\begin{aligned}\int_S|f(x)-m|\,dx &\le \int_{\{x\in S:f(x)\ge c\}}(f(x)-c)\,dx+(c-m)|\{x\in S:f(x)\ge m\}|\\ &\quad+\int_{\{x\in S:f(x)<c\}}(c-f(x))\,dx+(m-c)|\{x\in S:f(x)<m\}|\\ &=\int_S|f(x)-c|\,dx\\ &\quad+(m-c)\big[|\{x\in S:f(x)<m\}|-|\{x\in S:f(x)\ge m\}|\big]\\ &\le\int_S|f(x)-c|\,dx+(m-c)\left[\frac{1}{2}|S|-\frac{1}{2}|S|\right]\\ &=\int_S|f(x)-c|\,dx.\end{aligned}$$

In the case where $m<c$, we may apply the previous calculation to $-f$ and use the fact that $-m$ is a median of $-f$:

$$\int_S|f(x)-m|\,dx=\int_S|-f(x)-(-m)|\,dx\le\int_S|-f(x)-(-c)|\,dx=\int_S|f(x)-c|\,dx.$$

□

For $p=2$, we are able to do a few things at once. We are very simply able to obtain an equality that automatically determines the sharp constants for Proposition 4.5 and the constant $c$ for which the infimum is achieved.

PROPOSITION 4.9. *For any shape $S\in\mathscr{S}$,*

$$\inf_c\left(\fint_S|f-c|^2\right)^{1/2}=\left(\fint_S|f-f_S|^2\right)^{1/2},$$

*where the infimum is taken over all constants $c$.*

PROOF. Fix a shape $S\in\mathscr{S}$ and a constant $c$. As previously observed, $f-f_S$ is orthogonal to any constant in the sense of $L^2(S,\frac{dx}{|S|})$; in particular, $f-f_S$ is orthogonal to $f_S-c$. Thus,

$$\fint_S|f-c|^2=\fint_S|f-f_S|^2+\fint_S|f_S-c|^2\ge\fint_S|f-f_S|^2$$

with the minimum achieved when $c=f_S$. □

The following proposition shows that the action of Hölder continuous maps preserves the bound on the $p$-mean oscillation, up to a constant.

PROPOSITION 4.10. *Let $F:\mathbb{R}\to\mathbb{R}$ be $\alpha$-Hölder continuous for $0<\alpha\le 1$ with Hölder coefficient $L$. Fix any shape $S\in\mathscr{S}$ and suppose $f\in L^1(S)$ is real-valued. Then, for $1\le p<\infty$,*

$$\left(\fint_S|F\circ f-(F\circ f)_S|^p\right)^{\frac{1}{p}}\le 2L\left(\fint_S|f-f_S|^p\right)^{\frac{\alpha}{p}}. \tag{4.1}$$

*When* $p = 2$,

$$\left(\fint_S |F \circ f - (F \circ f)_S|^2\right)^{\frac{1}{2}} \leq L\left(\fint_S |f - f_S|^2\right)^{\frac{1}{2}}. \tag{4.2}$$

PROOF. Fix a shape $S \in \mathscr{S}$. By Proposition 4.5 and Jensen's inequality, we have that

$$\begin{aligned}\fint_S |F \circ f - (F \circ f)_S|^p &\leq 2^p \fint_S |F(f(x)) - F(f_S)|^p dx \\ &\leq 2L^p \fint_S |f(x) - f_S|^{\alpha p}\, dx \\ &\leq 2^p L^p \left(\fint_S |f - f_S|^p\right)^{\alpha}.\end{aligned}$$

When $p = 2$, Proposition 4.9 shows that the factor of $2^p$ in the first inequality can be dropped. □

As has been pointed out in [**17**], if one uses the equivalent norm defined by Proposition 4.5, the result of Proposition 4.10 holds with constant $L$ for any $p \geq 1$, since the factor of 2 comes from $(F \circ f)_S \neq F(f_S)$.

The following example demonstrates that the constants are sharp when $p = 1$ and $p = 2$.

EXAMPLE 4.11. Consider the function $F(x) = |x|$, so that $\alpha = 1 = L$, and fix a shape $S$ in $\Omega$. Taking the function $f = \chi_{E_1} - \chi_{E_3}$, as in Example 4.3, where $S$ is a disjoint union $E_1 \cup E_2 \cup E_3$ and $|E_1| = |E_3| = \beta|S|$ for some $0 < \beta < \frac{1}{2}$, we have $|f|_S = 2\beta$ and so

$$\fint_S ||f| - |f|_S| = 4(1 - 2\beta)\beta.$$

Since $\fint_S |f - f_S| = 2\beta$ and $\frac{4(1-2\beta)\beta}{2\beta} \to 2$ as $\beta \to 0^+$, the constant 2 is sharp for $p = 1$.

For $p = 2$, when $f \geq 0$ we have $F(f) = f$ and $F(f)_S = F(f_S)$, so equality holds in (4.2).

While we have shown that the shapewise inequalities (4.1), for $p = 1$, and (4.2) are sharp for $F(x) = |x|$, in the next section it will be shown that better constants can be obtained for norm inequalities.

Now we address how the $\mathrm{BMO}^p_{\mathscr{S}}(\Omega)$ spaces relate for different bases.

PROPOSITION 4.12. *For any shape* $S \in \mathscr{S}$, *if* $\tilde{S}$ *is another shape (from possibly another basis) such that* $\tilde{S} \subset S$ *and* $|\tilde{S}| \geq c|S|$ *for some constant* $c$, *then*

$$\left(\fint_{\tilde{S}} |f - f_{\tilde{S}}|^p\right)^{\frac{1}{p}} \leq 2c^{-1/p}\left(\fint_S |f - f_S|^p\right)^{\frac{1}{p}}.$$

PROOF. From Proposition 4.5,

$$\fint_{\tilde{S}} |f - f_{\tilde{S}}|^p \leq 2^p \fint_{\tilde{S}} |f - f_S|^p \leq 2^p c^{-1} \fint_S |f - f_S|^p.$$

□

An immediate consequence of this is that $\mathrm{BMO}^p_{\mathscr{S}}(\Omega) \subset \mathrm{BMO}^p_{\tilde{\mathscr{S}}}(\Omega)$ if $\tilde{\mathscr{S}} \trianglelefteq \mathscr{S}$. Moreover, if $\mathscr{S} \approx \tilde{\mathscr{S}}$ then $\mathrm{BMO}^p_{\mathscr{S}}(\Omega) \cong \mathrm{BMO}^p_{\tilde{\mathscr{S}}}(\Omega)$. In particular, it follows that $\mathrm{BMO}^p_{\mathcal{B}}(\Omega) \cong \mathrm{BMO}^p_{\mathcal{Q}}(\Omega)$. Since $\mathcal{Q} \subset \mathcal{R}$, it is automatic without passing through Proposition 4.12 that $\mathrm{BMO}^p_{\mathcal{R}}(\Omega) \subset \mathrm{BMO}^p(\Omega)$ with $\|f\|_{\mathrm{BMO}^p} \leq \|f\|_{\mathrm{BMO}^p_{\mathcal{R}}}$. The reverse inclusion is false. The following example of a function in $\mathrm{BMO}(\Omega)$ that is not in $\mathrm{BMO}_{\mathcal{R}}(\Omega)$ is taken from [**43**], where the calculations proving the claim can be found:

EXAMPLE 4.13. Consider $\Omega = (0,1) \times (0,1) \subset \mathbb{R}^2$. The function

$$f = \sum_{k=1}^{\infty} \chi_{(0,2^{-k+1}) \times (0,\frac{1}{k})}$$

belongs to $\mathrm{BMO}(\Omega) \setminus \mathrm{BMO}_{\mathcal{R}}(\Omega)$.

## 5. Rearrangements and the absolute value

Consider two measure spaces $(M,\mu)$ and $(N,\nu)$ such that $\mu(M) = \nu(N)$.

DEFINITION 5.1. We say that measurable functions $f : M \to \mathbb{R}$ and $g : N \to \mathbb{R}$ are equimeasurable if for all $s \in \mathbb{R}$ the quantities $\mu_f(s) = \mu(\{x \in M : f(x) > s\})$ and $\nu_g(s) = \nu(\{y \in N : g(y) > s\})$ coincide.

It is important to note that this is not the standard definition of equimeasurability. Typically (see, for example, [**3**]) equimeasurability means $\mu_{|f|}(s) = \mu_{|g|}(s)$ for all $s \geq 0$; however, for our purposes, it will be useful to distinguish between two functions being equimeasurable and the absolute value of two functions being equimeasurable. That said, it is true that

LEMMA 5.2. *Let $f$ and $g$ be measurable functions such that $\mu_f(s) = \nu_g(s) < \infty$ for all $s$. Then, $\mu_{|f|}(s) = \nu_{|g|}(s)$ for all $s$.*

PROOF. Fix $s \in \mathbb{R}$. Writing

$$\{x \in M : f(x) < -s\} = \bigcup_{n \in \mathbb{N}} \left\{x \in M : f(x) \leq -s - \frac{1}{n}\right\},$$

we have that

$$\mu(\{x \in M : f(x) < -s\}) = \lim_{n \to \infty} \mu(M) - \mu_f\left(-s - \frac{1}{n}\right).$$

Here we use the convention that infinity minus a finite number is infinity and use the fact that $\mu_f < \infty$. Since $\mu_f = \nu_g$, by assumption, it follows that

$$\mu(\{x \in M : f(x) < -s\}) = \nu(\{x \in N : g(x) < -s\}).$$

If $s \geq 0$, then

$$\begin{aligned}\mu(\{x \in M : |f(x)| > s\}) &= \mu(\{x \in M : f(x) > s\}) + \mu(\{x \in M : f(x) < -s\}) \\ &= \nu(\{y \in N : g(y) > s\}) + \nu(\{y \in N : g(y) < -s\}) \\ &= \nu(\{y \in N : |g(x)| > s\}).\end{aligned}$$

If $s < 0$, then

$$\mu(\{x \in M : |f(x)| > s\}) = \mu(M) = \nu(N) = \nu(\{y \in N : |g(x)| > s\}).$$

□

A useful tool is the following lemma. It is a consequence of Cavalieri's principle, also called the layer cake representation, which provides a way of expressing the integral of $\varphi(|f|)$ for a suitable transformation $\varphi$ in terms of a weighted integral of $\mu_{|f|}$. The simplest incarnation of this principal states that for any measurable set $A$,

$$\int_A |f|^p = \int_0^\infty p\alpha^{p-1}|\{x \in A : |f(x)| > \alpha\}|\, d\alpha,$$

where $0 < p < \infty$. A more general statement can be found in [**49**], Theorem 1.13 and its remarks.

LEMMA 5.3. *Let $M \subset \mathbb{R}^m$, $N \subset \mathbb{R}^n$ be Lebesgue measurable sets of equal measure, and $f : M \to \mathbb{R}$ and $g : N \to \mathbb{R}$ be measurable functions such that $|f|$ and $|g|$ are equimeasurable. Then, for $0 < p < \infty$,*

$$\int_M |f|^p = \int_N |g|^p \qquad \textit{and} \qquad \operatorname*{ess\,sup}_M |f| = \operatorname*{ess\,sup}_N |g|.$$

*Furthermore, under the hypothesis of Lemma 5.2, for any constant $c$,*

$$\int_M ||f| - c| = \int_N ||g| - c|.$$

Moving back to the setting of this paper, for this section we assume that $f$ is a measurable function on $\Omega$ that satisfies the condition

$$|\{x \in \Omega : |f(x)| > s\}| \to 0 \qquad \text{as} \qquad s \to \infty. \tag{5.1}$$

This guarantees that the rearrangements defined below are finite on their domains (see [**60**], V.3).

DEFINITION 5.4. Let $I_\Omega = (0, |\Omega|)$. The decreasing rearrangement of $f$ is the function

$$f^*(t) = \inf\{s \geq 0 : |\{x \in \Omega : |f(x)| > s\}| \leq t\}, \quad t \in I_\Omega.$$

This rearrangement is studied in the theory of interpolation and rearrangement-invariant function spaces. In particular, it can be used to define the Lorentz spaces, $L^{p,q}$, which are a refinement of the scale of Lebesgue spaces and can be used to strengthen certain inequalities such as those of Hardy-Littlewood-Sobolev and Hausdorff-Young. For standard references on these topics, see [**3**] or [**60**].

A related rearrangement is the following.

DEFINITION 5.5. The signed decreasing rearrangement of $f$ is defined as

$$f^\circ(t) = \inf\{s \in \mathbb{R} : |\{x \in \Omega : f(x) > s\}| \leq t\}, \quad t \in I_\Omega.$$

Clearly, $f^\circ$ coincides with $f^*$ when $f \geq 0$ and, more generally, $|f|^\circ = f^*$. Further information on this rearrangement can be found in [**12**, **43**].

Here we collect some of the basic properties of these rearrangements, the proofs for which are adapted from [**60**].

LEMMA 5.6. *Let $f : \Omega \to \mathbb{R}$ be measurable and satisfying* (5.1). *Then*

a) *its signed decreasing rearrangement $f^\circ : I_\Omega \to (-\infty, \infty)$ is decreasing and equimeasurable with $f$;*
b) *its decreasing rearrangement $f^* : I_\Omega \to [0, \infty)$ is decreasing and equimeasurable with $|f|$.*

PROOF. If $t_1 \geq t_2$, it follows that

$$\{s : |\{x \in \Omega : f(x) > s\}| \leq t_2\} \subset \{s : |\{x \in \Omega : f(x) > s\}| \leq t_1\}.$$

Since this is equally true for $|f|$ in place of $f$, it shows that both $f^*$ and $f^\circ$ are decreasing functions.

Fix $s$. For $t \in I_\Omega$, $f^\circ(t) > s$ if and only if $t < |\{x \in \Omega : f(x) > s\}|$, from where it follows that

$$|\{t \in I_\Omega : f^\circ(t) > s\}| = |\{x \in \Omega : f(x) > s\}|.$$

Again, applying this to $|f|$ in place of $f$ yields the corresponding statement for $f^*$. □

One may ask how the rearrangement $f^*$ behaves when additional conditions are imposed on $f$. In particular, is the map $f \mapsto f^*$ a bounded operator on various function spaces? A well-known result in this direction is that this map is an isometry on $L^p$, which follows immediately from Lemmas 5.3 and 5.6.

PROPOSITION 5.7. *For all $1 \leq p \leq \infty$, if $f \in L^p(\Omega)$ then $f^* \in L^p(I_\Omega)$ with $\|f^*\|_{L^p(I_\Omega)} = \|f\|_{L^p(\Omega)}$.*

Another well-known result is the Pólya-Szegő inequality, which asserts that the Sobolev norm decreases under the symmetric decreasing rearrangement ([**7**]), yet another kind of rearrangement. From this one can deduce the following (see, for instance, [**12**]).

THEOREM 5.8. *If $f \in W^{1,p}(\mathbb{R}^n)$ then*

$$n\omega_n^{1/n}\left(\int_0^\infty \left|\frac{d}{dt}f^*(t)\right|^p t^{p/n'}\,dt\right)^{\frac{1}{p}} \leq \left(\int_{\mathbb{R}^n} |\nabla f|^p\right)^{\frac{1}{p}},$$

*where $n'$ is the Hölder dual exponent of $n$ and $\omega_n$ denotes the volume of the unit ball in $\mathbb{R}^n$.*

Despite these positive results, there are some closely related spaces on which the operator $f \mapsto f^*$ is not bounded. One such example is the John-Nirenberg space $\mathrm{JN}_p(\Omega)$. We say that $f \in L^1_{\mathrm{loc}}(\Omega)$ is in $\mathrm{JN}_p(\Omega)$ if there exists a constant $K \geq 0$ such that

$$\sup \sum |Q_i| \left(\fint_{Q_i} |f - f_{Q_i}|\right)^p \leq K^p, \tag{5.2}$$

where the supremum is taken over all collections of pairwise disjoint cubes $Q_i$ in $\Omega$. We define the quantity $\|f\|_{\mathrm{JN}_p}$ as the smallest $K$ for which (5.2) holds. One can show that this is a norm on $\mathrm{JN}_p(\Omega)$ modulo constants. These spaces have been considered in the case where $\Omega$ is a cube in [**19, 37**] and a general Euclidean domain in [**33**], and generalised to a metric measure space in [**1, 52**].

While it is well known that $L^p(\Omega) \subset \mathrm{JN}_p(\Omega) \subset L^{p,\infty}(\Omega)$, the strictness of these inclusions has only recently been addressed ([**1, 19**]).

In the case where $\Omega = I$, a (possibly unbounded) interval, the following is obtained:

THEOREM 5.9 ([**19**]). *Let $f : I \to \mathbb{R}$ be a monotone function with $f \in L^1(I)$. Then there exists $c = c(p) > 0$ such that*

$$\|f\|_{\mathrm{JN}_p} \geq c\|f - C\|_{L^p}$$

*for some $C \in \mathbb{R}$.*

In other words, monotone functions are in $\mathrm{JN}_p(I)$ if and only if they are also in $L^p(I)$. In [**19**], an explicit example of a function $f \in \mathrm{JN}_p(I) \setminus L^p(I)$ is constructed when $I$ is a finite interval. This leads to the observation that the decreasing rearrangement is not bounded on $\mathrm{JN}_p(I)$.

COROLLARY 5.10. *If $I$ is a finite interval, there exists an $f \in \mathrm{JN}_p(I)$ such that $f^* \notin \mathrm{JN}_p(I_I)$.*

PROOF. Since $f \notin L^p(I)$, it follows from Proposition 5.7 that $f^* \notin L^p(I_I)$. As $f^*$ is monotone, it follows from the previous theorem that $f^* \notin \mathrm{JN}_p(I_I)$. □

We consider now the question of boundedness of rearrangements on $\mathrm{BMO}^p_{\mathscr{S}}(\Omega)$ spaces.

PROBLEM 5.11. Does there exist a constant $c$ such that for all $f \in \mathrm{BMO}^p_{\mathscr{S}}(\Omega)$, $\|f^*\|_{\mathrm{BMO}^p(I_\Omega)} \leq c\|f\|_{\mathrm{BMO}^p_{\mathscr{S}}(\Omega)}$? If so, what is the smallest constant, written $c_*(p, \mathscr{S})$, for which this holds?

PROBLEM 5.12. Does there exist a constant $c$ such that for all $f \in \mathrm{BMO}^p_{\mathscr{S}}(\Omega)$, $\|f^\circ\|_{\mathrm{BMO}^p(I_\Omega)} \leq c\|f\|_{\mathrm{BMO}^p_{\mathscr{S}}(\Omega)}$? If so, what is the smallest constant, written $c_\circ(p, \mathscr{S})$, for which this holds?

Clearly, if such constants exist, then they are at least equal to one. The work of Garsia-Rodemich and Bennett-DeVore-Sharpley implies an answer to the first problem and that $c_*(1, \mathcal{Q}) \leq 2^{n+5}$ when $\Omega = \mathbb{R}^n$:

THEOREM 5.13 ([**2**,**29**]). *If $f \in \mathrm{BMO}(\mathbb{R}^n)$, then $f^* \in \mathrm{BMO}((0,\infty))$ and*

$$\|f^*\|_{\mathrm{BMO}} \leq 2^{n+5}\|f\|_{\mathrm{BMO}}.$$

These results were obtained by a variant of the Calderón-Zygmund decomposition ([**57**]). Riesz' rising sun lemma, an analogous one-dimensional result that can often be used to obtain better constants, was then used by Klemes to obtain the sharp estimate that for $\Omega = I$, a finite interval, $c_\circ(1, \mathcal{I}) = 1$.

THEOREM 5.14 ([**40**]). *If $f \in \mathrm{BMO}(I)$, then $f^\circ \in \mathrm{BMO}(I_I)$ and*

$$\|f^\circ\|_{\mathrm{BMO}} \leq \|f\|_{\mathrm{BMO}}.$$

An elementary but key element of Klemes' proof that can be generalised to our context of general shapes is the following shapewise identity.

LEMMA 5.15. *For any shape $S$, if $f \in L^1(S)$ then*

$$\fint_S |f - f_S| = \frac{2}{|S|}\int_{\{f > f_S\}} (f - f_S) = \frac{2}{|S|}\int_{\{f < f_S\}} (f_S - f).$$

PROOF. Write

$$\int_S |f(x) - f_S|\,dx = \int_{\{x\in S: f(x) > f_S\}} (f(x) - f_S)\,dx + \int_{\{x \in S: f(x) < f_S\}} (f_S - f(x))\,dx.$$

Since

$$\int_{\{x\in S: f(x) > f_S\}} (f(x) - f_S)\,dx + \int_{\{x\in S: f(x) < f_S\}} (f(x) - f_S)\,dx = \int_S (f(x) - f_S)\,dx = 0,$$

it follows that

$$\int_{\{x\in S: f(x) > f_S\}} (f(x) - f_S)\,dx = \int_{\{x\in S: f(x) < f_S\}} (f_S - f(x))\,dx,$$

which gives the identity. □

The next sharp result concerning rearrangements is due to Korenovskii, showing that for $\Omega = I$, a finite interval, $c_*(1,\mathcal{I}) = 1$. The proof of this result makes direct use of Klemes' theorem.

THEOREM 5.16 ([**41**]). *If $f \in \mathrm{BMO}(I)$, then $f^* \in \mathrm{BMO}(I_I)$ and*

$$\|f^*\|_{\mathrm{BMO}} \leq \|f\|_{\mathrm{BMO}}.$$

Important in Korenovskii's transition from a sharp estimate for $c_\circ(1,\mathscr{S})$ to one for $c_*(1,\mathscr{S})$ is the fact that $|f|^\circ = f^*$, bringing us to consider the boundedness of the absolute value operator. Recall from Example 4.11 that $F(x) = |x|$ gives us the sharp shapewise inequality in Proposition 4.10 with $p = 1$, which implies that $\||f|\|_{\mathrm{BMO}_{\mathscr{S}}} \leq 2\|f\|_{\mathrm{BMO}_{\mathscr{S}}}$. However, this need not be sharp as a norm inequality, and so it is natural to ask

PROBLEM 5.17. What is the smallest constant $c_{|\cdot|}(p,\mathscr{S})$ such that $\||f|\|_{\mathrm{BMO}^p_{\mathscr{S}}} \leq c_{|\cdot|}(p,\mathscr{S})\|f\|_{\mathrm{BMO}^p_{\mathscr{S}}}$ holds for all $f \in \mathrm{BMO}^p_{\mathscr{S}}(\Omega)$?

It is clear that $c_{|\cdot|}(p,\mathscr{S}) \geq 1$ and Proposition 4.10 implies that $c_{|\cdot|}(p,\mathscr{S}) \leq 2$. Applying this estimate along with Klemes' theorem yields the non-sharp bound $c_*(1,\mathcal{I}) \leq 2$.

In order for Korenovskii to obtain a sharp result for $c_*(1,\mathcal{I})$, a more subtle argument was needed that allowed him to conclude that $c_{|\cdot|}(1,\mathcal{I}) = 1$ when $\Omega = I$:

THEOREM 5.18 ([**41**]). *If $f \in \mathrm{BMO}(I)$, then $\||f|\|_{\mathrm{BMO}} \leq \|f\|_{\mathrm{BMO}}$.*

The following is one of the essential parts of this argument. It demonstrates that the behaviour of the absolute value operator is more easily analyzed for decreasing functions.

THEOREM 5.19 ([**41**]). *Let $I$ be a finite interval and $f \in L^1(I)$ be a decreasing function. Then,*

$$\fint_I \big||f| - |f|_I\big| \leq \sup_{J\subset I} \fint_J |f - f_J|$$

*where the supremum is taken over all subintervals $J$ of $I$.*

Further sharp results were obtained by Korenovskii in the case where $\Omega = R$, a rectangle, and $\mathscr{S} = \mathcal{R}$: it was shown that, similar to the one-dimensional case just discussed, $c_{|\cdot|}(1,\mathcal{R}) = c_\circ(1,\mathcal{R}) = c_*(1,\mathcal{R}) = 1$.

THEOREM 5.20 ([**42**]). *If $f \in \mathrm{BMO}_{\mathcal{R}}(R)$, then $|f| \in \mathrm{BMO}_{\mathcal{R}}(R)$ and $f^\circ, f^* \in \mathrm{BMO}(I_R)$ with*

$$\||f|\|_{\mathrm{BMO}_{\mathcal{R}}} \leq \|f\|_{\mathrm{BMO}_{\mathcal{R}}}, \qquad \|f^\circ\|_{\mathrm{BMO}} \leq \|f\|_{\mathrm{BMO}_{\mathcal{R}}}, \qquad \|f^*\|_{\mathrm{BMO}} \leq \|f\|_{\mathrm{BMO}_{\mathcal{R}}}.$$

This demonstrates the paradigm that rectangles behave more similarly to one-dimensional intervals than cubes do. In particular, the generalization of Klemes' theorem to the higher-dimensional case of rectangles (the result that $c_\circ(1,\mathcal{R}) = 1$) employs a multidimensional analogue of Riesz' rising sun lemma using rectangles ([**44**]) when such a theorem could not exist for arbitrary cubes.

Following the techniques of [**41**], general relationships can be found between the constants $c_{|\cdot|}(1,\mathscr{S})$, $c_\circ(1,\mathscr{S})$, $c_*(1,\mathscr{S})$ for an arbitrary basis of shapes. First, we show that $c_{|\cdot|}(1,\mathscr{S}) \leq c_\circ(1,\mathscr{S})$.

PROPOSITION 5.21. *For any collection of shapes $\mathscr{S}$, if $f \in \mathrm{BMO}_{\mathscr{S}}(\Omega)$ and $\|f^\circ\|_{\mathrm{BMO}(I_\Omega)} \leq c\|f\|_{\mathrm{BMO}_{\mathscr{S}}(\Omega)}$, then $\||f|\|_{\mathrm{BMO}_{\mathscr{S}}(\Omega)} \leq c\|f\|_{\mathrm{BMO}_{\mathscr{S}}(\Omega)}$ for the same constant $c$.*

PROOF. Fix a shape $S \in \mathscr{S}$ and assume that $f \in \mathrm{BMO}_{\mathscr{S}}(\Omega)$ is supported on $S$.

Since $f$ is equimeasurable with $f^\circ$ by Lemma 5.6, it follows from Lemma 5.2 that $|f|$ is equimeasurable with $|f^\circ|$ (recall that $|S| < \infty$). Writing $E = (0, |S|)$, by Lemma 5.3 we have that $|f^\circ|_E = |f|_S$ and also, then, that

$$\int_0^{|S|} \big||f^\circ| - |f^\circ|_E\big| = \int_0^{|S|} \big||f^\circ| - |f|_S\big| = \int_S \big||f| - |f|_S\big|.$$

Thus, by Theorem 5.19,

$$\fint_S \big||f| - |f|_S\big| = \fint_{(0,|S|)} \big||f^\circ| - |f^\circ|_E\big| \leq \sup_{J \subset (0,|S|)} \fint_J |f^\circ - (f^\circ)_J|.$$

For all $J \subset (0, |S|)$, we have that

$$\fint_J |f^\circ - (f^\circ)_J| \leq \|f^\circ\|_{\mathrm{BMO}\left((0,|S|)\right)} \leq \|f^\circ\|_{\mathrm{BMO}(I_\Omega)} \leq c\|f\|_{\mathrm{BMO}_{\mathscr{S}}(\Omega)},$$

and, therefore,

$$\fint_S \big||f| - |f|_S\big| \leq c\|f\|_{\mathrm{BMO}_{\mathscr{S}}(\Omega)}.$$

Taking a supremum over all shapes $S \in \mathscr{S}$ yields the result. □

This result in turn allows us to prove the following relationship:

$$c_*(1, \mathscr{S}) \leq c_{|\cdot|}(1, \mathscr{S}) c_\circ(1, \mathscr{S}) \leq c_\circ(1, \mathscr{S})^2.$$

PROPOSITION 5.22. *For any collection of shapes $\mathscr{S}$, if $f \in \mathrm{BMO}_{\mathscr{S}}(\Omega)$ and $\|f^\circ\|_{\mathrm{BMO}(I_\Omega)} \leq c\|f\|_{\mathrm{BMO}_{\mathscr{S}}(\Omega)}$, then $\|f^*\|_{\mathrm{BMO}(I_\Omega)} \leq c^2\|f\|_{\mathrm{BMO}_{\mathscr{S}}(\Omega)}$.*

PROOF. By Proposition 5.21, it follows that

$$\||f|\|_{\mathrm{BMO}_{\mathscr{S}}(\Omega)} \leq c\|f\|_{\mathrm{BMO}_{\mathscr{S}}(\Omega)}.$$

Writing $f^* = |f|^\circ$, we have that

$$\|f^*\|_{\mathrm{BMO}(I_\Omega)} = \||f|^\circ\|_{\mathrm{BMO}(I_\Omega)} \leq c\||f|\|_{\mathrm{BMO}_{\mathscr{S}}(\Omega)} \leq c^2\|f\|_{\mathrm{BMO}_{\mathscr{S}}(\Omega)}.$$

□

From these results, we see that a sharp result of the form $c_\circ(1, \mathscr{S}) = 1$ would immediately imply two more sharp results, $c_{|\cdot|}(1, \mathscr{S}) = 1$ and $c_*(1, \mathscr{S}) = 1$.

Although the dyadic cubes do not, in general, cover a domain, the space dyadic BMO has been extensively studied in the literature (see [**28**] for an early work illustrating its connection to martingales). In fact, many of the results in this section hold for that space; as such, extending our notation to include $\mathscr{S} = \mathcal{Q}_d$ even though it does not form a basis, we provide here a sample of the known sharp results.

Klemes' theorem was extended to the higher-dimensional dyadic case by Nikolidakis, who shows, for $\Omega = Q$, that $c_\circ(1, \mathcal{Q}_d) \leq 2^n$:

THEOREM 5.23 ([**54**]). *If $f \in \mathrm{BMO}_{\mathcal{Q}_d}(Q)$, then $f^\circ \in \mathrm{BMO}(I_Q)$ and*

$$\|f^\circ\|_{\mathrm{BMO}} \leq 2^n\|f\|_{\mathrm{BMO}_{\mathcal{Q}_d}}.$$

As a corollary of Proposition 5.22 and Theorem 5.23 we have the following, which shows that $c_*(1, \mathcal{Q}_d) \leq 2^{n+1}$, an improvement on Theorem 5.13.

COROLLARY 5.24. *If $f \in \mathrm{BMO}_{\mathcal{Q}_d}(Q)$, then $f^* \in \mathrm{BMO}(I_Q)$ and*

$$\|f^*\|_{\mathrm{BMO}} \leq 2^{n+1}\|f\|_{\mathrm{BMO}_{\mathcal{Q}_d}}.$$

The previous discussion emphasized the situation when $p = 1$. For $p = 2$, even more powerful tools are available: using probabilistic methods, Stolyarov, Vasyunin and Zatitskiy prove the following sharp result.

THEOREM 5.25 ([**61**]). $c_*(2, \mathcal{Q}_d) = \frac{1+2^n}{2^{1+n/2}}$.

## 6. Truncations

An immediate consequence of the bounds for the absolute value is the following result demonstrating that BMO is a lattice.

PROPOSITION 6.1. *For any basis of shapes, if $f_1, f_2$ are real-valued functions in $\mathrm{BMO}^p_{\mathscr{S}}(\Omega)$, then $\max(f_1, f_2)$ and $\min(f_1, f_2)$ are in $\mathrm{BMO}^p_{\mathscr{S}}(\Omega)$, with*

$$\|\max(f_1, f_2)\|_{\mathrm{BMO}^p_{\mathscr{S}}} \leq \frac{1 + c_{|\cdot|}(p, \mathscr{S})}{2}\left(\|f_1\|_{\mathrm{BMO}^p_{\mathscr{S}}} + \|f_2\|_{\mathrm{BMO}^p_{\mathscr{S}}}\right).$$

*and*

$$\|\min(f_1, f_2)\|_{\mathrm{BMO}^p_{\mathscr{S}}} \leq \frac{1 + c_{|\cdot|}(p, \mathscr{S})}{2}\left(\|f_1\|_{\mathrm{BMO}^p_{\mathscr{S}}} + \|f_2\|_{\mathrm{BMO}^p_{\mathscr{S}}}\right).$$

PROOF. This follows from writing

$$\max(f_1, f_2) = \frac{(f_1 + f_2) + |f_1 - f_2|}{2} \quad \text{and} \quad \min(f_1, f_2) = \frac{(f_1 + f_2) - |f_1 - f_2|}{2}$$

and using the estimate for the absolute value:

$$\||f_1 - f_2|\|_{\mathrm{BMO}^p_{\mathscr{S}}} \leq c_{|\cdot|}(p, \mathscr{S})(\|f_1\|_{\mathrm{BMO}^p_{\mathscr{S}}} + \|f_2\|_{\mathrm{BMO}^p_{\mathscr{S}}}).$$

□

In particular, applying Theorems 5.18 and 5.20, this yields the sharp constant 1 for $p = 1$ when $\mathscr{S} = \mathcal{I}$ or $\mathcal{R}$.

We can also obtain the sharp constant 1 via a sharp shapewise inequality for the cases $p = 1$ and $p = 2$, regardless of the basis. The proof of the case $p = 2$ in the following result is given by Reimann and Rychener [**55**] for the basis $\mathscr{S} = \mathcal{Q}$.

PROPOSITION 6.2. *Let $p = 1$ or $p = 2$. Let $\mathscr{S}$ be any basis of shapes and fix a shape $S \in \mathscr{S}$. If $f_1, f_2 \in \mathrm{BMO}^p_{\mathscr{S}}(\Omega)$ and $f = \max(f_1, f_2)$ or $f = \min(f_1, f_2)$, we have*

$$\fint_S |f - f_S|^p \leq \fint_S |f_1 - (f_1)_S|^p + \fint_S |f_2 - (f_2)_S|^p. \tag{6.1}$$

*Consequently*

$$\|\max(f_1, f_2)\|_{\mathrm{BMO}^p_{\mathscr{S}}} \leq \|f_1\|_{\mathrm{BMO}^p_{\mathscr{S}}} + \|f_2\|_{\mathrm{BMO}^p_{\mathscr{S}}}$$

*and*

$$\|\min(f_1, f_2)\|_{\mathrm{BMO}^p_{\mathscr{S}}} \leq \|f_1\|_{\mathrm{BMO}^p_{\mathscr{S}}} + \|f_2\|_{\mathrm{BMO}^p_{\mathscr{S}}}.$$

PROOF. First, for $p = 1$, fix a shape $S \in \mathscr{S}$ and let $f = \min(f_1, f_2)$. Note that $f_S \le (f_i)_S$ as $f \le f_i$ on $S$ for $i = 1, 2$. Let

$$E_1 = \{x \in S : f_1(x) \le f_2(x)\} = \{x \in S : f(x) = f_1(x)\}, \quad E_2 = S \setminus E_1.$$

By Lemma 5.15,

$$\begin{aligned} \fint_S |f(x) - f_S|\, dx &= \frac{2}{|S|} \int_{\{x \in S : f(x) < f_S\}} (f_S - f(x))\, dx \\ &= \frac{2}{|S|} \sum_{i=1}^{2} \int_{\{x \in E_i : f_i(x) < f_S\}} (f_S - f_i(x))\, dx \\ &\le \frac{2}{|S|} \sum_{i=1}^{2} \int_{\{x \in S : f_i(x) < (f_i)_S\}} ((f_i)_S - f_i(x))\, dx \\ &= \fint_S |f_1 - (f_1)_S| + \fint_S |f_2 - (f_2)_S|. \end{aligned}$$

For $f = \max(f_1, f_2)$, the previous arguments follow in a similar way, except that we apply Lemma 5.15 to write the mean oscillation in terms of an integral over the set $\{x \in S : f(x) > f_S\}$.

For $p = 2$, we include, for the benefit of the reader, the proof from [**55**], with cubes replaced by shapes. Let $f = \max(f_1, f_2)$. We may assume without loss of generality that $(f_1)_S \ge (f_2)_S$. Consider the sets

$$S_1 = \{x \in S : f_2(x) < f_1(x)\}, \quad S_2 = \{x \in S : f_2(x) \ge f_1(x) \text{ and } f_2(x) \ge (f_1)_S\}$$

and

$$S_3 = S \setminus (S_1 \cup S_2) = \{x \in S : f_1(x) \le f_2(x) < (f_1)_S\}.$$

Then

$$\begin{aligned} \int_S |f - (f_1)_S|^2 &= \int_{S_1} |f_1(x) - (f_1)_S|^2\, dx + \int_{S_2} |f_2(x) - (f_1)_S|^2\, dx + \int_{S_3} |f_2(x) - (f_1)_S|^2\, dx \\ &\le \int_{S_1} |f_1(x) - (f_1)_S|^2\, dx + \int_{S_2} |f_2(x) - (f_2)_S|^2\, dx + \int_{S_3} |f_1(x) - (f_1)_S|^2\, dx \\ &\le \int_S |f_1(x) - (f_1)_S|^2\, dx + \int_S |f_2(x) - (f_2)_S|^2\, dx \end{aligned}$$

Using Proposition 4.9 and dividing by $|S|$ gives (6.1). Similarly, the result can be shown for $\min(f_1, f_2)$. □

For a real-valued measurable function $f$ on $\Omega$, define its truncation from above at level $k \in \mathbb{R}$ by

$$f^k = \min(f, k)$$

and its truncation from below at level $j \in \mathbb{R}$ as

$$f_j = \max(f, j).$$

We use the preceding propositions to prove boundedness of the upper and lower truncations on $\mathrm{BMO}^p_{\mathscr{S}}(\Omega)$.

PROPOSITION 6.3. *Let $\mathscr{S}$ be any basis of shapes and fix a shape $S \in \mathscr{S}$. If $f \in \mathrm{BMO}^p_{\mathscr{S}}(\Omega)$ then for all $k, j \in \mathbb{R}$,*

$$\max\left( \fint_S |f^k(x) - (f^k)_S|^p\, dx, \fint_S |f_j(x) - (f_j)_S|^p\, dx \right) \le c^p \fint_S |f(x) - f_S|^p\, dx \tag{6.2}$$

*where*

$$c = \begin{cases} 1, & p = 1 \text{ or } p = 2 \\ \min\left(2, \frac{1+c_{|\cdot|}(p,\mathscr{S})}{2}\right), & \text{otherwise.} \end{cases}$$

*Consequently*

$$\|f^k\|_{\mathrm{BMO}^p_{\mathscr{S}}} \leq c\|f\|_{\mathrm{BMO}^p_{\mathscr{S}}} \text{ and } \|f_j\|_{\mathrm{BMO}^p_{\mathscr{S}}} \leq c\|f\|_{\mathrm{BMO}^p_{\mathscr{S}}}.$$

PROOF. For the truncation from above, observing that $|f^k(x) - f^k(y)| \leq |f(x) - f(y)|$ almost everywhere and for any $k$, Proposition 4.1 gives (6.2) with $c = 2$. On the other hand, applying Proposition 6.1 to $f^k = \min(f, k)$ and using the fact that constant functions have zero mean oscillation, we get the constant $c = \frac{1+c_{|\cdot|}(p,\mathscr{S})}{2}$. These calculations hold for any $p \geq 1$.

For $p = 1$ and $p = 2$, we are able to strengthen this by deriving the sharp shapewise inequality, namely (6.2) with $c = 1$, from (6.1) in Proposition 6.2 and the fact that constants have zero mean oscillation. Alternatively, for $p = 2$, fix a shape $S \in \mathscr{S}$ and assume, without loss of generality, that $f_S = 0$. Then, by Proposition 4.9,

$$\fint_S |f^{(k)} - (f^{(k)})_S|^2 \leq \fint_S |f^{(k)} - f_S|^2 = \fint_S |f^{(k)}|^2 \leq \fint_S |f|^2 = \fint_S |f - f_S|^2.$$

The calculations for the truncation from below are analogous or can be derived by writing $f_j = -\min(-f, -j)$. □

For $p = 1$ and $p = 2$, this is clearly a sharp result: take any bounded function and either $k > \sup f$ or $j < \inf f$.

Also note that for $p = 1$, the sharp shapewise inequalities (6.1) and (6.2) give the sharp shapewise inequality (4.1) (with constant 2) for the absolute value by writing

$$|f| = f_+ + f_- = \max(f_+, f_-), \quad f_+ = f_0, \ f_- = -(f^0).$$

For a measurable function $f$ on $\Omega$, define its (full) truncation at level $k$ as

$$\mathrm{Tr}(f,k)(x) = \begin{cases} k, & f(x) > k \\ f(x), & -k \leq f(x) \leq k \\ -k, & f(x) < -k. \end{cases}$$

Note that $\mathrm{Tr}(f,k) = (f^k)_{-k} = (f_{-k})^k$ and that $\mathrm{Tr}(f,k) \in L^\infty(\Omega)$ for each $k$. Moreover, $\mathrm{Tr}(f,k) \to f$ pointwise and in $L^1_{\mathrm{loc}}(\Omega)$ as $k \to \infty$ ([**39**]).

For a general function $f \in \mathrm{BMO}(\Omega)$, it is not true that $\mathrm{Tr}(f,k) \to f$ in $\mathrm{BMO}(\Omega)$, unless $\Omega$ is bounded and $f \in \mathrm{VMO}(\Omega)$, the space of functions of vanishing mean oscillation ([**6**]). Nonetheless, as shown in Corollary 6.5 below, $\|\mathrm{Tr}(f,k)\|_{\mathrm{BMO}} \to \|f\|_{\mathrm{BMO}}$ (as mentioned without proof in [**58, 59**]), and in fact for any basis $\mathscr{S}$.

PROBLEM 6.4. Does there exist a constant $c$ such that, for all $k$ and for all $f \in \mathrm{BMO}^p_{\mathscr{S}}(\Omega)$, $\|\mathrm{Tr}(f,k)\|_{\mathrm{BMO}^p_{\mathscr{S}}} \leq c\|f\|_{\mathrm{BMO}^p_{\mathscr{S}}}$, and if so, what is the smallest constant $c_T(p,\mathscr{S})$ for which this holds?

If $c_T(p,\mathscr{S})$ exists, then clearly $c_T(p,\mathscr{S}) \geq 1$. The results above provide a positive answer to the first question and some information about $c_T(p,\mathscr{S})$.

COROLLARY 6.5. *For any choice of shapes, if* $f \in \mathrm{BMO}^p_{\mathscr{S}}(\Omega)$ *then* $\mathrm{Tr}(f,k) \in \mathrm{BMO}^p_{\mathscr{S}}(\Omega)$ *for all* $k$ *and* $\|\mathrm{Tr}(f,k)\|_{\mathrm{BMO}^p_{\mathscr{S}}} \leq c\|f\|_{\mathrm{BMO}^p_{\mathscr{S}}}$*, where*

$$c = \begin{cases} 1, & p = 1 \text{ or } p = 2 \\ \min\left(2, \left(\frac{1+c_{|\cdot|}(p,\mathscr{S})}{2}\right)^2\right), & \text{otherwise} \end{cases}.$$

*Moreover, for* $p = 1$ *or* $p = 2$*,* $\|\mathrm{Tr}(f,k)\|_{\mathrm{BMO}^p_{\mathscr{S}}} \to \|f\|_{\mathrm{BMO}^p_{\mathscr{S}}}$ *as* $k \to \infty$*.*

PROOF. Writing $\mathrm{Tr}(f,k) = (f^k)_{-k}$ and applying inequality (6.2) in Proposition 6.3 gives us the shapewise inequality

$$\fint_S |\mathrm{Tr}(f,k) - (\mathrm{Tr}(f,k))_S|^p \leq c \fint_S |f - f_S|^p \tag{6.3}$$

with constant $c = 1$ for $p = 1$ and $p = 2$, and $c \leq \left(\min\left(2, \frac{1+c_{|\cdot|}(p,\mathscr{S})}{2}\right)\right)^2$. We get $c \leq 2$ (as opposed to $2^2$) by applying Proposition 4.1 directly with the estimate $|\mathrm{Tr}(f,k)(x) - \mathrm{Tr}(f,k)(y)| \leq |f(x) - f(y)|$.

For the convergence of the norms in the case $p = 1$ and $p = 2$, we have that $|\mathrm{Tr}(f,k)| \leq |f| \in L^1(S)$ implies $(\mathrm{Tr}(f,k))_S \to f_S$, and since $\mathrm{Tr}(f,k) \to f$ pointwise a.e. on $S$, we can apply Fatou's lemma and (6.3) with $c = 1$ to get

$$\begin{aligned} \fint_S |f - f_S|^p &\leq \liminf_{k\to\infty} \fint_S |\mathrm{Tr}(f,k) - (\mathrm{Tr}(f,k))_S|^p \\ &\leq \limsup_{k\to\infty} \fint_S |\mathrm{Tr}(f,k) - (\mathrm{Tr}(f,k))_S|^p \leq \fint_S |f - f_S|^p. \end{aligned}$$

□

This result gives the sharp constant for $p = 1, 2$ and an upper bound for $c_T(p,\mathscr{S})$. Of course, the known upper bound $c_{|\cdot|}(p,\mathscr{S}) \leq 2$ implies

$$\left(\frac{1 + c_{|\cdot|}(p,\mathscr{S})}{2}\right)^2 \leq \frac{9}{4}$$

(which appears, for example, in Exercise 3.1.4 in [**31**]), but this is worse than the truncation bound $c_T(p,\mathscr{S}) \leq 2$. On the other hand, if $c_{|\cdot|}(p,\mathscr{S}) \leq 2\sqrt{2} - 1$, then the bound depends on $c_{|\cdot|}(p,\mathscr{S})$. In particular, a result of $c_{|\cdot|}(p,\mathscr{S}) = 1$ would imply that $c_T(p,\mathscr{S}) = 1$.

## 7. The John-Nirenberg inequality

We now come to the most important inequality in the theory of BMO, originating in the paper of John and Nirenberg [**37**].

DEFINITION 7.1. Let $X \subset \mathbb{R}^n$ be a set of finite Lebesgue measure. We say that $f \in L^1(X)$ satisfies the John-Nirenberg inequality on $X$ if there exist constants $c_1, c_2 > 0$ such that

$$|\{x \in X : |f(x) - f_X| > \alpha\}| \leq c_1|X|e^{-c_2\alpha}, \qquad \alpha > 0. \tag{7.1}$$

The following is sometimes referred to as the John-Nirenberg Lemma.

THEOREM 7.2 ([**37**]). *If* $X = Q$*, a cube in* $\mathbb{R}^n$*, then there exist constants* $c$ *and* $C$ *such that for all* $f \in \mathrm{BMO}(Q)$*, (7.1) holds with* $c_1 = c$*,* $c_2 = C/\|f\|_{\mathrm{BMO}}$*.*

More generally, given a basis of shapes $\mathscr{S}$ on a domain $\Omega \subset \mathbb{R}^n$, $|\Omega| < \infty$, one can pose the following problem.

PROBLEM 7.3. Does $f \in \mathrm{BMO}^p_{\mathscr{S}}(\Omega)$ imply that $f$ satisfies the John-Nirenberg inequality on $\Omega$? That is, do there exist constants $c, C > 0$ such that

$$|\{x \in \Omega : |f(x) - f_\Omega| > \alpha\}| \leq c|\Omega| \exp\left(-\frac{C}{\|f\|_{\mathrm{BMO}^p_{\mathscr{S}}}}\alpha\right), \qquad \alpha > 0$$

holds for all $f \in \mathrm{BMO}^p_{\mathscr{S}}(\Omega)$? If so, what is the smallest constant $c_{\Omega,\mathrm{JN}}(p, \mathscr{S})$ and the largest constant $C_{\Omega,\mathrm{JN}}(p, \mathscr{S})$ for which this inequality holds?

When $n = 1$, $\Omega = I$, a finite interval, and $\mathscr{S} = \mathcal{I}$, the positive answer is a special case of Theorem 7.2. Sharp constants are known for the cases $p = 1$ and $p = 2$. For $p = 1$, it is shown in [**48**] that $c_{I,\mathrm{JN}}(1, \mathcal{I}) = \frac{1}{2}e^{4/e}$ and in [**41**] that $C_{I,\mathrm{JN}}(1, \mathcal{I}) = 2/e$. For $p = 2$, Bellman function techniques are used in [**62**] to give $c_{I,\mathrm{JN}}(2, \mathcal{I}) = 4/e^2$ and $C_{I,\mathrm{JN}}(2, \mathcal{I}) = 1$.

When $n \geq 2$, $\Omega = R$, a rectangle, and $\mathscr{S} = \mathcal{R}$, a positive answer is provided by a less well-known result due to Korenovskii in [**42**], where he also shows the sharp constant $C_{R,\mathrm{JN}}(1, \mathcal{R}) = 2/e$.

Dimension-free bounds on these constants are also of interest. In [**18**], Cwikel, Sagher, and Shvartsman conjecture a geometric condition on cubes and prove dimension-free bounds for $c_{\Omega,\mathrm{JN}}(1, \mathcal{Q})$ and $C_{\Omega,\mathrm{JN}}(1, \mathcal{Q})$ conditional on this hypothesis being true.

Rather than just looking at $\Omega$, we can also consider whether the John-Nirenberg inequality holds for all shapes $S$.

DEFINITION 7.4. We say that a function $f \in L^1_{\mathrm{loc}}(\Omega)$ has the John-Nirenberg property with respect to a basis $\mathscr{S}$ of shapes on $\Omega$ if there exist constants $c_1, c_2 > 0$ such that for all $S \in \mathscr{S}$, (7.1) holds for $X = S$.

We can now formulate a modified problem.

PROBLEM 7.5. For which bases $\mathscr{S}$ and $p \in [1, \infty)$ does $f \in \mathrm{BMO}^p_{\mathscr{S}}(\Omega)$ imply that $f$ has the John-Nirenberg property with respect to $\mathscr{S}$? If this is the case, what is the smallest constant $c = c_{\mathrm{JN}}(p, \mathscr{S})$ and the largest constant $C = C_{\mathrm{JN}}(p, \mathscr{S})$ for which (7.1) holds for all $f \in \mathrm{BMO}_{\mathscr{S}}(\Omega)$ and $S \in \mathscr{S}$ with $c_1 = c$, $c_2 = C/\|f\|_{\mathrm{BMO}^p_{\mathscr{S}}}$?

Since Theorem 7.2 holds for any cube $Q$ in $\mathbb{R}^n$ with constants independent of $Q$, it follows that for a domain $\Omega \subset \mathbb{R}^n$, any $f \in \mathrm{BMO}(\Omega)$ has the John-Nirenberg property with respect to $\mathcal{Q}$, and equivalently $\mathcal{B}$. Similarly, every $f \in \mathrm{BMO}_{\mathcal{R}}(\Omega)$ has the John-Nirenberg property with respect to $\mathcal{R}$.

In the negative direction, $f \in \mathrm{BMO}^p_{\mathcal{C}}(\mathbb{R}^n)$ does not necessarily have the John-Nirenberg property with respect to $\mathcal{C}$ ([**45, 50**]). This space, known in the literature as CMO for central mean oscillation or CBMO for central bounded mean oscillation, was originally defined with the additional constraint that the balls have radius at least 1 ([**11, 26**]).

We now state the converse to Theorem 7.2, namely that the John-Nirenberg property is sufficient for BMO, in more generality.

THEOREM 7.6. *If $f \in L^1_{loc}(\Omega)$ and $f$ has the John-Nirenberg property with respect to $\mathscr{S}$, then $f \in \mathrm{BMO}^p_{\mathscr{S}}(\Omega)$ for all $p \in [1, \infty)$, with*

$$\|f\|_{\mathrm{BMO}^p_{\mathscr{S}}} \leq \frac{(c_1 p \Gamma(p))^{1/p}}{c_2}.$$

PROOF. Take $S \in \mathscr{S}$. By Cavalieri's principle and (7.1),

$$\begin{aligned}\fint_S |f - f_S|^p &= \frac{p}{|S|}\int_0^\infty \alpha^{p-1}|\{x \in S : |f(x) - f_S| > \alpha\}|\, d\alpha \\ &\leq pc_1 \int_0^\infty \alpha^{p-1}\exp(-c_2\alpha)\, d\alpha \\ &= \frac{c_1 p\Gamma(p)}{c_2{}^p},\end{aligned}$$

from where the result follows. □

By Lemma 3.3, this theorem shows that every $f$ with the John-Nirenberg property is in $L^p_{\mathrm{loc}}(\Omega)$.

A consequence of the John-Nirenberg Lemma, Theorem 7.2, is that

$$\mathrm{BMO}^{p_1}(\mathbb{R}^n) \cong \mathrm{BMO}^{p_2}(\mathbb{R}^n)$$

for all $1 \leq p_1, p_2 < \infty$. This can be stated in more generality as a corollary of the preceding theorem.

COROLLARY 7.7. *If there exists $p_0 \in [1,\infty)$ such that every $f \in \mathrm{BMO}^{p_0}_{\mathscr{S}}(\Omega)$ has the John-Nirenberg property with respect to $\mathscr{S}$, then*

$$\mathrm{BMO}^{p_1}_{\mathscr{S}}(\Omega) \cong \mathrm{BMO}^{p_2}_{\mathscr{S}}(\Omega), \quad p_0 \leq p_1, p_2 < \infty.$$

PROOF. The hypothesis means that there are constants $c_{\mathrm{JN}}(p_0,\mathscr{S})$, $C_{\mathrm{JN}}(p_0,\mathscr{S})$ such that if $f \in \mathrm{BMO}^{p_0}_{\mathscr{S}}(\Omega)$ then $f$ satisfies (7.1) for all $S \in \mathscr{S}$, with $c_1 = c_{\mathrm{JN}}(p_0,\mathscr{S})$, $c_2 = C_{\mathrm{JN}}(p_0,\mathscr{S})/\|f\|_{\mathrm{BMO}^{p_0}_{\mathscr{S}}}$.

From the preceding theorem, this implies that $\mathrm{BMO}^{p_0}_{\mathscr{S}}(\Omega) \subset \mathrm{BMO}^{p}_{\mathscr{S}}(\Omega)$ for all $p \in [1,\infty)$, with

$$\|f\|_{\mathrm{BMO}^p_{\mathscr{S}}} \leq \frac{(c_{\mathrm{JN}}(p_0,\mathscr{S})p\Gamma(p))^{1/p}}{C_{\mathrm{JN}}(p_0,\mathscr{S})}\|f\|_{\mathrm{BMO}^{p_0}_{\mathscr{S}}}.$$

Conversely, Proposition 3.2 gives us that $\mathrm{BMO}^{p}_{\mathscr{S}}(\Omega) \subset \mathrm{BMO}^{p_0}_{\mathscr{S}}(\Omega)$ whenever $p_0 \leq p < \infty$, with $\|f\|_{\mathrm{BMO}^{p_0}_{\mathscr{S}}} \leq \|f\|_{\mathrm{BMO}^{p}_{\mathscr{S}}}$.

Thus all the spaces $\mathrm{BMO}^{p}_{\mathscr{S}}(\Omega)$, $p_0 \leq p < \infty$, are congruent to $\mathrm{BMO}^{p_0}_{\mathscr{S}}(\Omega)$. □

The John-Nirenberg Lemma gives the hypothesis of Corollary 7.7 for the bases $\mathcal{Q}$ and $\mathcal{B}$ on $\mathbb{R}^n$ with $p_0 = 1$. As pointed out, by results of [**42**] this also applies to the basis $\mathcal{R}$, showing that $\mathrm{BMO}^{p_1}_{\mathcal{R}}(\Omega) \cong \mathrm{BMO}^{p_2}_{\mathcal{R}}(\Omega)$ for all $1 \leq p_1, p_2 < \infty$.

PROBLEM 7.8. If the hypothesis of Corollary 7.7 is satisfied with $p_0 = 1$, what is the smallest constant $c(p,\mathscr{S})$ such that $\|f\|_{\mathrm{BMO}^p_{\mathscr{S}}} \leq c(p,\mathscr{S})\|f\|_{\mathrm{BMO}_{\mathscr{S}}}$ holds for all $f \in \mathrm{BMO}^p_{\mathscr{S}}(\Omega)$?

The proof of Corollary 7.7 gives a well-known upper bound on $c(p,\mathscr{S})$:

$$c(p,\mathscr{S}) \leq \frac{\left(c_{\mathrm{JN}}(1,\mathscr{S})p\Gamma(p)\right)^{1/p}}{C_{\mathrm{JN}}(1,\mathscr{S})}.$$

## 8. Product decomposition

In this section, we assume that the domain $\Omega$ can be decomposed as

$$\Omega = \Omega_1 \times \Omega_2 \times \cdots \times \Omega_k \tag{8.1}$$

for $2 \leq k \leq n$, where each $\Omega_i$ is a domain in $\mathbb{R}^{m_i}$ for $1 \leq m_i \leq n-1$, having its own basis of shapes $\mathscr{S}_i$.

We will require some compatibility between the basis $\mathscr{S}$ on all of $\Omega$ and these individual bases.

DEFINITION 8.1. We say that $\mathscr{S}$ satisfies the weak decomposition property with respect to $\{\mathscr{S}_i\}_{i=1}^k$ if for every $S \in \mathscr{S}$ there exist $S_i \in \mathscr{S}_i$ for each $i = 1, \ldots, k$ such that $S = S_1 \times S_2 \times \ldots S_k$. If, in addition, for every $\{S_i\}_{i=1}^k$, $S_i \in \mathscr{S}_i$, the set $S_1 \times S_2 \times \ldots S_k \in \mathscr{S}$, then we say that the basis $\mathscr{S}$ satisfies the strong decomposition property with respect to $\{\mathscr{S}_i\}_{i=1}^k$.

Using $\mathcal{R}_i$ to denote the basis of rectangles in $\Omega_i$ (interpreted as $\mathcal{I}_i$ if $m_i = 1$), note that the basis $\mathcal{R}$ on $\Omega$ satisfies the strong decomposition property with respect to $\{\mathcal{R}_i\}_{i=1}^k$ and for any $k$. Meanwhile, the basis $\mathcal{Q}$ on $\Omega$ satisfies the weak decomposition property with respect to $\{\mathcal{R}_i\}_{i=1}^k$ for any $k$, but not the strong decomposition property.

We now turn to the study of the spaces $\mathrm{BMO}^p_{\mathscr{S}_i}(\Omega)$, first defined using different notation in the context of the bidisc $\mathbb{T} \times \mathbb{T}$ in [**16**]. For simplicity, we only define $\mathrm{BMO}^p_{\mathscr{S}_1}(\Omega)$, as the other $\mathrm{BMO}^p_{\mathscr{S}_i}(\Omega)$ for $i = 2, \ldots, k$ are defined analogously. We write a point in $\Omega$ as $(x, y)$, where $x \in \Omega_1$ and $y \in \widetilde{\Omega} = \Omega_2 \times \cdots \times \Omega_k$. Writing $f_y(x) = f(x, y)$, functions in $\mathrm{BMO}^p_{\mathscr{S}_1}(\Omega)$ are those for which $f_y$ is in $\mathrm{BMO}^p_{\mathscr{S}_1}(\Omega_1)$, uniformly in $y$:

DEFINITION 8.2. A function $f \in L^1_{\mathrm{loc}}(\Omega)$ is said to be in $\mathrm{BMO}^p_{\mathscr{S}_1}(\Omega)$ if

$$\|f\|_{\mathrm{BMO}^p_{\mathscr{S}_1}(\Omega)} = \sup_{y \in \tilde{\Omega}} \|f_y\|_{\mathrm{BMO}^p_{\mathscr{S}_1}(\Omega_1)} < \infty,$$

where $f_y(x) = f(x, y)$.

Although this norm combines a supremum with a BMO norm, we are justified in calling $\mathrm{BMO}^p_{\mathscr{S}_1}(\Omega)$ a BMO space as it inherits many properties from $\mathrm{BMO}^p_{\mathscr{S}}(\Omega)$. In particular, for each $i = 1, 2, \ldots, k$, $L^\infty(\Omega) \subset \mathrm{BMO}^p_{\mathscr{S}_i}(\Omega)$ and $\mathrm{BMO}^{p_1}_{\mathscr{S}_i}(\Omega) \cong \mathrm{BMO}^{p_2}_{\mathscr{S}_i}(\Omega)$ for all $1 \leq p_1, p_2 < \infty$ if the hypothesis of Corollary 7.7 is satisfied with $p_0 = 1$ for $\mathscr{S}_i$ on $\Omega_i$. Moreover, under certain conditions, the spaces $\mathrm{BMO}^p_{\mathscr{S}_i}(\Omega)$ can be quite directly related to $\mathrm{BMO}^p_{\mathscr{S}}(\Omega)$, revealing the product nature of $\mathrm{BMO}^p_{\mathscr{S}}(\Omega)$. This depends on the decomposition property of the basis $\mathscr{S}$, as well as some differentiation properties of the $\mathscr{S}_i$.

Before stating the theorem, we briefly recall the main definitions related to the theory of differentiation of the integral; see the survey [**32**] for a standard reference. For a basis of shapes $\mathscr{S}$, denote by $\mathscr{S}(x)$ the subcollection of shapes that contain $x \in \Omega$. We say that $\mathscr{S}$ is a differentiation basis if for each $x \in \Omega$ there exists a sequence of shapes $\{S_k\} \subset \mathscr{S}(x)$ such that $\delta(S_k) \to 0$ as $k \to \infty$. Here, $\delta(\cdot)$ is the Euclidean diameter. For $f \in L^1_{\mathrm{loc}}(\Omega)$, we define the upper derivative of $\int f$ with respect to $\mathscr{S}$ at $x \in \Omega$ by

$$\overline{D}\Big(\int f, x\Big) = \sup\left\{\limsup_{k\to\infty} \fint_{S_k} f : \{S_k\} \subset \mathscr{S}(x),\ \delta(S_k) \to 0 \text{ as } k \to \infty\right\}$$

and the lower derivative of $\int f$ with respect to $\mathscr{S}$ at $x \in \Omega$ by

$$\underline{D}(\int f, x) = \inf \left\{ \liminf_{k\to\infty} \fint_{S_k} f : \{S_k\} \subset \mathscr{S}(x),\ \delta(S_k) \to 0 \text{ as } k \to \infty \right\}.$$

We say, then, that a differentiation basis $\mathscr{S}$ differentiates $L^1_{\text{loc}}(\Omega)$ if for every $f \in L^1_{\text{loc}}(\Omega)$ and for almost every $x \in \Omega$, $\overline{D}(\int f, x) = \underline{D}(\int f, x) = f(x)$. The classical Lebesgue differentiation theorem is a statement that the basis $\mathcal{B}$ (equivalently, $\mathcal{Q}$) differentiates $L^1_{\text{loc}}(\Omega)$. It is known, however, that the basis $\mathcal{R}$ does not differentiate $L^1_{\text{loc}}(\Omega)$, but does differentiate the Orlicz space $L(\log L)^{n-1}(\Omega)$ ([**35**]).

THEOREM 8.3. *Let $\Omega$ be a domain satisfying* (8.1), *$\mathscr{S}$ be a basis of shapes for $\Omega$ and $\mathscr{S}_i$ be a basis of shapes for $\Omega_i$, $1 \le i \le k$.*

a) *Let $f \in \bigcap_{i=1}^k \mathrm{BMO}^p_{\mathscr{S}_i}(\Omega)$. If $\mathscr{S}$ satisfies the weak decomposition property with respect to $\{\mathscr{S}_i\}_{i=1}^k$, then $f \in \mathrm{BMO}^p_{\mathscr{S}}(\Omega)$ with*

$$\|f\|_{\mathrm{BMO}^p_{\mathscr{S}}(\Omega)} \le \sum_{i=1}^k \|f\|_{\mathrm{BMO}^p_{\mathscr{S}_i}(\Omega)}.$$

b) *Let $f \in \mathrm{BMO}^p_{\mathscr{S}}(\Omega)$. If $\mathscr{S}$ satisfies the strong decomposition property with respect to $\{\mathscr{S}_i\}_{i=1}^k$ and each $\mathscr{S}_i$ contains a differentiation basis that differentiates $L^1_{loc}(\Omega_i)$, then $f \in \bigcap_{i=1}^k \mathrm{BMO}^p_{\mathscr{S}_i}(\Omega)$ with*

$$\max_{i=1,\dots,k} \{\|f\|_{\mathrm{BMO}^p_{\mathscr{S}_i}(\Omega)}\} \le 2^{k-1} \|f\|_{\mathrm{BMO}^p_{\mathscr{S}}(\Omega)}.$$

*When $p = 2$, the constant $2^{k-1}$ can be replaced by 1.*

This theorem was first pointed out in [**16**] in the case of $\mathcal{R}$ in $\mathbb{T} \times \mathbb{T}$. Here, we prove it in the setting of more general shapes and domains, clarifying the role played by the theory of differentiation and keeping track of constants.

PROOF. We first present the proof in the case of $k = 2$. We write $\Omega = X \times Y$, denoting by $(x, y)$ an element in $\Omega$ with $x \in X$ and $y \in Y$. The notations $\mathscr{S}_X$ and $\mathscr{S}_Y$ will be used for the basis in $X$ and $Y$, respectively. Similarly, the measures $dx$ and $dy$ will be used for Lebesgue measure on $X$ and $Y$, respectively, while $dA$ will be used for the Lebesgue measure on $\Omega$.

To prove (a), assume that $\mathscr{S}$ satisfies the weak decomposition property with respect to $\{\mathscr{S}_X, \mathscr{S}_Y\}$ and let $f \in \mathrm{BMO}^p_{\mathscr{S}_X}(\Omega) \cap \mathrm{BMO}^p_{\mathscr{S}_Y}(\Omega)$. Fixing a shape $R \in \mathscr{S}$, write $R = S \times T$ for $S \in \mathscr{S}_X$ and $T \in \mathscr{S}_Y$. Then, by Minkowski's inequality,

$$\left(\fint_R |f(x,y) - f_R|^p dA\right)^{\frac{1}{p}} \le \left(\fint_R |f(x,y) - (f_y)_S|^p dA\right)^{\frac{1}{p}} + \left(\fint_T |(f_y)_S - f_R|^p dy\right)^{\frac{1}{p}}.$$

For the first integral, we estimate

$$\fint_R |f(x,y) - (f_y)_S|^p \, dA \le \fint_T \|f_y\|^p_{\mathrm{BMO}^p_{\mathscr{S}_X}(X)} \, dy \le \|f\|^p_{\mathrm{BMO}^p_{\mathscr{S}_X}(\Omega)}. \tag{8.2}$$

For the second integral, Jensen's inequality gives

$$\begin{aligned}\left(\fint_T |(f_y)_S - f_R|^p\,dy\right)^{1/p} &= \left(\fint_T \left|\fint_S f_y(x)\,dx - \fint_S \fint_T f(x,y)\,dy\,dx\right|^p dy\right)^{1/p}\\ &= \left(\fint_T \left|\fint_S (f(x,y) - (f_x)_T)\,dx\right|^p dy\right)^{1/p}\\ &\le \left(\fint_R |f(x,y) - (f_x)_T|^p\,dA\right)^{1/p}\\ &\le \|f\|_{\mathrm{BMO}^p_{\mathscr{S}_Y}(\Omega)},\end{aligned}$$

where the last inequality follows in the same way as (8.2). Therefore, we may conclude that $f \in \mathrm{BMO}^p_{\mathscr{S}}(\Omega)$ with $\|f\|_{\mathrm{BMO}^p_{\mathscr{S}}(\Omega)} \le \|f\|_{\mathrm{BMO}^p_{\mathscr{S}_X}(\Omega)} + \|f\|_{\mathrm{BMO}^p_{\mathscr{S}_Y}(\Omega)}$.

We now come to the proof of (b). To simplify the notation, we use $\mathcal{O}_p(f,S)$ for the $p$-mean oscillation of the function $f$ on the shape $S$, i.e.

$$\mathcal{O}_p(f,S) := \fint_S |f - f_S|^p.$$

Assume that $\mathscr{S}$ satisfies the strong decomposition property with respect to $\{\mathscr{S}_X, \mathscr{S}_Y\}$, and that $\mathscr{S}_X$ and $\mathscr{S}_Y$ each contain a differentiation basis that differentiates $L^1_{\mathrm{loc}}(X)$ and $L^1_{\mathrm{loc}}(Y)$, respectively.

Let $f \in \mathrm{BMO}^p_{\mathscr{S}}(\Omega)$. Fix a shape $S_0 \in \mathscr{S}_X$ and consider the $p$-mean oscillation of $f_y$ on $S_0$, $\mathcal{O}_p(f_y, S_0)$, as a function of $y$. For any $T \in \mathscr{S}_Y$, writing $R = S_0 \times T \in \mathscr{S}$, we have that $R \in \mathscr{S}$ by the strong decomposition property of $\mathscr{S}$, so $f \in \mathrm{BMO}^p_{\mathscr{S}}(\Omega)$ implies $f \in L^1(R)$ and therefore

$$\int_T \mathcal{O}_p(f_y, S_0)\,dy = \frac{1}{|S_0|}\int_T\int_{S_0} |f_y(x) - (f_y)_{S_0}|^p\,dx\,dy \le \frac{2^p}{|S_0|}\int_R |f(x,y)|^p\,dx\,dy < \infty.$$

By Lemma 3.3, this is enough to guarantee that $\mathcal{O}_p(f_y, S_0) \in L^1_{\mathrm{loc}}(Y)$.

Let $\varepsilon > 0$. Since $\mathscr{S}_Y$ contains a differentiation basis, for almost every $y_0 \in Y$ there exists a shape $T_0 \in \mathscr{S}_Y$ containing $y_0$ such that

$$\left|\fint_{T_0} \mathcal{O}_p(f_y, S_0)\,dy - \mathcal{O}_p(f_{y_0}, S_0)\right| < \varepsilon. \tag{8.3}$$

Fix such an $x_0$ and a $T_0$ and let $R_0 = S_0 \times T_0$. Then the strong decomposition property implies that $R_0 \in \mathscr{S}$, and by Proposition 4.5 applied to the mean oscillation of $f_y$ on $S_0$, we have

$$\begin{aligned}\fint_{T_0} \mathcal{O}_p(f_y, S_0)\,dy &= \fint_{T_0}\fint_{S_0} |f_y(x) - (f_y)_{S_0}|^p\,dx\,dy\\ &\le 2^p \fint_{T_0}\fint_{S_0} |f(x,y) - f_{R_0}|^p\,dx\,dy\\ &= 2^p \fint_{R_0} |f(x,y) - f_{R_0}|^p\,dA\\ &\le 2^p \|f\|^p_{\mathrm{BMO}^p_{\mathscr{S}}(\Omega)}.\end{aligned}$$

Note that when $p = 2$, Proposition 4.9 implies that the factor of $2^p$ can be dropped.

Combining this with (8.3), it follows that

$$\mathcal{O}_p(f_{y_0}, S_0) < \varepsilon + 2^p \|f\|^p_{\mathrm{BMO}^p_{\mathscr{S}}(\Omega)}.$$

Taking $\varepsilon \to 0^+$, since $S_0$ is arbitrary, this implies that $f_{y_0} \in \mathrm{BMO}^p_{\mathscr{S}_X}(X)$ with

$$\|f_{y_0}\|_{\mathrm{BMO}^p_{\mathscr{S}_X}(X)} \leq 2\|f\|_{\mathrm{BMO}^p_{\mathscr{S}}(\Omega)}. \tag{8.4}$$

The fact that this is true for almost every $y_0 \in Y$ implies that $\|f\|_{\mathrm{BMO}^p_{\mathscr{S}_X}(\Omega)} \leq 2\|f\|_{\mathrm{BMO}^p_{\mathscr{S}}(\Omega)}$.

Similarly, one can show that $\|f\|_{\mathrm{BMO}^p_{\mathscr{S}_Y}(\Omega)} \leq 2\|f\|_{\mathrm{BMO}^p_{\mathscr{S}}(\Omega)}$. Thus we have shown $f \in \mathrm{BMO}^p_{\mathscr{S}_X}(\Omega) \cap \mathrm{BMO}^p_{\mathscr{S}_Y}(\Omega)$ with

$$\max\{\|f\|_{\mathrm{BMO}^p_{\mathscr{S}_X}(\Omega)}, \|f\|_{\mathrm{BMO}^p_{\mathscr{S}_Y}(\Omega)}\} \leq 2\|f\|_{\mathrm{BMO}^p_{\mathscr{S}}(\Omega)}.$$

Again, when $p = 2$ the factor of 2 disappears.

For $k > 2$, let us assume the result holds for $k-1$. Write $X = \Omega_1 \times \Omega_2 \times \ldots \times \Omega_{k-1}$ and $Y = \Omega_k$. Set $\mathscr{S}_Y = \mathscr{S}_k$. By the weak decomposition property of $\mathscr{S}$, we can define the projection of the basis $\mathscr{S}$ onto $X$, namely

$$\mathscr{S}_X = \{S_1 \times S_2 \times \ldots \times S_{k-1} : S_i \in \mathscr{S}_i, \exists S_k \in \mathscr{S}_k, \prod_{i=1}^{k} S_i \in \mathscr{S}\}, \tag{8.5}$$

and this is a basis of shapes on $X$ which by definition also has the weak decomposition property. Moreover, $\mathscr{S}$ has the weak decomposition property with respect to $\mathscr{S}_X$ and $\mathscr{S}_Y$.

To prove part (a) for $k$ factors, we first apply the result of part (a) proved above for $k = 2$, followed by the definitions and part (a) applied again to $X$, since we are assuming it is valid with $k-1$ factors. This gives us the inclusion $\bigcap_{i=1}^k \mathrm{BMO}^p_{\mathscr{S}_i}(\Omega) \subset \mathrm{BMO}^p_{\mathscr{S}}(\Omega)$ with the following estimates on the norms (we use the notation $\widehat{x_i}$ for the $k-2$ tuple of variables obtained from $(x_1, \ldots, x_{k-1})$ by removing $x_i$):

$$\begin{aligned}
\|f\|_{\mathrm{BMO}^p_{\mathscr{S}}(\Omega)} &\leq \|f\|_{\mathrm{BMO}^p_{\mathscr{S}_X}(\Omega)} + \|f\|_{\mathrm{BMO}^p_{\mathscr{S}_Y}(\Omega)} \\
&= \sup_{y\in Y} \|f_y\|_{\mathrm{BMO}^p_{\mathscr{S}_X}(X)} + \|f\|_{\mathrm{BMO}^p_{\mathscr{S}_Y}(\Omega)} \\
&\leq \sup_{y\in Y} \sum_{i=1}^{k-1} \|f_y\|_{\mathrm{BMO}^p_{\mathscr{S}_i}(X)} + \|f\|_{\mathrm{BMO}^p_{\mathscr{S}_Y}(\Omega)} \\
&= \sup_{y\in Y} \sum_{i=1}^{k-1} \sup_{\widehat{x_i}} \|(f_y)_{\widehat{x_i}}\|_{\mathrm{BMO}^p_{\mathscr{S}_i}(\Omega_i)} + \|f\|_{\mathrm{BMO}^p_{\mathscr{S}_Y}(\Omega)} \\
&\leq \sum_{i=1}^{k-1} \sup_{(\widehat{x_i}, y)} \|f_{(\widehat{x_i}, y)}\|_{\mathrm{BMO}^p_{\mathscr{S}_i}(\Omega_i)} + \|f\|_{\mathrm{BMO}^p_{\mathscr{S}_Y}(\Omega)} \\
&= \sum_{i=1}^{k-1} \|f\|_{\mathrm{BMO}^p_{\mathscr{S}_i}(\Omega)} + \|f\|_{\mathrm{BMO}^p_{\mathscr{S}_k}(\Omega)} \\
&= \sum_{i=1}^{k} \|f\|_{\mathrm{BMO}^p_{\mathscr{S}_i}(\Omega)}.
\end{aligned}$$

To prove part (b) for $k > 2$, we have to be more careful. First note that if $\mathscr{S}$ has the strong decomposition property, then so does $\mathscr{S}_X$ defined by (8.5). We repeat the first part of the proof of (b) for the case $k = 2$ above, with $X = \Omega_1 \times \Omega_2 \times \ldots \times \Omega_{k-1}$ and $Y = \Omega_k$, leading up to the estimate (8.4) for the function $f_{y_0}$ for some $y_0 \in Y$.

Note that in this part we only used the differentiation properties of $Y$, which hold by hypothesis in this case since $Y = \Omega_k$. Now we repeat the process for the function $f_{y_0}$ instead of $f$, with $X_1 = \Omega_1 \times \Omega_2 \times \ldots \times \Omega_{k-2}$ and $Y_1 = \Omega_{k-1}$. This gives

$$\|(f_{y_0})_{y_1}\|_{\mathrm{BMO}^p_{\mathscr{S}_{X_1}}(X_1)} \leq 2\|f_{y_0}\|_{\mathrm{BMO}^p_{\mathscr{S}}(X)} \leq 4\|f\|_{\mathrm{BMO}^p_{\mathscr{S}}(\Omega)} \quad \forall y_1 \in \Omega_{k-1}, y_0 \in \Omega_k.$$

We continue until we get to $X_{k-1} = \Omega_1$, for which $\mathscr{S}_{X_k} = \mathscr{S}_1$, yielding the estimate

$$\|f_{(y_{k-2},\ldots,y_0)}\|_{\mathrm{BMO}^p_{\mathscr{S}_1}(\Omega_1)} \leq \cdots \leq 2^{k-2}\|f_{y_0}\|_{\mathrm{BMO}^p_{\mathscr{S}}(X)} \leq 2^{k-1}\|f\|_{\mathrm{BMO}^p_{\mathscr{S}}(\Omega)}$$

for all $k-1$-tuples $y = (y_{k-2}, \ldots, y_0) \in \widetilde{\Omega} = \Omega_2 \times \ldots \Omega_k$. Taking the supremum over all such $y$, we have, by Definition 8.2, that $f \in \mathrm{BMO}^p_{\mathscr{S}_1}(\Omega)$ with

$$\|f\|_{\mathrm{BMO}^p_{\mathscr{S}_1}(\Omega)} = \sup_{y\in\widetilde{\Omega}}\|f_y\|_{\mathrm{BMO}^p_{\mathscr{S}_1}(\Omega_1)} \leq 2^{k-1}\|f\|_{\mathrm{BMO}^p_{\mathscr{S}}(\Omega)}.$$

A similar process for $i = 2, \ldots k$ shows that $f \in \mathrm{BMO}^p_{\mathscr{S}_i}(\Omega)$ with $\|f\|_{\mathrm{BMO}^p_{\mathscr{S}_i}(\Omega)} \leq 2^{k-1}\|f\|_{\mathrm{BMO}^p_{\mathscr{S}}(\Omega)}$. As the factor of 2 appears in the proof for $k = 2$ only when $p \neq 2$, the same will happen here. □

Since $\mathcal{Q}$ satisfies the weak decomposition property, the claim of part (a) holds for BMO, a fact pointed out in [**58**] without proof. Also, it is notable that there was no differentiation assumption required for this direction.

In the proof of part (b), differentiation is key and the strong decomposition property of the basis cannot be eliminated as there would be no guarantee that arbitrary $S$ and $T$ would yield a shape $R$ in $\Omega$. In fact, if the claim were true for bases with merely the weak decomposition property, this would imply that BMO and $\mathrm{BMO}_{\mathcal{R}}$ are congruent, which is not true (see Example 4.13).

## Acknowledgments

The authors would like to thank the editors, Donatella Danielli and Irina Mitrea, as well as the anonymous referee. They would also like to thank Almut Burchard for discussions that initiated much of the investigation that lead to this paper, as well as Andrew Lavigne and Hong Yue for their useful comments.

## References

[1] D. Aalto, L. Berkovits, O. E. Kansanen, and H. Yue, *John-Nirenberg lemmas for a doubling measure*, Studia Math. **204** (2011), no. 1, 21–37, DOI 10.4064/sm204-1-2. MR2794938

[2] C. Bennett, R. A. DeVore, and R. Sharpley, *Weak-$L^\infty$ and BMO*, Ann. of Math. (2) **113** (1981), no. 3, 601–611, DOI 10.2307/2006999. MR621018

[3] C. Bennett and R. Sharpley, *Interpolation of operators*, Pure and Applied Mathematics, vol. 129, Academic Press, Inc., Boston, MA, 1988. MR928802

[4] O. Beznosova and A. Reznikov, *Dimension-free properties of strong Muckenhoupt and reverse Hölder weights for radon measures*, J. Geom. Anal. **29** (2019), no. 2, 1109–1115, DOI 10.1007/s12220-018-0028-0. MR3935251

[5] H. Brezis and L. Nirenberg, *Degree theory and BMO. I. Compact manifolds without boundaries*, Selecta Math. (N.S.) **1** (1995), no. 2, 197–263, DOI 10.1007/BF01671566. MR1354598

[6] H. Brezis and L. Nirenberg, *Degree theory and BMO. II. Compact manifolds with boundaries*, Selecta Math. (N.S.) **2** (1996), no. 3, 309–368, DOI 10.1007/BF01587948. With an appendix by the authors and Petru Mironescu. MR1422201

[7] A. Burchard. A Short Course on Rearrangement Inequalities. Lecture notes. http://www.math. utoronto.ca/almut/rearrange.pdf, 2009.

[8] A. Butaev, *On some refinements of the embedding of critical Sobolev spaces into BMO*, Pacific J. Math. **298** (2019), no. 1, 1–26, DOI 10.2140/pjm.2019.298.1. MR3910046

[9] S.-Y. A. Chang and R. Fefferman, *A continuous version of duality of $H^1$ with BMO on the bidisc*, Ann. of Math. (2) **112** (1980), no. 1, 179–201, DOI 10.2307/1971324. MR584078

[10] S.-Y. A. Chang and R. Fefferman, *Some recent developments in Fourier analysis and $H^p$-theory on product domains*, Bull. Amer. Math. Soc. (N.S.) **12** (1985), no. 1, 1–43, DOI 10.1090/S0273-0979-1985-15291-7. MR766959

[11] Y. Z. Chen and K.-S. Lau, *Some new classes of Hardy spaces*, J. Funct. Anal. **84** (1989), no. 2, 255–278, DOI 10.1016/0022-1236(89)90097-9. MR1001460

[12] A. Cianchi and L. Pick, *Sobolev embeddings into BMO, VMO, and $L_\infty$*, Ark. Mat. **36** (1998), no. 2, 317–340, DOI 10.1007/BF02384772. MR1650446

[13] R. R. Coifman and G. Weiss, *Extensions of Hardy spaces and their use in analysis*, Bull. Amer. Math. Soc. **83** (1977), no. 4, 569–645, DOI 10.1090/S0002-9904-1977-14325-5. MR447954

[14] A. Córdoba and R. Fefferman, *On differentiation of integrals*, Proc. Nat. Acad. Sci. U.S.A. **74** (1977), no. 6, 2211–2213, DOI 10.1073/pnas.74.6.2211. MR476977

[15] A. Córdoba, *Maximal functions: a proof of a conjecture of A. Zygmund*, Bull. Amer. Math. Soc. (N.S.) **1** (1979), no. 1, 255–257, DOI 10.1090/S0273-0979-1979-14576-2. MR513753

[16] M. Cotlar and C. Sadosky, *Two distinguished subspaces of product BMO and Nehari-AAK theory for Hankel operators on the torus*, Integral Equations Operator Theory **26** (1996), no. 3, 273–304, DOI 10.1007/BF01306544. MR1415032

[17] M. Cwikel, Y. Sagher, and P. Shvartsman, *A new look at the John-Nirenberg and John-Strömberg theorems for BMO*. Lecture notes. arXiv:1011.0766v1 [math.FA], 2010.

[18] M. Cwikel, Y. Sagher, and P. Shvartsman, *A new look at the John-Nirenberg and John-Strömberg theorems for BMO*, J. Funct. Anal. **263** (2012), no. 1, 129–166, DOI 10.1016/j.jfa.2012.04.003. MR2920844

[19] G. Dafni, T. Hytönen, R. Korte, and H. Yue, *The space $JN_p$: nontriviality and duality*, J. Funct. Anal. **275** (2018), no. 3, 577–603, DOI 10.1016/j.jfa.2018.05.007. MR3806819

[20] G. Dafni and H. Yue, *Some characterizations of local bmo and $h^1$ on metric measure spaces*, Anal. Math. Phys. **2** (2012), no. 3, 285–318, DOI 10.1007/s13324-012-0034-5. MR2958361

[21] X. T. Duong, J. Li, Y. Ou, B. D. Wick, and D. Yang, *Product BMO, little BMO, and Riesz commutators in the Bessel setting*, J. Geom. Anal. **28** (2018), no. 3, 2558–2601, DOI 10.1007/s12220-017-9920-2. MR3833807

[22] X. T. Duong, J. Li, B. D. Wick, and D. Yang, *Commutators, little BMO and weak factorization* (English, with English and French summaries), Ann. Inst. Fourier (Grenoble) **68** (2018), no. 1, 109–129. MR3795472

[23] C. Fefferman, *Characterizations of bounded mean oscillation*, Bull. Amer. Math. Soc. **77** (1971), 587–588, DOI 10.1090/S0002-9904-1971-12763-5. MR280994

[24] C. Fefferman and E. M. Stein, *$H^p$ spaces of several variables*, Acta Math. **129** (1972), no. 3-4, 137–193, DOI 10.1007/BF02392215. MR0447953

[25] S. H. Ferguson and C. Sadosky, *Characterizations of bounded mean oscillation on the polydisk in terms of Hankel operators and Carleson measures*, J. Anal. Math. **81** (2000), 239–267, DOI 10.1007/BF02788991. MR1785283

[26] J. García-Cuerva, *Hardy spaces and Beurling algebras*, J. London Math. Soc. (2) **39** (1989), no. 3, 499–513, DOI 10.1112/jlms/s2-39.3.499. MR1002462

[27] J. B. Garnett, *Bounded analytic functions*, 1st ed., Graduate Texts in Mathematics, vol. 236, Springer, New York, 2007. MR2261424

[28] A. M. Garsia, *Martingale inequalities: Seminar notes on recent progress*, W. A. Benjamin, Inc., Reading, Mass.-London-Amsterdam, 1973. Mathematics Lecture Notes Series. MR0448538

[29] A. M. Garsia and E. Rodemich, *Monotonicity of certain functionals under rearrangement* (English, with French summary), Ann. Inst. Fourier (Grenoble) **24** (1974), no. 2, vi, 67–116. MR414802

[30] D. Goldberg, *A local version of real Hardy spaces*, Duke Math. J. **46** (1979), no. 1, 27–42. MR523600

[31] L. Grafakos, *Modern Fourier analysis*, 3rd ed., Graduate Texts in Mathematics, vol. 250, Springer, New York, 2014. MR3243741

[32] M. de Guzmán, *Differentiation of integrals in $R^n$*, Lecture Notes in Mathematics, Vol. 481, Springer-Verlag, Berlin-New York, 1975. With appendices by Antonio Córdoba, and Robert Fefferman, and two by Roberto Moriyón. MR0457661

[33] R. Hurri-Syrjänen, N. Marola, and A. V. Vähäkangas, *Aspects of local-to-global results*, Bull. Lond. Math. Soc. **46** (2014), no. 5, 1032–1042, DOI 10.1112/blms/bdu061. MR3262204

[34] B. Jawerth and A. Torchinsky, *The strong maximal function with respect to measures*, Studia Math. **80** (1984), no. 3, 261–285, DOI 10.4064/sm-80-3-261-285. MR783994

[35] B. Jessen, J. Marcinkiewicz, A. Zygmund, *Note on the differentiability of multiple integrals*, Fund. Math. **25** (1935), 217–234.

[36] F. John, *Rotation and strain*, Comm. Pure Appl. Math. **14** (1961), 391–413, DOI 10.1002/cpa.3160140316. MR0138225

[37] F. John and L. Nirenberg, *On functions of bounded mean oscillation*, Comm. Pure Appl. Math. **14** (1961), 415–426, DOI 10.1002/cpa.3160140317. MR0131498

[38] P. W. Jones, *Extension theorems for BMO*, Indiana Univ. Math. J. **29** (1980), no. 1, 41–66, DOI 10.1512/iumj.1980.29.29005. MR554817

[39] J.-L. Journé, *Calderón-Zygmund operators, pseudodifferential operators and the Cauchy integral of Calderón*, Lecture Notes in Mathematics, vol. 994, Springer-Verlag, Berlin, 1983. MR706075

[40] I. Klemes, *A mean oscillation inequality*, Proc. Amer. Math. Soc. **93** (1985), no. 3, 497–500, DOI 10.2307/2045621. MR774010

[41] A. A. Korenovskiĭ, *The connection between mean oscillations and exact exponents of summability of functions* (Russian), Mat. Sb. **181** (1990), no. 12, 1721–1727, DOI 10.1070/SM1992v071n02ABEH001409; English transl., Math. USSR-Sb. **71** (1992), no. 2, 561–567. MR1099524

[42] A. A. Korenovskiĭ, *The Riesz "rising sun" lemma for several variables, and the John-Nirenberg inequality* (Russian, with Russian summary), Mat. Zametki **77** (2005), no. 1, 53–66, DOI 10.1007/s11006-005-0005-3; English transl., Math. Notes **77** (2005), no. 1-2, 48–60. MR2158697

[43] A. Korenovskii, *Mean oscillations and equimeasurable rearrangements of functions*, Lecture Notes of the Unione Matematica Italiana, vol. 4, Springer, Berlin; UMI, Bologna, 2007. MR2363526

[44] A. A. Korenovskyy, A. K. Lerner, and A. M. Stokolos, *On a multidimensional form of F. Riesz's "rising sun" lemma*, Proc. Amer. Math. Soc. **133** (2005), no. 5, 1437–1440, DOI 10.1090/S0002-9939-04-07653-1. MR2111942

[45] J. D. Lakey, *Constructive decomposition of functions of finite central mean oscillation*, Proc. Amer. Math. Soc. **127** (1999), no. 8, 2375–2384, DOI 10.1090/S0002-9939-99-04806-6. MR1486741

[46] E. Yu. Leonchik. Oscillations of functions and differential-difference properties of singular integrals (in Russian). Diss., Odessa, 2003.

[47] A. K. Lerner, *BMO-boundedness of the maximal operator for arbitrary measures*, Israel J. Math. **159** (2007), 243–252, DOI 10.1007/s11856-007-0045-3. MR2342480

[48] A. K. Lerner, *The John-Nirenberg inequality with sharp constants* (English, with English and French summaries), C. R. Math. Acad. Sci. Paris **351** (2013), no. 11-12, 463–466, DOI 10.1016/j.crma.2013.07.007. MR3090130

[49] E. H. Lieb and M. Loss, *Analysis*, 2nd ed., Graduate Studies in Mathematics, vol. 14, American Mathematical Society, Providence, RI, 2001. MR1817225

[50] S. Lu and D. Yang, *The central BMO spaces and Littlewood-Paley operators*, Approx. Theory Appl. (N.S.) **11** (1995), no. 3, 72–94. MR1370776

[51] T. Luque, C. Pérez, and E. Rela, *Reverse Hölder property for strong weights and general measures*, J. Geom. Anal. **27** (2017), no. 1, 162–182, DOI 10.1007/s12220-016-9678-y. MR3606549

[52] N. Marola and O. Saari, *Local to global results for spaces of BMO type*, Math. Z. **282** (2016), no. 1-2, 473–484, DOI 10.1007/s00209-015-1549-x. MR3448391

[53] U. Neri, *Some properties of functions with bounded mean oscillation*, Studia Math. **61** (1977), no. 1, 63–75, DOI 10.4064/sm-61-1-63-75. MR445210

[54] E. N. Nikolidakis, *Dyadic-BMO functions, the dyadic Gurov-Reshetnyak condition on* $[0,1]^n$ *and equimeasurable rearrangements of functions*, Ann. Acad. Sci. Fenn. Math. **42** (2017), no. 2, 551–562, DOI 10.5186/aasfm.2017.4228. MR3701634

[55] H. M. Reimann and T. Rychener, *Funktionen beschränkter mittlerer Oszillation* (German), Lecture Notes in Mathematics, Vol. 487, Springer-Verlag, Berlin-New York, 1975. MR0511997

[56] S. Saks. On the strong derivatives of functions of intervals. *Fund. Math.*, 25:235-252, 1935.

[57] E. M. Stein, *Singular integrals and differentiability properties of functions*, Princeton Mathematical Series, No. 30, Princeton University Press, Princeton, N.J., 1970. MR0290095

[58] E. M. Stein, *Harmonic analysis: real-variable methods, orthogonality, and oscillatory integrals*, Princeton Mathematical Series, vol. 43, Princeton University Press, Princeton, NJ, 1993. With the assistance of Timothy S. Murphy; Monographs in Harmonic Analysis, III. MR1232192

[59] E. M. Stein and R. Shakarchi, *Functional analysis*, Princeton Lectures in Analysis, vol. 4, Princeton University Press, Princeton, NJ, 2011. Introduction to further topics in analysis. MR2827930

[60] E. M. Stein and G. Weiss, *Introduction to Fourier analysis on Euclidean spaces*, Princeton University Press, Princeton, N.J., 1971. Princeton Mathematical Series, No. 32. MR0304972

[61] D. M. Stolyarov, V. I. Vasyunin, and P. B. Zatitskiy, *Monotonic rearrangements of functions with small mean oscillation*, Studia Math. **231** (2015), no. 3, 257–267. MR3471053

[62] V. Vasyunin and A. Volberg, *Sharp constants in the classical weak form of the John-Nirenberg inequality*, Proc. Lond. Math. Soc. (3) **108** (2014), no. 6, 1417–1434, DOI 10.1112/plms/pdt063. MR3218314

[63] D. Yang, *Local Hardy and BMO spaces on non-homogeneous spaces*, J. Aust. Math. Soc. **79** (2005), no. 2, 149–182, DOI 10.1017/S1446788700010430. MR2176342

[64] A. Zygmund, *A note on the differentiability of integrals*, Colloq. Math. **16** (1967), 199–204, DOI 10.4064/cm-16-1-199-204. MR210847

Concordia University, Department of Mathematics and Statistics, Montréal, Québec, H3G-1M8, Canada
*Email address*: galia.dafni@concordia.ca

Concordia University, Department of Mathematics and Statistics, Montréal, Québec, H3G-1M8, Canada
*Email address*: ryan.gibara@concordia.ca

Contemporary Mathematics
Volume **748**, 2020
https://doi.org/10.1090/conm/748/15055

# Applications of harmonic analysis techniques to regularity problems of dissipative equations

Mimi Dai and Han Liu

Abstract. We discuss recent advances in the regularity problem of a variety of fluid equations and systems. The purpose is to illustrate the advantage of harmonic analysis techniques in obtaining sharper conditional regularity results when compared to classical energy methods.

## 1. Overview

In this paper we would like to draw readers' attention to recent progress in the regularity problems of a variety of dissipative equations that describe the motion of certain fluids or complex fluids. The emphasis is to introduce a type of low modes regularity criteria for the Navier-Stokes equations (NSE) and several other partial differential equation models akin to it. In other words, the existence of global regular solutions to these equations can be achieved under the condition that some norm of the low frequency parts of the solutions are controlled. These low modes regularity criteria were established with tools from harmonic analysis under a wavenumber splitting regime, which we shall present in the following sections.

We first recall some background of the NSE and various fluid equations.

**1.1. From the NSE to complex fluid systems.** The 3D incompressible NSE, a prototype of a series of fluid equations to appear in this paper, are given by

$$
\begin{cases} u_t + (u\cdot\nabla)u - \nu\Delta u = -\nabla p, \quad x\in\mathbb{R}^3,\ t\geq 0,\\ \nabla\cdot u = 0, \end{cases} \tag{1.1}
$$

where $u$ is the fluid velocity, $p$ the pressure, and $\nu$ the constant viscosity coefficient.

The global regularity for the NSE in dimension three remains an outstanding open problem. Nevertheless, the Leray-Hopf solutions, a class of weak solutions to the NSE constructed by Leray [**39**] and Hopf [**31**], are known to exist globally in time. To introduce the definition of Leray-Hopf solutions, we denote by $C_w$ the space of weakly continuous functions, and by $\dot{C}_0^\infty$ the space of divergence-free smooth functions with compact support.

2010 *Mathematics Subject Classification.* Primary 76D03, 35Q35.

*Key words and phrases.* Navier-Stokes equations; complex fluids; Littlewood-Paley decomposition theory; regularity/blow-up criteria.

The work of the authors was partially supported by NSF Grant DMS-1815069.

DEFINITION 1.1 (Leray-Hopf weak solution). Let $\langle , \rangle$ be the usual $L^2$-inner product. A $L^2$-weakly continuous vector field $u(t,x) \in C_w(0,T; L^2(\mathbb{R}^3))$ is called a weak solution of the NSE on $[0,T]$ if it satisfies the following weak formulation of (1.1)

$$\begin{cases} \langle u(s), \varphi(s)\rangle\big|_0^t = \int_0^t (\langle u(s), \partial_s \varphi(s)\rangle + \langle (u(s)\cdot\nabla)\varphi(s), u(s)\rangle + \nu\langle u(s), \Delta\varphi(s)\rangle) \mathrm{d}s, \\ \int_0^t \langle u(s), \nabla\varphi(s)\rangle \mathrm{d}s = 0, \quad \forall \varphi \in \dot{C}_0^\infty([0,T]\times\mathbb{R}^3) \end{cases}$$

for any $t \in [0,T]$.

Let $\nu > 0$. A weak solution $u$ to system (1.1) on $[0,T]$ is a Leray-Hopf solution provided that $u \in L^\infty(0,T; L^2(\mathbb{R}^3)) \cap L^2(0,T; H^1(\mathbb{R}^3))$ and that the following energy inequality

$$\|u(t)\|_2^2 + \nu \int_{t_0}^t \|\nabla u(s)\|_2^2 \mathrm{d}s \le \|u(t_0)\|_2^2 \tag{1.2}$$

holds for almost every $t_0 \in [0,T]$ and $t \in (t_0, T]$.

Leray, in his pioneering work [**39**], established the global existence of Leray-Hopf solutions for initial data with finite energy.

THEOREM 1.2. *Let $u_0 \in L^2_\sigma(\mathbb{R}^3)$, with $L^2_\sigma$ being the space of divergence free $L^2$-functions. There exists a global in time Leray-Hopf solution $u$ to system (1.1).*

It is worthwhile to point out that the uniqueness of Leray-Hopf solutions is unknown. The question that whether a Leray-Hopf solution develops singularity at finite time is yet to be resolved. However, there have been several important partial regularity results in forms of conditional uniqueness and regularity criteria. Although there is a vast literature on this topic, we shall only allude to the ones that are relevant to the main purpose of this paper. Below we summarize the works of Prodi [**44**], Serrin [**47**] and Ladyzhenskaya [**38**], as well as that of Escauriaza, Seregin and Šverak [**26**].

THEOREM 1.3 (Prodi-Serrin-Ladyzhenskaya). *Let $u$ be a Leray-Hopf solution to system (1.1) with $u_0 \in L^2_\sigma(\mathbb{R}^3)$. If $u \in L^\alpha(0,T; L^\beta(\mathbb{R}^3))$, with $\frac{2}{\alpha} + \frac{3}{\beta} = 1$ and $\beta \in (3, +\infty]$, then $u$ is smooth on $[0,T]$.*

THEOREM 1.4 (Escauriaza-Seregin-Šverak). *Let $u$ be a Leray-Hopf solution to system (1.1) with $u_0 \in L^2_\sigma(\mathbb{R}^3)$. If $u \in L^\infty(0,T; L^3(\mathbb{R}^3))$, then $u$ is smooth on $[0,T]$.*

We note that system (1.1) is invariant under the following scaling:

$$u_\lambda(x,t) = \lambda u(\lambda x, \lambda^2 t), \ p_\lambda(t,x) = \lambda^2 p(\lambda x, \lambda^2 t).$$

The norms of critical spaces i.e., function spaces invariant under the scaling, are of particular importance for the study of the global regularity problem. Indeed, if $(u,p)$ is a solution to the NSE on $[0,T)$, then $(u_\lambda, p_\lambda)$ is a solution to the NSE on $[0, \lambda^{-2}T)$ with initial data $\lambda u_0(\lambda\ \cdot\ )$; hence, it is natural that conditions that guarantee global well-posedness of the system are scaling-invariant. Some examples of critical spaces for the NSE are

$$\dot{H}^{\frac{1}{2}}(\mathbb{R}^3) \hookrightarrow L^3(\mathbb{R}^3) \hookrightarrow \dot{B}_{p,\infty}^{-1+\frac{3}{p}}(\mathbb{R}^3) \hookrightarrow BMO^{-1}(\mathbb{R}^3) \hookrightarrow \dot{B}_{\infty,\infty}^{-1}(\mathbb{R}^3), \quad p < \infty.$$

The time-space Lebesgue spaces in Theorem 1.3 are invariant with respect to the scaling, thus critical. However, a Leray-Hopf solution belongs to $L^\alpha(0,T,L^\beta(\mathbb{R}^3))$ with $\frac{2}{\alpha}+\frac{3}{\beta}=\frac{3}{2}$, which are supercritical. In fact, one of the essential reasons why the NSE regularity problem is particularly challenging is that all the quantities that are known to be controlled to produce a priori bounds e.g., the energy $\|u\|_{L^2}^2$ etc., are supercritical. In view of the scaling-invariance and the notion of criticality, we can define regular solutions as follows.

DEFINITION 1.5. Let $\nu > 0$. A weak solution of system (1.1) is regular on a time interval $\mathcal{I}$ if $\|u\|_{H^s(\mathbb{R}^3)}$ is continuous on $\mathcal{I}$ for some $s > \frac{1}{2}$.

In the case $\nu = 0$, system (1.1) reduces to the Euler equations for incompressible, inviscid flows, whose solvability in 3D remains more of a mystery than that of the NSE, since not even the global existence of Leray-Hopf type solutions is known, while local existence of mild solutions was established by Kato [**33**]. A classical result concerning the Euler equations is the extensibility criterion in terms of the time-integrability of the vorticity, obtained by Beale, Kato and Majda [**4**].

THEOREM 1.6 (Beale-Kato-Majda). *Let $\nu = 0$. Suppose that $u_0 \in H^s(\mathbb{R}^3)$, $s \geq 3$, then there exists a time $T$ depending on $\|u_0\|_{H^3}$, so that system (1.1) has a solution $u \in C(0,T;H^s(\mathbb{R}^3)) \cap C^1(0,T;H^{s-1}(\mathbb{R}^3))$. Moreover, $u$ can be extended beyond $T$ if and only if*

$$\int_0^T \|\nabla \times u(t)\|_\infty \mathrm{d}t < +\infty.$$

Besides the NSE and the surface quasi-geostrophic equation (SQG), in this paper we also include several complex fluid models such as the magneto-hydrodynamics system (MHD), the Hall-magneto-hydrodynamics system (Hall-MHD), the nematic LCD system with Q-tensor, and the chemotaxis-Navier-Stokes system. These systems have many features (e.g., scaling properties) analogous to those of the NSE. In particular, for each of the aforementioned systems, certain Prodi-Serrin-Ladyzhenskaya or Beale-Kato-Majda type regularity criterion is known and more recently, improvements in the form of low modes regularity criteria have been made. We shall provide more detailed reviews of these fluid equations in the upcoming sections.

**1.2. Kolmogorov's theory of turbulence.** Seemingly fluctuating and chaotic behaviors that fluid flows can exhibit at times pose a major difficulty to obtaining a complete mathematical theory for the fluid equations such as the NSE. A turbulent flow may be characterized by eddies of different sizes where the energy cascade takes place. While eddies of larger scales break up into those of smaller scales, kinetic energy is also transferred from larger to smaller scales successively, and finally converted into heat by viscosity, as L. F. Richardson depicted, "Big whirls have little whirls that feed on their velocity, and little whirls have lesser whirls and so on to viscosity in the molecular sense."

In 1941, A. N. Kolmogorov formulated a mathematical theory of turbulence [**36**]. For a fluid with sufficiently high Reynolds number, Kolmogorov suggested that the turbulent flow is statistically isotropic at small scales, in other words, the statistics of the turbulent flow at small scales is independent of directional conditions. Morevoer, it is postulated that at small scales, the statistics of a turbulent flow is universally and uniquely determined by the rate of energy suppy $\epsilon$ and viscosity $\nu$, while determined solely by $\epsilon$ at large scales. The heuristic is that in the NSE, the norm of $\Delta u$ increases as the scales decreases in the energy cascade, thus eventually the molecular dissipation term $\Delta u$ overwhelms the inertial term $u \cdot \nabla u$.

Via a dimensional argument, one can infer that the length scale at which the turbulence switches from the inertial range to the dissipation range should be $\ell_{\rm d} \sim (\frac{\nu^3}{\epsilon})^{\frac{1}{4}}$. Kolmogorov's dissipation wavenumber $\kappa_{\rm d}$, below which is the range where viscous effects in a turbulent flow are negligible, is then defined as

$$\kappa_{\rm d} = \epsilon^{\frac{1}{4}} \nu^{-\frac{3}{4}} \sim \frac{1}{\ell_{\rm d}}.$$

Taking into account the effect of spatial intermittency on $\epsilon$, as observed in experiments, we are suggested to adjust the above formula into

$$\kappa_{\rm d} = \left( \frac{\epsilon}{\nu^3} \right)^{\frac{1}{d+1}},$$

where the parameter $d \in [0, 3]$ is the physical space dimension of the set in which dissipation occurs.

The analysis above forms a very import conjecture: the low frequency part below $\kappa_{\rm d}$ are essential to describe the flow, while the high frequency part above $\kappa_{\rm d}$ are asymptotically controlled by low modes. The wavenumber splitting approach to be introduced later is inspired by this conjecture.

## 2. Harmonic analysis tools

**2.1. Littlewood-Paley theory.** We recall the Littlewood-Paley decomposition, a tool extensively used in the mathematical study of fluids and waves. To start, we define a family of functions with annular support, $\{\varphi_q(\xi)\}_{q=-1}^{\infty}$, which forms a dyadic partition of unity in the frequency domain. Let $\lambda_q = 2^q, q \in \mathbb{Z}$. We choose a radial function $\chi \in C_0^\infty(\mathbb{R}^n)$ satisfying

$$\chi(\xi) = \begin{cases} 1, \text{ for } |\xi| \le \frac{3}{4} \\ 0, \text{ for } |\xi| \ge 1, \end{cases}$$

and define $\varphi(x) = \chi(\frac{\xi}{2}) - \chi(\xi)$ and $\varphi_q(\xi) = \begin{cases} \varphi(\lambda_q^{-1}\xi), \text{for } q \ge 0, \\ \chi(\xi), \text{ for } q = -1. \end{cases}$

Given a vector field $u \in \mathcal{S}'$, its Littlewood-Paley projections are defined as

$$\begin{cases} \Delta_{-1} u = u_{-1} =: \mathcal{F}^{-1}(\chi(\xi)\hat{u}(\xi)) = \int \tilde{h}(y) u(x-y) dy, \\ \Delta_q u = u_q =: \mathcal{F}^{-1}(\varphi_q(\xi)\hat{u}(\xi)) = \lambda_q^n \int h(\lambda_q y) u(x-y) dy, \end{cases}$$

with $\tilde{h} = \check{\chi}$ and $h = \check{\varphi}$.

Thus, the following identity holds in the distributional sense

$$u = \sum_{q=-1}^{\infty} u_q.$$

For convenience, we introduce the following notations

$$u_{\le Q} = \sum_{q=-1}^{Q} u_q, \quad u_{(P,Q]} = \sum_{q=P+1}^{Q} u_q, \quad \tilde{u}_q = \sum_{|p-q| \le 1} u_p.$$

The projections $\{\Delta_q\}_{q=-1}^{\infty}$ provide us with an elegant tool for frequency localization. An observation is that for a function whose Fourier transform is supported in an annulus e.g., $u_q$, a derivative costs exactly one $\lambda_q$. More precisely, we have the following norm equivalence

$$\|u\|_{H^s} \sim \Big(\sum_{q\geq -1}^{\infty} \lambda_q^{2s}\|u_q\|_2^2\Big)^{\frac{1}{2}}.$$

For functions with annular support in the frequency domain, e.g., $u_q$, the following Bernstein's inequality holds, which is applied extensively.

LEMMA 2.1. *Let $s \geq r \geq 1$. We have, in $n$-dimensional space*

$$\|u_q\|_r \lesssim \lambda_q^{n(\frac{1}{r}-\frac{1}{s})}\|u_q\|_s.$$

Finally, the Besov space $B^s_{p,\infty}$ can be defined in a straightforward manner with the Littlewood-Paley projections.

DEFINITION 2.2. The Besov space $B^s_{p,\infty}$ consists of functions $u \in \mathcal{S}'$ such that

$$\|u\|_{B^s_{p,\infty}} =: \sup_{q\geq -1} \lambda_q^s\|u_q\|_p < \infty.$$

Here $\|\cdot\|_{B^s_{p,\infty}}$ is the norm of the Besov space.

We refer readers to the work of Bahouri, Chemin and Danchin [**3**], as well as that of Grafakos [**28**] for more details about the Littlewood-Paley theory and its applications.

**2.2. The commutator and Bony's paraproduct.** To deal with the quadratic nonlinear term $(u\cdot\nabla)u$ which makes the NSE intriguing, we seek a way to decompose the product of two functions within the Littlewood-Paley framework. Formally, the product of two distributions $u$ and $v$ can be written as

$$uv = \sum_{p,q\geq -1} u_p v_q.$$

In 1981 the para-differential calculus was introduced by J. M. Bony, soon finding its applications in many fields such as the study of nonlinear hyperbolic systems. Here we just introduce Bony's paraproduct, which distinguishes three parts in the product $uv$, that is,

$$uv = \sum_{q\geq -1}\big(u_{\leq q-2}v_q + u_q v_{\leq q-2} + \tilde{u}_q v_q\big).$$

Another tool to facilitate the estimation of the convection or inertial terms is the commutator notation

$$[\Delta_q, u_{\leq p-2}\cdot\nabla]v_p = \Delta_q(u_{\leq p-2}\cdot\nabla v_p) - u_{\leq p-2}\cdot\nabla\Delta_q v_p,$$

which enjoys the following estimate.

LEMMA 2.3. *For $\frac{1}{r_1} = \frac{1}{r_2} + \frac{1}{r_3}$, we have*

$$\|[\Delta_q, u_{\leq p-2}\cdot\nabla]v_p\|_{r_1} \lesssim \|v_p\|_{r_2}\sum_{p'\leq p-2}\lambda_{p'}\|u_{p'}\|_{r_3}.$$

PROOF. By definition of $\Delta_q$,

$$
\begin{aligned}
[\Delta_q, u_{\le p-2}\cdot\nabla]v_p =&\lambda_q^3\int_{\mathbb{R}^3} h(\lambda_q(x-y))(u_{\le p-2}(y)-u_{\le p-2}(x))\nabla_y v_p(y)\mathrm{d}y\\
=&-\lambda_q^3\int_{\mathbb{R}^3} \nabla_y h(\lambda_q(x-y))(u_{\le p-2}(y)-u_{\le p-2}(x))v_p(y)\mathrm{d}y\\
=&\int_{\mathbb{R}^3}\lambda_q^3|x-y|\nabla_y h(\lambda_q(x-y))\frac{u_{\le p-2}(x)-u_{\le p-2}(y)}{|x-y|}v_p(y)\mathrm{d}y.
\end{aligned}
$$

By Young's inequality for convolutions,

$$
\begin{aligned}
\|[\Delta_q, u_{\le p-2}\cdot\nabla]v_p\|_{r_1}\le&\|v_p\|_{r_2}\|u_{\le p-2}\|_{r_3}\int_{\mathbb{R}^3}|z||\nabla h(z)|\mathrm{d}z\\
\lesssim&\|v_p\|_{r_2}\|u_{\le p-2}\|_{r_3}.
\end{aligned}
$$

□

As we shall see in the next section, the commutator, together with the divergence free condition, reveals certain cancellations within the nonlinear interactions.

## 3. Low modes regularity criteria for fluid equations

**3.1. The NSE.** The results of interest in this section are low modes regularity criteria for the NSE, which have improved previously known regularity criteria. In this section, more detailed analysis shall be included as the NSE is the prototypical case for other fluid systems. We shall also discuss more about the results' connection to Kolmogorov's theory of turbulence.

The following result, due to Cheskidov and Shvydkoy [**16**], is the foremost among a series of low modes regularity criteria of interest in this paper.

THEOREM 3.1 (Cheskidov-Shvydkoy). *Let $u$ be a weak solution to system (1.1) on $[0,T]$. If $u(t)$ is regular on $[0,T)$ and*

$$
\int_0^T \|(\nabla\times u)_{\le Q(t)}\|_{B^0_{\infty,\infty}}\mathrm{d}t<\infty,
$$

*then $u(t)$ is regular on $[0,T]$.*

*Here $Q(t)=\log_2\Lambda(t)$, with the wavenumber $\Lambda(t)$ for the NSE defined as*

$$
\Lambda(t)=\min\{\lambda_q:\lambda_p^{-1+\frac{3}{r}}\|u_p\|_r<c_r\nu,\ 2\le r\le\infty,\forall p>q,q\in\mathbb{N}\}.
$$

This result gives a unified regularity criterion for the NSE and the Euler equations, since in the case of $\nu=0$, Theorem (3.1 reduces to the Beale-Kato-Majda regularity criterion for the Euler equations.

Later, Cheskidov and Dai [**14**] further weakened the above regularity criterion.

THEOREM 3.2 (Cheskidov-Dai). *Let $u$ be a weak solution to system (1.1) on $[0,T]$ such that $u(t)$ is regular on $[0,T)$. If for certain constant $c_r$, $2\le r\le\infty$*

$$
\limsup_{q\to\infty}\int_{\frac{T}{2}}^T 1_{q\le Q(t)}\lambda_q\|u_q\|_\infty\mathrm{d}t<c_r,
$$

*then $u(t)$ is regular on $[0,T]$.*

The preludes to the above two results are notable results of regularity criteria in terms of Besov norms. Theorem 3.1 has improved the following Prodi-Serrin-Ladyzhenskaya criterion extended to Besov spaces, obtained by Kozono, Ogawa and Taniuchi [**37**].

THEOREM 3.3 (Kozono-Ogawa-Taniuchi). *Let $u$ be a weak solution to system (1.1) such that it is regular on $[0,T)$. If $u \in L^1(0,T;B^{-1}_{\infty,\infty}(\mathbb{R}^3))$, then $u$ can be extended beyond time $T$.*

In fact, it can be shown that the condition in Theorem 3.1 is weaker than any Prodi-Serrin-Ladyzhenskaya type condition $u \in L^r(0,T;B^{2/r-1}_{\infty,\infty})$. Meanwhile, Theorem 3.2 has improved not only Theorem 3.1 but also Planchon's refined Beale-Kato-Majda criterion (see [**43**]), stated as follows.

THEOREM 3.4 (Planchon). *Let $u \in C(0,T;B^s_{p,q}(\mathbb{R}^3)), s \geq 1+\frac{3}{p}, 1 \leq p,q \leq \infty$ be a solution to the Euler equations, that is, system (1.1) with $\nu = 0$. There exists a constant $M_0$ such that $T$ is the maximal existence time iff*

$$\lim_{\varepsilon\to 0}\sup_{q\geq -1}\int_{T-\varepsilon}^{T}\|\Delta_q(\nabla\times u)\|_\infty \mathrm{d}t \geq M_0.$$

A sketch of the proofs of Theorem 3.1 and Theorem 3.2 starts with considering the Littlewood-Paley projections of the NSE in higher order energy spaces.

$$\begin{aligned}\frac{1}{2}\frac{\mathrm{d}}{\mathrm{d}t}\sum_{q\geq -1}^{\infty}\lambda_q^{2s}\|u_q\|_2^2 \leq& -\nu\sum_{q\geq -1}^{\infty}\lambda_q^{2s+2}\|u_q\|_2^2 - \sum_{q\geq -1}^{\infty}\int_{\mathbb{R}^3}\Delta_q(u\cdot\nabla u)u_q \mathrm{d}x\\ =:& -\nu\sum_{q\geq -1}^{\infty}\lambda_q^{2s+2}\|u_q\|_2^2 + I.\end{aligned} \tag{3.3}$$

It then becomes clear that the essential step is to analyze the nonlinear term $(u\cdot\nabla)u$, which translates into term $I$ in (3.3). As we shall see, the wavenumber splitting approach yields, for $s > \frac{1}{2}$,

$$|I| \lesssim c_r\nu\sum_{q\geq -1}\lambda_q^{2s+2}\|u_q\|_2^2 + Q(t)\|u_{\leq Q(t)}\|_{B^1_{\infty,\infty}}\sum_{q\geq -1}\lambda_q^{2s}\|u_q\|_2^2.$$

To proceed, $I$ is split into many terms, which allows one to analyze the interactions between different frequencies. Bony's paraproduct decomposition yields

$$\begin{aligned}I =& -\sum_{q\geq -1}\sum_{|p-q|\leq 2}\int_{\mathbb{R}^3}\Delta_q(u_p\cdot\nabla u_{\leq p-2})u_q \mathrm{d}x\\ &-\sum_{q\geq -1}\sum_{|p-q|\leq 2}\int_{\mathbb{R}^3}\Delta_q(u_{\leq p-2}\cdot\nabla u_p)u_q \mathrm{d}x\\ &-\sum_{q\geq -1}\sum_{|p-q|\leq 2}\int_{\mathbb{R}^3}\Delta_q(u_p\cdot\nabla \tilde{u}_p)u_q \mathrm{d}x\\ =:& I_1+I_2+I_3.\end{aligned}$$

Re-writing the terms using the commutator further reveals certain cancellations. We notice that in the following expression $I_{22}$ vanishes as a consequence of the facts $\sum_{|p-q|\le 2}\Delta_q u_p = u_q$ and $\nabla\cdot u_{\le p-2}=0$.

$$\begin{aligned}
I_2 =& -\sum_{q\ge -1}\sum_{|p-q|\le 2}\int_{\mathbb{R}^3}[\Delta_q, u_{\le p-2}\cdot\nabla]u_p u_q \mathrm{d}t\\
&-\sum_{q\ge -1}\sum_{|p-q|\le 2}\int_{\mathbb{R}^3} u_{\le q-2}\nabla\Delta_q u_p u_q \mathrm{d}t\\
&-\sum_{q\ge -1}\sum_{|p-q|\le 2}\int_{\mathbb{R}^3}(u_{\le p-2}-u_{\le q-2})\nabla\Delta_q u_p u_q \mathrm{d}t\\
=:& I_{21}+I_{22}+I_{23}.
\end{aligned}$$

One then utilizes $Q(t)$ to split all the terms above into low modes and high modes, for example

$$\begin{aligned}
I_1 =& -\sum_{-1\le q\le Q}\sum_{|p-q|\le 2}\int_{\mathbb{R}^3}\Delta_q(u_p\cdot\nabla u_{\le p-2})u_q \mathrm{d}x\\
&-\sum_{q>Q}\sum_{|p-q|\le 2}\int_{\mathbb{R}^3}\Delta_q(u_p\cdot\nabla u_{\le Q})u_q \mathrm{d}x\\
&-\sum_{q>Q}\sum_{|p-q|\le 2}\int_{\mathbb{R}^3}\Delta_q(u_p\cdot\nabla u_{(Q,p-2]})u_q \mathrm{d}x\\
=:& I_{11}+I_{12}+I_{13},
\end{aligned}$$

and

$$\begin{aligned}
I_{21} =& -\sum_{-1\le p\le Q+2}\sum_{|p-q|\le 2}\int_{\mathbb{R}^3}[\Delta_q, u_{\le p-2}\cdot\nabla]u_p u_q \mathrm{d}t\\
&-\sum_{p>Q+2}\sum_{|p-q|\le 2}\int_{\mathbb{R}^3}[\Delta_q, u_{\le Q}\cdot\nabla]u_p u_q \mathrm{d}t\\
&-\sum_{p>Q+2}\sum_{|p-q|\le 2}\int_{\mathbb{R}^3}[\Delta_q, u_{(Q,p-2]}\cdot\nabla]u_p u_q \mathrm{d}t\\
=:& I_{211}+I_{212}+I_{213},
\end{aligned}$$

where $I_{11}, I_{12}, I_{211}$ and $I_{212}$ are low modes, while $I_{13}$ and $I_{213}$ are high modes. To estimate terms involving the commutator, Lemma 2.3 is used, while Hölder's and Young's inequalities are used to estimate terms such as $I_{11}, I_{12}$ and $I_{13}$. It turns out that for low modes and high modes terms the following estimates holds true, respectively.

$$\begin{aligned}
|I_{\text{low modes}}| \lesssim& Q(t)\|(\nabla\times u)_{\le Q(t)}\|_{B^0_{\infty,\infty}}\sum_{q\ge -1}^{\infty}\lambda_q^{2s}\|u_q\|_2^2,\\
|I_{\text{high modes}}| \lesssim& c_r\nu\sum_{q\ge -1}^{\infty}\lambda_q^{2s+2}\|u_q\|_2^2.
\end{aligned}$$

Hence the eventual outcome is a Grönwall type inequality

$$\frac{\mathrm{d}}{\mathrm{d}t}\sum_{q\geq -1}^{\infty}\lambda_q^{2s}\|u_q\|_2^2 \lesssim (-1+c_r)\nu\sum_{q\geq -1}^{\infty}\lambda_q^{2s+2}\|u_q\|_2^2 + Q(t)\|u_{\leq Q(t)}\|_{B^1_{\infty,\infty}}\sum_{q\geq -1}^{\infty}\lambda_q^{2s}\|u_q\|_2^2, \tag{3.4}$$

where $s > \frac{1}{2}$, which already puts $u$ in a subcritical Sobolev space and $c_r$ can be chosen such that $c_r < 1$. By the definition of $\Lambda(t)$ and Bernstein's inequality, one can infer that

$$\Lambda \lesssim \|u_Q\|_\infty \lesssim \Lambda\lambda_Q^{\frac{1}{2}}\|u_Q\|_2 \lesssim \Lambda\lambda_Q^{s}\|u_Q\|_2.$$

Therefore $Q(t) \lesssim (1+\log\|u\|_{H^s})$, and inequality (3.4) becomes

$$\frac{\mathrm{d}}{\mathrm{d}t}\|u\|_{H^s}^2 \lesssim \|(\nabla\times u)_{\leq Q(t)}\|_{B^0_{\infty,\infty}}(1+\log\|u\|_{H^s})\|u\|_{H^s}^2,$$

implying that $u \in L^\infty(0,T;H^s)\cap L^2(0,T;H^{s+1})$ is regular on $[0,T]$ provided that $\|(\nabla\times u)_{\leq Q(t)}\|_{B^0_{\infty,\infty}} \in L^1(0,T)$, which is the condition in Theorem 3.1.

In the proof of Theorem 3.2 the condition $\|(\nabla\times u)_{\leq Q(t)}\|_{B^0_{\infty,\infty}} \in L^1(0,T)$ is weakened via a more delicate analysis. First, by the same procedure as above, one arrives at the following Grönwall type inequality, which slightly differs from (3.4).

$$\frac{\mathrm{d}}{\mathrm{d}t}\sum_{p\geq -1}^{\infty}\lambda_p^{2s}\|u_p\|_2^2 \leq C(\nu,r,s)\sum_{q\leq Q(t)}\lambda_q\|u_q\|_\infty\sum_{p\geq -1}^{\infty}\lambda_p^{2s}\|u_p\|_2^2. \tag{3.5}$$

Assuming that the condition in Theorem 3.2 holds, one can introduce an index

$$q^* = \left\{q : \int_{T/2}^{T} 1_{q\leq Q(t)}\lambda_q\|u_q\|_\infty \mathrm{d}t < c_r,\ \forall q > q^*\right\},$$

with $c_r = \epsilon\ln 2/C(\nu,r,s)$ for some small $\epsilon > 0$. Splitting the sum yields

$$\sum_{q\leq Q(t)}\lambda_q\|u_q\|_\infty = \sum_{q\leq q^*}\lambda_q\|u_q\|_\infty + \sum_{q^*<q\leq Q(t)}\lambda_q\|u_q\|_\infty =: f_{\leq q^*} + f_{>q^*}.$$

It is then not difficult to see that, for $t\in[\frac{T}{2},T]$

$$\int_{\frac{T}{2}}^{t}\sum_{q\leq Q(\tau)}\lambda_q\|u_q\|_\infty \mathrm{d}\tau = \int_{\frac{T}{2}}^{t} f_{\leq q^*}(\tau)\mathrm{d}\tau + \int_{\frac{T}{2}}^{t} f_{>q^*}(\tau)\mathrm{d}\tau \leq q^*\lambda_{q^*}^{\frac{5}{2}}\|u_0\|_2 T + \bar{Q}(t)c_r,$$

where $\bar{Q}(t) = \sup_{\frac{T}{2}\leq\tau\leq t} Q(\tau)$.

By the definition of $\Lambda(t)$ and Bernstein's inequality, one has, for $\Lambda(t) > 1$

$$c_r\nu\Lambda^{1-\frac{3}{r}} \leq \|u_Q\|_r \lesssim \Lambda^{\frac{3}{2}-\frac{3}{r}}\|u_Q\|_2,$$

from which one can infer

$$c_r\nu\Lambda^{s-\frac{1}{2}} \lesssim \Lambda^s\|u_Q\|_2 \lesssim \|u\|_{H^s}.$$

Let $s = \frac{1}{2} + \epsilon$, the above inequality yields $2^{\epsilon \bar{Q}(t)} \lesssim \frac{1}{\nu}\|u\|_{H^s}$. Hence, from inequality (3.5) one has

$$\begin{aligned} \|u(t)\|_{H^s}^2 \leq & \exp\Big(C(\nu,r,s)\int_{\frac{T}{2}}^{t}(f_{\leq q^*}(\tau)+f_{>q^*}(\tau))\mathrm{d}\tau\Big)\|u(T/2)\|_{H^s}^2 \\ \leq & \frac{1}{\nu}\exp(C(\nu,r,s)q^*\lambda {q^*}^{\frac{5}{2}}\|u_0\|_2 T) \sup_{\frac{T}{2}\leq\tau\leq t}\|u(\tau)\|_{H^s}\|u(T/2)\|_{H^s}^2. \end{aligned} \tag{3.6}$$

Noticing that the right hand side of (3.6) is bounded, one can thus conclude Theorem 3.2, which assumes a condition weaker than all the ones listed below.

$$\limsup_{q\to\infty}\int_{\mathcal{T}_q}^{T}\|\Delta_q(\nabla\times u)\|_\infty \mathrm{d}t < c_r, \tag{i}$$

$$\lim_{\varepsilon\to 0}\limsup_{q\to\infty}\int_{T-\varepsilon}^{T}\|\Delta_q(\nabla\times u)\|_\infty \mathrm{d}t < c_r, \tag{ii}$$

$$\int_0^T \sup_{q\leq Q(t)}\|\Delta_q(\nabla\times u)\|_\infty \mathrm{d}t < \infty, \tag{iii}$$

$$\limsup_{q\to\infty}\int_0^T 1_{q\leq Q(t)}\big(\lambda_q^{-1+\frac{3}{r}+\frac{2}{\ell}}\|u\|_r\big)^\ell \mathrm{d}t < \nu^{\ell-1}c_r^\ell, \tag{iv}$$

$$\limsup_{q\to\infty}\int_{\mathcal{T}_q}^{T}\big(\lambda_q^{-1+\frac{3}{r}+\frac{2}{\ell}}\|u\|_r\big)^\ell \mathrm{d}t < \nu^{\ell-1}c_r^\ell, \tag{v}$$

$$\lim_{\varepsilon\to 0}\limsup_{q\to\infty}\int_{T-\varepsilon}^{T}\big(\lambda_q^{-1+\frac{3}{r}+\frac{2}{\ell}}\|u\|_r\big)^\ell \mathrm{d}t < \nu^{\ell-1}c_r^\ell, \tag{vi}$$

$$\int_0^T \big(\|u_{\leq Q(t)}\|_{B_{r,\infty}^{-1+\frac{3}{r}+\frac{2}{\ell}}}\big)^\ell \mathrm{d}t < \nu^{\ell-1}c_r^\ell, \tag{vii}$$

$$\limsup_{t\to T-}\|u(t)-u(T)\|_{B_{r,\infty}^{-1+\frac{3}{r}}} \leq \frac{c_r}{2}, \tag{viii}$$

where $2\leq r\leq\infty, 1\leq\ell\leq\infty$, and $\mathcal{T}_q := \sup\{t\in(\frac{T}{2},T) : Q(\tau) < q, \forall\tau\in(\frac{T}{2},t)\}$. Here the conditions (vi), (viii) and (iii), (vii) can be found in Cheskidov and Shvydkoy's works [**15**] and [**16**], respectively.

We recall the wavenumber with intermittency correction $\kappa_{\mathrm{d}} = (\epsilon/\nu^3)^{\frac{1}{4-s}}$ as well as Kolmogorov's theory, according to which the rate of energy dissipation/supply $\epsilon$, defined as

$$\epsilon := \frac{1}{T}\int_0^T \|\nabla u\|_2^2 \mathrm{d}t$$

plays a role in both the inertial range and the dissipation range, whereas $\nu$ the viscosity is significant only in the dissipation range. A remarkable feature of $\Lambda(t)$ is that it divides the dissipation range and the inertial range more precisely than $\kappa_{\mathrm{d}}$ in the sense that $\langle\Lambda\rangle \lesssim \kappa_{\mathrm{d}}$, where $\langle\Lambda\rangle$ denotes the time average of $\Lambda(t)$. To demonstrate this, we denote $\langle\Lambda\rangle_U =: \frac{1}{T}\int_U \Lambda(t)\mathrm{d}t$, with $U := \{t\in[0,T] : \Lambda(t) > 1\}$, and calculate as follows:

$$\begin{aligned}\langle\Lambda\rangle - 1 \leq &\langle\Lambda\rangle_U \leq \left(\frac{\langle\Lambda(\nu c_\infty)^{\frac{2}{4-\sigma}}\rangle_U^{4-\sigma}}{\nu^2 c_\infty^2}\right)^{\frac{1}{4-\sigma}} \\ \leq &\left(\frac{\langle\|u_Q\|_\infty^{\frac{2}{4-\sigma}}\Lambda^{\frac{2-\sigma}{4-\sigma}}\rangle_U^{4-\sigma}}{\nu^2 c_\infty^2}\right)^{\frac{1}{4-\sigma}} \leq \left(\frac{\langle\|u_Q\|_\infty^2\Lambda^{2-\sigma}\rangle_U}{\nu^2 c_\infty^2}\right)^{\frac{1}{4-\sigma}}.\end{aligned}$$

Here the parameter $\sigma \in [0,3]$ is the dimension of the set in the Fourier space where dissipation occurs, whose dual $\mathrm{d} = 3-\sigma$ is the dimension of the set in the physical space where dissipation occurs, as mentioned in section (1.2). Choosing any $\sigma$ such that $\Lambda^{2-\sigma}\|u_Q\|_\infty^2 \lesssim \Lambda^2\|u_Q\|_2^2$ is satisfied, it follows that

$$\langle\Lambda\rangle - 1 \lesssim \left(\frac{1}{\nu^3}\frac{\nu}{T}\int_U \Lambda(t)^2\|u_Q\|_2^2 \mathrm{d}t\right)^{\frac{1}{4-\sigma}} \lesssim \left(\frac{\epsilon}{\nu^3}\right)^{\frac{1}{4-\sigma}} = \kappa_{\mathrm{d}}.$$

This observation indicates that the effects of viscosity start to manifest earlier than predicted by the dimensional argument. Furthermore, it can be shown that if $\sigma < \frac{3}{2}$, then $u$ is in fact regular on $[0,T]$.

**3.2. The MHD system.** The MHD equations describe the evolution of a system consisting of an electrically conducting fluid and an external magnetic field, influencing each other. It is a model vital to plasma physics, geophysics and several branches of engineering. The following form of the MHD system, in which $u$ is the fluid velocity, $p$ the pressure and $b$ the magnetic field, can be derived from the coupling of the NSE with the Maxwell's equations of electromagnetism.

$$\begin{cases} u_t + u\cdot\nabla u - b\cdot\nabla b + \nabla p = \nu\Delta u, \\ b_t + u\cdot\nabla b - b\cdot\nabla u = \mu\Delta b \\ \nabla\cdot u = 0, \ \nabla\cdot b = 0, \end{cases} \tag{3.7}$$

Theorem with $x \in \mathbb{R}^3$ and $t \geq 0$. The constants $\nu$ and $\mu$ are the fluid viscosity coefficient and magnetic resistivity coefficient, respectively. The MHD system and the NSE share many similar features; in fact, the former reduces to the latter if $b \equiv 0$. System (3.7) enjoys the following scaling-invariance -

$$u_\lambda(t,x) = \lambda u(\lambda x, \lambda^2 t), \ b_\lambda(t,x) = \lambda b(\lambda x, \lambda^2 t), \ p_\lambda(t,x) = \lambda^2 p(\lambda x, \lambda^2 t)$$

solve system (3.7) with initial data $(\lambda u_0(\lambda x), \lambda b_0(\lambda x))$, provided that $(u(t,x), b(t,x), p(t,x))$ is a solution corresponding to the initial data $(u_0(x), b_0(x))$. Global existence of Leray-Hopf type weak solutions was established by Sermange and Temam [**46**], as well as by Duvaut and Lions [**25**]. As the MHD system and the NSE have the same scaling properties, the definitions of regular solutions are naturally analogous.

DEFINITION 3.5. A weak solution of system (3.7) is regular on a time interval $\mathcal{I}$ if $\left(\|u\|_{H^s(\mathbb{R}^3)} + \|b\|_{H^s(\mathbb{R}^3)}\right)$ is continuous on $\mathcal{I}$ for some $s > \frac{1}{2}$.

For the inviscid case $\nu = 0$, a Beale-Kato-Majda type regularity criterion was due to Caflisch, Klapper and Steele [**6**].

THEOREM 3.6. *Let $(u_0, b_0) \in H^s, s > 3$. Then there exists a solution $(u,b) \in C(0,T;H^s(\mathbb{R}^3)) \cap C^1(0,T;H^{s-1}(\mathbb{R}^3))$ to system (3.7), which blows up at time $T^*$ iff*

$$\int_0^{T*} (\|\nabla\times u\|_\infty + \|\nabla\times b\|_\infty)\mathrm{d}t = +\infty.$$

A Planchon-type regularity criterion, which has extended the above regularity criterion, was due to Cannone, Miao and Chen [**7**].

THEOREM 3.7. *Let* $(u_0, b_0) \in B^s_{p,q}(\mathbb{R}^3), s > \frac{3}{p} + 1, 1 \le p, q < \infty$. *Suppose that* $(u, b) \in C(0, T; B^s_{p,q}(\mathbb{R}^3)) \cap C^1(0, T; B^{s-1}_{p,q}(\mathbb{R}^3))$ *is a regular solution to system* (3.7) *on* $[0, T)$. *Then* $(u, b)$ *can be extended beyond time* $T$ *iff there exists a constant* $M_0$ *such that*

$$\lim_{\epsilon\to 0} \sup_{q\ge -1} \int_{T-\epsilon}^{T} (\|\Delta_q(\nabla \times u)\|_\infty + \|\Delta_q(\nabla \times b)\|_\infty)dt < M_0.$$

Consistent with numerical and theoretical observations that the velocity $u$ plays the dominant role in the interactions of $u$ and $b$, Prodi-Serrin type regularity criteria in terms of only $u$ or $\nabla u$ can be found in the works of He and Xin [**30**], as well as, He and Wang [**29**], which can be summarized as follows.

THEOREM 3.8. *Let* $(u_0, b_0) \in L^2_\sigma(\mathbb{R}^3)$. *Suppose that* $(u, b) \in L^\infty(0, T; H^1(\mathbb{R}^3)) \cap L^2(0, T; H^2(\mathbb{R}^3))$ *is a regular solution to system* (3.7). *Then* $(u, b)$ *can be extended beyond time* $T$ *iff*

$$u \in L^\alpha(0, T; L^\beta(\mathbb{R}^3)), \ \frac{2}{\alpha} + \frac{3}{\beta} = 1, \ 3 < \beta \le \infty,$$
$$\text{or} \ \ \nabla u \in L^\alpha(0, T; L^\beta(\mathbb{R}^3)), \ \frac{2}{\alpha} + \frac{3}{\beta} = 2, \ 3 < \beta \le \infty.$$

The above result was later extended to Besov spaces by Chen, Miao and Zhang [**12, 13**], who also obtained a Beale-Kato-Majda type condition in terms of $u$ only. The following theorems summarize their results.

THEOREM 3.9. *Let* $(u_0, b_0) \in L^2_\sigma(\mathbb{R}^3)$. *Suppose that* $(u, b)$ *is a weak solution to system* (3.7) *on* $[0, T)$. *Then* $(u, b)$ *is regular on* $[0, T]$ *iff*

$$u \in L^q(0, T; B^s_{p,\infty}(\mathbb{R}^3)), \ \frac{2}{q} + \frac{3}{p} = 1 + s, \ \frac{3}{1+s} < p \le \infty.$$

THEOREM 3.10. *Let* $(u_0, b_0) \in H^s, s > \frac{1}{2}(\mathbb{R}^3)$. *Suppose that* $(u, b) \in C(0, T; H^s(\mathbb{R}^3)) \cap C^1(0, T; H^{s+1}(\mathbb{R}^3))$ *is a regular solution to system* (3.7) *on* $[0, T)$. *Then* $(u, b)$ *can be extended beyond time* $T$ *iff there exists a constant* $M_0$ *such that*

$$\lim_{\epsilon\to 0} \sup_{q\ge -1} \int_{T-\epsilon}^{T} \|\Delta_q(\nabla \times u)\|_\infty dt < M_0.$$

Finally, by the same wavenumber splitting approach as that applied to the NSE, Cheskidov and Dai [**14**] proved a regularity criterion in terms of the low modes of $u$, which poses an improvement to all the previous results listed above.

THEOREM 3.11. *Let* $(u, b)$ *be a weak solution to system* (3.7) *that is regular on* $[0, T)$. *Then* $(u, b)$ *is regular on* $[0, T]$ *iff*

$$\limsup_{q\to\infty} \int_{\frac{T}{2}}^{T} 1_{q\le Q(t)} \lambda_q \|u_q\|_\infty dt < c_r, \ 2 \le r \le 6.$$

*With* $Q(t) = \log_2 \Lambda(t)$, *where the wavenumber* $\Lambda(t)$ *is defined as*

$$\Lambda(t) := \min\{\lambda_q : \lambda_p^{-1+\frac{3}{r}} \|u_p(t)\|_r < c_r\nu, \ 2 \le r \le 6, \forall p > q, q \in \mathbb{N}\}.$$

Proof of Theorem 3.11 is essentially along the same line as that of Theorem 3.2. As the energy for system (3.7) is $\|u(t)\|_2^2+\|b(t)\|_2^2$, analyzing the projected equations in higher order energy spaces leads to a Grönwall-type inequality. Compared to the NSE, the MHD system has more terms to control and fewer cancellations to utilize, so the above result is somehow surprising. While the magnetic field $b$ didn't appear in the regularity condition, its presence did restrict the choice of index $r$, reducing the range of $r$ to $[2,6]$ from the $[2,\infty]$ in the case of the NSE.

**3.3. The Hall-MHD system.** The following Hall-MHD system differs from the MHD system from the previous section by the extra Hall term $-\nabla\times((\nabla\times b)\times b)$ resulting from replacing the resistive Ohm's law by a more generalized Ohm's law which takes the Hall effect into account.

$$
\begin{cases} u_t + u\cdot\nabla u - b\cdot\nabla b + \nabla p = \nu\Delta u, \\ b_t + u\cdot\nabla b - b\cdot\nabla u - \nabla\times((\nabla\times b)\times b) = \mu\Delta b, \\ \nabla\cdot u = 0,\ \nabla\cdot b = 0, \end{cases} \tag{3.8}
$$

with $x\in\mathbb{R}^3$ and $t\geq 0$. The Hall-MHD system accurately models plasma with large magnetic gradients and is useful in the study of the magnetic reconnection process. Leray-Hopf type weak solutions to system (3.8) satisfy the same energy inequality as that for system (3.7), as the additional Hall term vanishes at energy level. Yet, the Hall term annuls the scaling property of the MHD system and renders the equation for the magnetic field $b$ nonlinear. As a result, the low modes regularity criterion for system (3.8), due to Dai [**20**], has to involve both $u$ and $b$, in contrast to that for system (3.7).

THEOREM 3.12. *Let $(u,b)$ be a weak solution to system (3.8) on $[0,T]$. Assume that $(u,b)$ is regular on $[0,T)$, that is, $(u,b)\in C(0,T;H^s(\mathbb{R}^3))$, $s>\frac{5}{2}$. If*

$$
\int_0^T \|u_{\leq Q_u(t)}(t)\|_{B^1_{\infty,\infty}} + \Lambda_b(t)\|b_{\leq Q_b(t)}(t)\|_{B^0_{\infty,\infty}}\,dt < \infty,
$$

*then $(u,b)$ is regular on $[0,T]$.*

Here $2^{Q_u(t)}=\Lambda_u(t)$ and $2^{Q_b(t)}=\Lambda_b(t)$, with

$$
\begin{aligned}
\Lambda_u(t) &:= \min\{\lambda_q\geq -1 : \lambda_p^{-1}\|u_p(t)\|_\infty < c_0\min\{\nu,\mu\}, \forall p>q\}, \\
\Lambda_b(t) &:= \min\{\lambda_q\geq -1 : \lambda_{p-q}^{\delta}\|b_p(t)\|_\infty < c_0\min\{\nu,\mu\}, \forall p>q\},
\end{aligned}
$$

where $\lambda_{p-q}^{\delta}$ represents a kernel with $\delta\geq s$.

Theorem 3.12 improves known regularity criteria for system (3.8), notably the following Prodi-Serrin-Ladyzhenskaya type regularity criterion and its extension into the BMO space, proved by Chae and Lee [**9**].

THEOREM 3.13. *Suppose that $(u,b)$ is a weak solution to system (3.8) on $[0,T]$ that is regular on $[0,T)$. If $(u,b)$ satisfy*

$$
\begin{cases} u\in L^q(0,T;L^p(\mathbb{R}^3)),\ \frac{3}{p}+\frac{2}{q}\leq 1, p\in(3,\infty], \\ \nabla b\in L^r(0,T;L^s(\mathbb{R}^3)),\ \frac{3}{s}+\frac{2}{r}\leq 1, p\in(3,\infty]; \end{cases}
$$
$$
\text{or } u,\nabla b\in L^2(0,T;BMO(\mathbb{R}^3)),
$$

*then $(u,b)$ is regular on $[0,T]$.*

In order to prove Theorem 3.12, it is essential to tame the Hall term, which involves the strongest nonlinearity in system (3.8). This could be done thanks to the following new commutators and their corresponding estimates, found in [**20**].

LEMMA 3.14. *Given vector valued functions $F$ and $G$, define the commutators*

$$\begin{aligned}[\Delta_q, F\times\nabla\times]G =&\Delta_q(F\times(\nabla\times G))-F\times(\nabla\times G_q),\\ [\Delta_q,(\nabla\times F)\times]G =&\Delta_q(\nabla\times(F\times G))-(\nabla\times F)\times G_q.\end{aligned}$$

*The following estimates hold true*

$$\begin{aligned}\|[\Delta_q, F\times\nabla\times]G\|_r &\lesssim \|\nabla F\|_\infty\|G\|_r,\\ \|[\Delta_q,(\nabla\times F)\times]G\|_r &\lesssim \|\nabla F\|_\infty\|G\|_r,\end{aligned}$$

*provided that $\nabla\cdot F=0$ and $G$ vanishes for large $|x|$, with $x\in\mathbb{R}^3$.*

Indeed, the Hall term is translated into the following term

$$H=-\lambda_q^{2s}\int_{\mathbb{R}^3}\Delta_q((\nabla\times b)\times b)\cdot(\nabla\times b_q)\mathrm{d}x,$$

which can be decomposed using Bony's paraproduct as

$$\begin{aligned}H=&-\sum_{q\ge -1}\sum_{|p-q|\le 2}\lambda_q^{2s}\int_{\mathbb{R}^3}\Delta_q((\nabla\times b_p)\times b_{\le p-2})\cdot(\nabla\times b_q)\mathrm{d}x\\ &-\sum_{q\ge -1}\sum_{|p-q|\le 2}\lambda_q^{2s}\int_{\mathbb{R}^3}\Delta_q((\nabla\times b_{\le p-2})\times b_p)\cdot(\nabla\times b_q)\mathrm{d}x\\ &-\sum_{q\ge -1}\sum_{p\ge q-2}\lambda_q^{2s}\int_{\mathbb{R}^3}\Delta_q((\nabla\times b_p)\times \tilde b_p)\cdot(\nabla\times b_q)\mathrm{d}x\\ =:&H_1+H_2+H_3.\end{aligned}$$

Applying the commutators to the above terms reveals cancellations, for example:

$$\begin{aligned}H_1=&\sum_{q\ge -1}\sum_{|p-q|\le 2}\lambda_q^{2s}\int_{\mathbb{R}^3}[\Delta_q,(b_{\le p-2}\times\nabla\times]b_p\cdot(\nabla\times b_q)\mathrm{d}x\\ &+\sum_{q\ge -1}\lambda_q^{2s}\int_{\mathbb{R}^3}b_{\le q-2}\times(\nabla\times b_q)\cdot(\nabla\times b_q)\mathrm{d}x\\ &+\sum_{q\ge -1}\sum_{|p-q|\le 2}\lambda_q^{2s}\int_{\mathbb{R}^3}(b_{\le p-2}-b_{\le q-2})\times(\nabla\times(b_p)_q)\cdot(\nabla\times b_q)\mathrm{d}x\\ =:&H_{11}+H_{12}+H_{13},\end{aligned}$$

where $H_{12}=0$ due to the property of cross product.

One then estimates the various terms with the help of commutator estimates, Hölder's, Jensen's and Young's inequalities, as well as the definitions of the wavenumbers, ultimately reaching a Grönwall type inequality which yields Theorem 3.12.

**3.4. The supercritical SQG equation.** In [**21**], the wavenumber spittling method was applied to the supercritical SQG equation, written as follows.

$$\begin{cases}\theta_t+u\cdot\nabla\theta+\kappa\Lambda^\alpha\theta=0, & x\in\mathbb{R}^2,\ t\ge 0,\\ u=R^\perp\theta,\end{cases}\tag{3.9}$$

with $0 < \alpha < 1, \kappa > 0, \Lambda = (-\Delta)^{\frac{1}{2}}$ and $R^{\perp}\theta = \Lambda^{-1}(-\partial_2\theta, \partial_1\theta)$. In system (3.9), the scalar function $\theta$ is the potential temperature and the vector valued function $u$ is the fluid velocity. System (3.9), which arises from modeling geophysical flows in atmospheric sciences and oceanography, is also a toy model for the 3D Euler equations due to many parallels between the two in their forms and solutions' behaviors. While the subcritical ($1 < \alpha \leq 2$) and critical ($\alpha = 1$) SQG equations are known to be globally well-posed [**5, 18, 34, 35**], the regularity problem for the supercritical SQG equation (3.9) remains an intriguing unresolved question.

Constantin, Majda and Tabak [**17**] proved a regularity criterion analogous to the one for the Euler equations.

THEOREM 3.15. *Given $\theta_0 \in H^s(\mathbb{R}^2), s \geq 3$, there exists a unique smooth solution to system (3.9) $\theta \in L^\infty(0, T^*; H^s(\mathbb{R}^2))$, and $T^*$ is the maximal existence time iff*

$$\int_0^{T^*} \|\nabla^{\perp}\theta(t)\|_\infty \mathrm{d}t = +\infty.$$

A Prodi-Serrin-Ladyzhenskaya type criterion was established by Chae [**8**].

THEOREM 3.16. *Let $\theta$ be a solution to system (3.9). If*

$$\nabla^{\perp}\theta \in L^q(0, T; L^p(\mathbb{R}^2)), \ \frac{2}{p} + \frac{\alpha}{q} = \alpha, \ p \in (\frac{2}{\alpha}, \infty),$$

*then singularity does not occur up to time $T$.*

The Besov space version of the Prodi-Serrin-Ladyzhenskaya type regularity criterion was due to Dong and Pavlović [**24**].

THEOREM 3.17. *Let $\theta$ be a weak solution to system (3.9). If*

$$\theta \in L^r(0, T; B^s_{p,\infty}(\mathbb{R}^2)), \ s = \frac{2}{p} + 1 - \alpha + \frac{\alpha}{r}, \ 2 \leq p, r < \infty,$$

*then $\theta$ is a smooth solution, i.e., $\theta \in C^\infty([0, T] \times \mathbb{R}^2)$.*

Notice that system (3.9) enjoys invariance under the scaling transform

$$\theta(x, t) \mapsto \theta_\lambda(x, t) = \lambda^{\alpha-1}\theta(\lambda x, \lambda^\alpha t),$$

thus some critical spaces are $H^{2-\alpha}(\mathbb{R}^2)$, $C^{1-\alpha}(\mathbb{R}^2)$, $L^\infty(\mathbb{R}^2)$ and $B^{1-\alpha}_{\infty,\infty}(\mathbb{R}^2)$, the last being the largest critical space for system (3.9). In particular, the conditions in the theorems listed above are all in terms of critical quantities.

While Leray-Hopf type weak solutions to system (3.9) are weak solutions that satisfy the energy inequality

$$\|\theta(t)\|_2^2 + 2\kappa \int_{t_0}^t \|\nabla\theta(s)\|_2^2 \mathrm{d}s \leq \|u(t_0)\|_2^2,$$

the notion of viscosity solutions also comes into play. A weak solution to system (3.9) is a viscosity solution if it is the weak limit of of a sequence of solutions to the following systems with $\epsilon \to 0$,

$$\begin{cases} \theta^\epsilon_t + R^{\perp}\theta^\epsilon \cdot \nabla\theta^\epsilon + \kappa\Lambda^\alpha\theta^\epsilon = \epsilon\Delta\theta^\epsilon, \\ \theta^\epsilon(x, 0) = \theta_0, \end{cases}$$

where $\theta_0 \in H^s(\mathbb{R}^2), s > 1$.

Define the wavenumber $\Lambda(t) = 2^{Q(t)}$ as

$$\Lambda(t) = \min\{\lambda_q : \lambda_p^{1-\alpha}\|\theta_p(t)\|_\infty < c_0\kappa, \forall p > q \geq 1\},$$

where $c_0$ is some constant. The low modes regularity criterion for the SQG equation states as follows.

THEOREM 3.18. *Let $\theta$ be a viscosity solution to system (3.9) on $[0,T]$. Assume that*

$$\int_0^T \|\nabla\theta_{\leq Q(t)}(t)\|_{B^0_{\infty,\infty}} \mathrm{d}t < +\infty,$$

*then $\theta$ is a smooth solution on $[0,T]$.*

It can be shown that Theorem 3.18 has improved all of the regularity criteria for system (3.9) mentioned before. A key ingredient of the proof is the following regularity result due to Constantin and Wu [**19**].

THEOREM 3.19. *Let $\theta$ be a Leray-Hopf weak solution to system (3.9). If*

$$\theta \in L^\infty(t_0, t; C^\delta(\mathbb{R}^2)),\ \delta > 1-\alpha,\ 0 < t_0 < t < \infty,$$

*then $\theta \in C^\infty([t_0,t]\times\mathbb{R}^2)$.*

Instead of considering the projected equation in $L^2$-based Sobolev spaces, one approaches the problem via the Besov space $B^s_{l,l}$. Frequency localization yield the following projected equation.

$$\frac{\mathrm{d}}{\mathrm{d}t}\sum_{q\geq -1}\lambda_q^{sl}\|\theta_q\|_l^l \leq C\kappa\sum_{q\geq -1}\lambda_q^{sl+\alpha}\|\theta_q\|_l^l + l\sum_{q\geq -1}\lambda_q^{sl}\int_{\mathbb{R}^3}\Delta_q(u\cdot\nabla\theta)|\theta_q|^{l-2}\theta_q \mathrm{d}x.$$

Analyzing using the same tools as those used for the NSE, one obtains a Grönwall type inequality which implies that $\theta \in L^\infty(0,T; B^s_{l,l}(\mathbb{R}^2))$ provided that the condition in Theorem 3.18 holds. Choosing $s, l$ such that $0 < s < 1, sl > 2$ and $\frac{\alpha}{l} < 1 - s < \alpha - \frac{2}{l}$, one infers from Theorem 3.19) and the embedding $B^s_{l,l} \subset C^{0,s-\frac{2}{l}}$ that $\theta$ is in fact a smooth solution.

**3.5. The nematic LCD system with Q-tensor.** Low modes regularity criteria have been established for the following nematic LCD system with Q-tensor in [**22**].

$$\begin{cases} u_t + (u\cdot\nabla)u + \nabla p = \nu\Delta u + \nabla\cdot\Sigma(\mathbb{Q}), \\ \mathbb{Q}_t + (u\cdot\nabla)\mathbb{Q} - \mathbb{S}(\nabla u, \mathbb{Q}) = \mu\Delta\mathbb{Q} - \mathcal{L}[\partial F(\mathbb{Q})], \\ \nabla\cdot u = 0, \end{cases} \tag{3.10}$$

with $x \in \mathbb{R}^3$ and $t \geq 0$. System (3.10) is a model for nematic liquid crystal flows. Liquid crystals are matters in an intermediate state between the conventionally observed solid and liquid. The nematic phase, in which the rod-like molecules possess no positional order but long-range directional order through self-aligning in an almost parallel manner, is one of the most common liquid crystal phases. Nematics, due to its fluidity and optical properties, are extremely important to liquid crystal displays (LCD). In system (3.10), $u$ is the fluid velocity, $p$ the fluid pressure, while the local configuration of the crystal and the ordering of the molecules are represented by the symmetric and traceless Q-tensor $\mathbb{Q}(t,x) \in \mathbb{R}^{3\times 3}_{\mathrm{sym},0}$. The constants $\nu$

and $\mu$ stand for the fluid viscosity and the elasticity of the molecular orientation field, respectively. In the simplified case tensors $\Sigma$ and $\mathbb{S}$ are given by

$$\Sigma(\mathbb{Q}) = \Delta\mathbb{Q}\mathbb{Q} - \mathbb{Q}\Delta\mathbb{Q} - \nabla\mathbb{Q} \otimes \nabla\mathbb{Q}, \ \mathbb{S}(\nabla u, \mathbb{Q}) = \Omega(u)\mathbb{Q} - \mathbb{Q}\Omega(u)$$

where $(\nabla\mathbb{Q} \otimes \nabla\mathbb{Q})_{ij} = \partial_i \mathbb{Q}_{\alpha\beta}\partial_j\mathbb{Q}_{\alpha\beta}$ and the skew-symmetric parts of the rate of the stress tensor $\Omega(u) = \frac{1}{2}(\nabla u - \nabla^t u)$. The operator $\mathcal{L}$, which projects onto the space of traceless matrices, is defined as

$$\mathcal{L}[\mathbb{A}] = \mathbb{A} - \frac{1}{3}\operatorname{tr}[\mathbb{A}]\mathbb{I},$$

and the bulk potential function $F(\mathbb{Q})$ takes the Landau-de Gennes form

$$F(\mathbb{Q}) = \frac{a}{2}|\mathbb{Q}|^2 + \frac{b}{3}\operatorname{tr}|\mathbb{Q}|^3 + \frac{c}{4}|\mathbb{Q}|^4.$$

Paicu and Zarnescu [**41**,**42**] proved existence of weak solutions to system (3.10). On bounded domains, Abels, Dolzmann and Liu [**1**,**2**] proved existence and uniqueness of local-in-time strong solutions subject to various boundary conditions. A Prodi-Serrin-Ladyzhenskaya type regularity condition

$$\nabla u \in L^p(0,T;L^q), \ \frac{2}{p} + \frac{3}{q} = 2, \ 2 \le p \le 3$$

can be found in [**27**].

For system (3.10), a solution $(u, \mathbb{Q})$ is regular on time interval $\mathcal{I}$ if $\|u(t)\|_{H^s}$ and $\|\nabla\mathbb{Q}(t)\|_{H^s}$ are continuous on $\mathcal{I}$ for some $s \ge \frac{1}{2}$. Dai [**22**] established the corresponding low modes regularity criteria through the wavenumber splitting scheme. The wavenumber $\Lambda(t) = 2^{Q(t)}$ can be defined in an almost identical fashion as that for the MHD system:

$$\Lambda(t) = \min\{\lambda_q : \lambda_p^{-1+\frac{3}{r}}\|u_p(t)\|_r < c_r \min\{\nu,\mu\}, \forall p > q, q \in \mathbb{N}\}$$

where $c_r$ is a constant depending only on $r \in [2,6)$. Moreover, it turns out that the low modes regularity criteria for the two systems are almost identical as well.

THEOREM 3.20. ([**22**]) *Let $(u,\mathbb{Q})$ be a weak solution to system (3.10) on $[0,T]$ that is also regular on $[0,T)$. Assume that*

$$\int_0^T \|\nabla u_{\le Q(t)}(t)\|_{B^0_{\infty,\infty}} dt < +\infty,$$

*then $(u,\mathbb{Q})$ is regular on $[0,T]$.*

THEOREM 3.21. ([**22**]) *Let $(u,\mathbb{Q})$ be a weak solution to system (3.10) that is regular on $[0,T)$. Assume that*

$$\limsup_{q\to\infty} \int_{\frac{T}{2}}^T 1_{q\le Q(t)}\lambda_q\|u_q\|_\infty dt < c$$

*for some small constant $c$, then $(u,\mathbb{Q})$ is regular on $[0,T]$.*

The analysis shares much in common with that of system (3.7), while additional new commutator estimates were introduced to handle the terms involving $\mathbb{Q}$. Similar to Theorem 3.10 in the case of the MHD equations, Theorem 3.21 is a refinement of Theorem 3.20. An obvious yet meaningful implication of Theorem 3.20 and Theorem 3.21 is given by the following corollary, also found in [**22**].

COROLLARY 3.22. *Let $(u, \mathbb{Q})$ be a weak solution to system (3.10). Suppose that either the Prodi-Serrin-Ladyzhenskaya type condition*

$$u \in L^s(0,T;L^r(\mathbb{R}^3)), \quad \frac{2}{s}+\frac{3}{r}=1, \ 3<r<6$$

*or the Beale-Kato-Majda type condition*

$$\int_0^T \|\nabla \times u(t)\|_\infty \mathrm{d}t < +\infty$$

*holds true, then $(u, \mathbb{Q})$ is regular on $[0,T]$.*

**3.6. The chemotaxis-Navier-Stokes system.** The following chemotaxis-Navier-Stokes system is a coupling of the NSE with the parabolic Keller-Segel system, emerging from the study of aggregation behaviors of chemotactic cells.

$$\begin{cases} n_t + u\cdot\nabla n = \kappa\Delta n - \nabla\cdot(n\nabla c), \\ c_t + u\cdot\nabla c = \mu\Delta c - nc, \\ u_t + (u\cdot\nabla)u + \nabla p = \nu\Delta u - n\nabla\Phi, \\ \nabla\cdot u = 0, \ (t,x)\in\mathbb{R}^+\times\mathbb{T}^3, \end{cases} \tag{3.11}$$

where $n$ is the cell density, $c$ the concentration of the attractant, $u$ the fluid velocity, $p$ the fluid pressure, and $\nabla\Phi$ a constant vector field. System (3.11) models the situation in which the cells e.g., *Bacillus subtilis*, which are attracted to certain chemical substance e.g., oxygen, swim in sessile drops of water. Lorz [**40**] established the existence of local-in-time weak solutions to system (3.11) on bounded domains in 3D. Global existence of weak solutions under more general assumptions was proved via entropy-energy estimates by Winkler [**48**].

In particular, Chae, Kang and Lee [**10, 11**] proved Prodi-Serrin-Ladyzhenskaya type regularity criteria in the following theorem.

THEOREM 3.23. *Let $(n,c,u)$ be a weak solution to system (3.11) which is regular on $[0,T)$, that is, $(n,c,u)\in L^\infty(0,T;H^{m-1}\times H^m\times H^m(\mathbb{R}^3)), m\ge 3$. If*

$$\|u\|_{L^q(0,T;L^p(\mathbb{R}^3))} + \|\nabla c\|_{L^2(0,T;L^\infty(\mathbb{R}^3))} < \infty,$$
$$\text{or } \|u\|_{L^q(0,T;L^p(\mathbb{R}^3))} + \|n\|_{L^r(0,T;L^s(\mathbb{R}^3))} < \infty,$$

*where $\frac{3}{p}+\frac{2}{q}=1$, $3<p\le\infty$ and $\frac{3}{s}+\frac{2}{r}=2$, $\frac{3}{2}<s\le\infty$, then $(n,c,u)$ is regular on $[0,T]$.*

System (3.11) satisfies the following scaling property: suppose that $(n,c,u)(t,x)$ is a solution to system (3.11) with initial data $(n_0,c_0,u_0)(x)$, then

$$n_\lambda(t,x)=\lambda^2 n(\lambda^2 t,\lambda x), \ c_\lambda(t,x)=c(\lambda^2 t,\lambda x), \ u_\lambda(t,x)=\lambda u(\lambda^2 t,\lambda x)$$

also solves system (3.11) with initial data

$$n_{\lambda,0}=\lambda^2 n(\lambda x), \quad c_{\lambda,0}=c(\lambda x), \quad u_{\lambda,0}=\lambda u(\lambda x).$$

Therefore, the Sobolev space $\dot H^{-\frac12}\times\dot H^{\frac12}\times\dot H^{\frac32}(\mathbb{R}^3)$ is critical according to the above scaling of the system and the notion of regular solutions can thus be understood via the subcritical spaces $\dot H^s\times\dot H^{s+1}\times\dot H^{s+2}(\mathbb{R}^3)$ with $s>-\frac12$. Notice that the conditions in Theorem 3.23 are in terms of scaling-invariant quantities. In addition, a weak solution $(n,c,u)$ to system (3.11) possesses the following properties of mass conservation and monotonicity of oxygen concentration:

$$\|n(t)\|_1 = \|n_0\|_1, \ \|c(t)\|_\infty \le \|c_0\|_\infty.$$

The following low modes regularity criterion, proved in [**23**], seems somehow surprising, given that the wavenumber splitting method was applied to the $c$ equation, which differs from fluid equations in nature.

THEOREM 3.24. *Let $(n(t), c(t), u(t))$ be a weak solution to (3.11) on $[0, T]$. Assume that $(n(t), c(t), u(t))$ is regular on $[0, T)$ and*

$$\int_0^T \|\nabla c_{\le Q_c(t)}(t)\|_{L^\infty}^2 + \|u_{\le Q_u(t)}(t)\|_{B^1_{\infty,\infty}} dt < \infty,$$

*then $(n(t), c(t), u(t))$ is regular on $[0, T]$.*

While the definition of the wavenumber $\Lambda_u(t)$ and the condition on $u$ are the same as those for the NSE in [**14**], to weaken the condition on $\nabla c$ to low modes $\nabla c_{\le Q_c}$ is rather challenging due to the lack of divergence free condition. This difficulty is reflected by the rather narrow range for the parameter $r$ in the following definition of the wavenumber $\Lambda_c(t)$ :

$$\Lambda_c(t) = \min\{\lambda_q : \lambda_p^{\frac{3}{r}} \|c_p(t)\|_r < C_0 \min\{\kappa, \mu, \nu\}, \forall p > q, q \in \mathbb{N}\}, \; 3 < r < \frac{3}{1-\varepsilon},$$

where $C_0$ and $\varepsilon$ are both positive small constants.

Considering the frequency localized equations of system (3.11) in Sobolev space $(\dot{H}^s \times \dot{H}^{s+1} \times \dot{H}^{s+1})(\mathbb{T}^3)$ and analyzing using the same set of tools as before lead to a Grönwall type inequality from which one can conclude that

$$(n, c, u) \in L^\infty(0, T; \dot{H}^s \times \dot{H}^{s+1} \times \dot{H}^{s+1}(\mathbb{T}^3)),$$
$$(\nabla n, \nabla c, \nabla u) \in L^2(0, T; \dot{H}^s \times \dot{H}^{s+1} \times \dot{H}^{s+1}(\mathbb{T}^3))$$

for certain $s = -\varepsilon < 0$. While $n$ and $u$ are already in subcritical spaces, one must lift $c$ to a subcritical space by a parabolic bootstrap argument, which could be carried out thanks to the mixed derivative theorem found in the works of Prüss and Simonett [**45**], the fact that $c \in L^\infty(0, T; L^\infty(\mathbb{T}^3))$ as well as the following lemma on the derivative gain for the heat equation.

LEMMA 3.25. *Let $u$ be a solution to the heat equation $u_t - \Delta u = f$ on $\mathbb{T}^n, n \ge 2$. Assume that $u_0 \in H^{\alpha+1}$ and $f \in L^2(0, T; H^\alpha)$ for $\alpha \in \mathbb{R}$, then*

$$u \in L^2(0, T; H^{\alpha+2}(\mathbb{T}^n)) \cap H^1(0, T; H^\alpha(\mathbb{T}^n)).$$

It is noteworthy that the first two equations of system (3.11) i.e., the Keller-Segel model, arise from a discipline quite distant from hydrodynamics. The lack of divergence free conditions, along with the entropy functional $\int_{\mathbb{T}^3} n \ln n \, dx$ further dissociates the $n$ and $c$ equations from the family of equations of incompressible fluid flows. The fact that the wavenumber splitting method finds its application in a model of population dynamics motivates one to ask the question whether or how one can apply the same harmonic analysis techniques that bear fruits in the realm of fluids to some other utterly different models.

## Acknowledgments

The authors would like to thank the anonymous referee for valuable suggestions, which led to an improved version of the article.

## References


[1] Helmut Abels, Georg Dolzmann, and YuNing Liu, *Well-posedness of a fully coupled Navier-Stokes/Q-tensor system with inhomogeneous boundary data*, SIAM J. Math. Anal. **46** (2014), no. 4, 3050–3077, DOI 10.1137/130945405. MR3252809

[2] Helmut Abels, Georg Dolzmann, and YuNing Liu, *Strong solutions for the Beris-Edwards model for nematic liquid crystals with homogeneous Dirichlet boundary conditions*, Adv. Differential Equations **21** (2016), no. 1-2, 109–152. MR3449332

[3] Hajer Bahouri, Jean-Yves Chemin, and Raphaël Danchin, *Fourier analysis and nonlinear partial differential equations*, Grundlehren der Mathematischen Wissenschaften [Fundamental Principles of Mathematical Sciences], vol. 343, Springer, Heidelberg, 2011. MR2768550

[4] J. T. Beale, T. Kato, and A. Majda, *Remarks on the breakdown of smooth solutions for the 3-D Euler equations*, Comm. Math. Phys. **94** (1984), no. 1, 61–66. MR763762

[5] Luis A. Caffarelli and Alexis Vasseur, *Drift diffusion equations with fractional diffusion and the quasi-geostrophic equation*, Ann. of Math. (2) **171** (2010), no. 3, 1903–1930, DOI 10.4007/annals.2010.171.1903. MR2680400

[6] Russel E. Caflisch, Isaac Klapper, and Gregory Steele, *Remarks on singularities, dimension and energy dissipation for ideal hydrodynamics and MHD*, Comm. Math. Phys. **184** (1997), no. 2, 443–455, DOI 10.1007/s002200050067. MR1462753

[7] Marco Cannone, Qionglei Chen, and Changxing Miao, *A losing estimate for the ideal MHD equations with application to blow-up criterion*, SIAM J. Math. Anal. **38** (2007), no. 6, 1847–1859, DOI 10.1137/060652002. MR2299432

[8] Dongho Chae, *On the regularity conditions for the dissipative quasi-geostrophic equations*, SIAM J. Math. Anal. **37** (2006), no. 5, 1649–1656, DOI 10.1137/040616954. MR2215601

[9] Dongho Chae and Jihoon Lee, *On the blow-up criterion and small data global existence for the Hall-magnetohydrodynamics*, J. Differential Equations **256** (2014), no. 11, 3835–3858, DOI 10.1016/j.jde.2014.03.003. MR3186849

[10] Myeongju Chae, Kyungkeun Kang, and Jihoon Lee, *Existence of smooth solutions to coupled chemotaxis-fluid equations*, Discrete Contin. Dyn. Syst. **33** (2013), no. 6, 2271–2297, DOI 10.3934/dcds.2013.33.2271. MR3007686

[11] Myeongju Chae, Kyungkeun Kang, and Jihoon Lee, *Global existence and temporal decay in Keller-Segel models coupled to fluid equations*, Comm. Partial Differential Equations **39** (2014), no. 7, 1205–1235, DOI 10.1080/03605302.2013.852224. MR3208807

[12] Qionglei Chen, Changxing Miao, and Zhifei Zhang, *On the regularity criterion of weak solution for the 3D viscous magneto-hydrodynamics equations*, Comm. Math. Phys. **284** (2008), no. 3, 919–930, DOI 10.1007/s00220-008-0545-y. MR2452599

[13] Qionglei Chen, Changxing Miao, and Zhifei Zhang, *The Beale-Kato-Majda criterion for the 3D magneto-hydrodynamics equations*, Comm. Math. Phys. **275** (2007), no. 3, 861–872, DOI 10.1007/s00220-007-0319-y. MR2336368

[14] A. Cheskidov and M. Dai. *Regularity criteria for the 3D Navier-Stokes and MHD equations.* arXiv:1507.06611, 2015.

[15] A. Cheskidov and R. Shvydkoy, *The regularity of weak solutions of the 3D Navier-Stokes equations in $B^{-1}_{\infty,\infty}$*, Arch. Ration. Mech. Anal. **195** (2010), no. 1, 159–169, DOI 10.1007/s00205-009-0265-2. MR2564471

[16] A. Cheskidov and R. Shvydkoy, *A unified approach to regularity problems for the 3D Navier-Stokes and Euler equations: the use of Kolmogorov's dissipation range*, J. Math. Fluid Mech. **16** (2014), no. 2, 263–273, DOI 10.1007/s00021-014-0167-4. MR3208714

[17] Peter Constantin, Andrew J. Majda, and Esteban Tabak, *Formation of strong fronts in the 2-D quasigeostrophic thermal active scalar*, Nonlinearity **7** (1994), no. 6, 1495–1533. MR1304437

[18] Peter Constantin and Vlad Vicol, *Nonlinear maximum principles for dissipative linear nonlocal operators and applications*, Geom. Funct. Anal. **22** (2012), no. 5, 1289–1321, DOI 10.1007/s00039-012-0172-9. MR2989434

[19] Peter Constantin and Jiahong Wu, *Regularity of Hölder continuous solutions of the supercritical quasi-geostrophic equation*, Ann. Inst. H. Poincaré Anal. Non Linéaire **25** (2008), no. 6, 1103–1110, DOI 10.1016/j.anihpc.2007.10.001. MR2466323

[20] Mimi Dai, *Regularity criterion for the 3D Hall-magneto-hydrodynamics*, J. Differential Equations **261** (2016), no. 1, 573–591, DOI 10.1016/j.jde.2016.03.019. MR3487269

[21] Mimi Dai, *Regularity criterion and energy conservation for the supercritical quasi-geostrophic equation*, J. Math. Fluid Mech. **19** (2017), no. 2, 191–202, DOI 10.1007/s00021-017-0320-y. MR3649436

[22] Mimi Dai, *Regularity problem for the nematic LCD system with Q-tensor in $\mathbb{R}^3$*, SIAM J. Math. Anal. **49** (2017), no. 6, 5007–5030, DOI 10.1137/16M109137X. MR3735290

[23] M. Dai and H. Liu. *Low modes regularity criterion for a chemotaxis-Navier-Stokes system* arXiv:1805.08117

[24] Hongjie Dong and Nataša Pavlović, *A regularity criterion for the dissipative quasi-geostrophic equations*, Ann. Inst. H. Poincaré Anal. Non Linéaire **26** (2009), no. 5, 1607–1619, DOI 10.1016/j.anihpc.2008.08.001. MR2566702

[25] G. Duvaut and J.-L. Lions, *Inéquations en thermoélasticité et magnétohydrodynamique* (French), Arch. Rational Mech. Anal. **46** (1972), 241–279, DOI 10.1007/BF00250512. MR346289

[26] L. Iskauriaza, G. A. Serëgin, and V. Shverak, *$L_{3,\infty}$-solutions of Navier-Stokes equations and backward uniqueness* (Russian, with Russian summary), Uspekhi Mat. Nauk **58** (2003), no. 2(350), 3–44, DOI 10.1070/RM2003v058n02ABEH000609; English transl., Russian Math. Surveys **58** (2003), no. 2, 211–250. MR1992563

[27] Francisco Guillén-González and María Ángeles Rodríguez-Bellido, *Weak time regularity and uniqueness for a Q-tensor model*, SIAM J. Math. Anal. **46** (2014), no. 5, 3540–3567, DOI 10.1137/13095015X. MR3272623

[28] Loukas Grafakos, *Modern Fourier analysis*, 2nd ed., Graduate Texts in Mathematics, vol. 250, Springer, New York, 2009. MR2463316

[29] Cheng He and Yun Wang, *On the regularity criteria for weak solutions to the magnetohydrodynamic equations*, J. Differential Equations **238** (2007), no. 1, 1–17, DOI 10.1016/j.jde.2007.03.023. MR2334589

[30] Cheng He and Zhouping Xin, *On the regularity of weak solutions to the magnetohydrodynamic equations*, J. Differential Equations **213** (2005), no. 2, 235–254, DOI 10.1016/j.jde.2004.07.002. MR2142366

[31] Eberhard Hopf, *The partial differential equation $u_t + uu_x = \mu u_{xx}$*, Comm. Pure Appl. Math. **3** (1950), 201–230, DOI 10.1002/cpa.3160030302. MR0047234

[32] Jie Jiang, Hao Wu, and Songmu Zheng, *Blow-up for a three dimensional Keller-Segel model with consumption of chemoattractant*, J. Differential Equations **264** (2018), no. 8, 5432–5464, DOI 10.1016/j.jde.2018.01.004. MR3760180

[33] Tosio Kato, *Nonstationary flows of viscous and ideal fluids in $\mathbf{R}^3$*, J. Functional Analysis **9** (1972), 296–305, DOI 10.1016/0022-1236(72)90003-1. MR0481652

[34] A. Kiselev and F. Nazarov, *A variation on a theme of Caffarelli and Vasseur* (English, with English and Russian summaries), Zap. Nauchn. Sem. S.-Peterburg. Otdel. Mat. Inst. Steklov. (POMI) **370** (2009), no. Kraevye Zadachi Matematicheskoĭ Fiziki i Smezhnye Voprosy Teorii Funktsiĭ. 40, 58–72, 220, DOI 10.1007/s10958-010-9842-z; English transl., J. Math. Sci. (N.Y.) **166** (2010), no. 1, 31–39. MR2749211

[35] A. Kiselev, F. Nazarov, and A. Volberg, *Global well-posedness for the critical 2D dissipative quasi-geostrophic equation*, Invent. Math. **167** (2007), no. 3, 445–453, DOI 10.1007/s00222-006-0020-3. MR2276260

[36] A. Kolmogoroff, *The local structure of turbulence in incompressible viscous fluid for very large Reynold's numbers*, C. R. (Doklady) Acad. Sci. URSS (N.S.) **30** (1941), 301–305. MR0004146

[37] Hideo Kozono, Takayoshi Ogawa, and Yasushi Taniuchi, *The critical Sobolev inequalities in Besov spaces and regularity criterion to some semi-linear evolution equations*, Math. Z. **242** (2002), no. 2, 251–278, DOI 10.1007/s002090100332. MR1980623

[38] O. A. Ladyženskaja, *Uniqueness and smoothness of generalized solutions of Navier-Stokes equations* (Russian), Zap. Naučn. Sem. Leningrad. Otdel. Mat. Inst. Steklov. (LOMI) **5** (1967), 169–185. MR0236541

[39] Jean Leray, *Sur le mouvement d'un liquide visqueux emplissant l'espace* (French), Acta Math. **63** (1934), no. 1, 193–248, DOI 10.1007/BF02547354. MR1555394

[40] Alexander Lorz, *Coupled chemotaxis fluid model*, Math. Models Methods Appl. Sci. **20** (2010), no. 6, 987–1004, DOI 10.1142/S0218202510004507. MR2659745

[41] Marius Paicu and Arghir Zarnescu, *Global existence and regularity for the full coupled Navier-Stokes and Q-tensor system*, SIAM J. Math. Anal. **43** (2011), no. 5, 2009–2049, DOI 10.1137/10079224X. MR2837493

[42] Marius Paicu and Arghir Zarnescu, *Energy dissipation and regularity for a coupled Navier-Stokes and Q-tensor system*, Arch. Ration. Mech. Anal. **203** (2012), no. 1, 45–67, DOI 10.1007/s00205-011-0443-x. MR2864407

[43] Fabrice Planchon, *An extension of the Beale-Kato-Majda criterion for the Euler equations*, Comm. Math. Phys. **232** (2003), no. 2, 319–326, DOI 10.1007/s00220-002-0744-x. MR1953068

[44] Giovanni Prodi, *Un teorema di unicità per le equazioni di Navier-Stokes* (Italian), Ann. Mat. Pura Appl. (4) **48** (1959), 173–182, DOI 10.1007/BF02410664. MR126088

[45] J. Prüss and G. Simonett. *J. Prüss and G. Simonett.* Monographs in Mathematics, Birkhäuser, Basel, 2016.

[46] Michel Sermange and Roger Temam, *Some mathematical questions related to the MHD equations*, Comm. Pure Appl. Math. **36** (1983), no. 5, 635–664, DOI 10.1002/cpa.3160360506. MR716200

[47] James Serrin, *The initial value problem for the Navier-Stokes equations*, Nonlinear Problems (Proc. Sympos., Madison, Wis., 1962), Univ. of Wisconsin Press, Madison, Wis., 1963, pp. 69–98. MR0150444

[48] Michael Winkler, *Global large-data solutions in a chemotaxis-(Navier-)Stokes system modeling cellular swimming in fluid drops*, Comm. Partial Differential Equations **37** (2012), no. 2, 319–351, DOI 10.1080/03605302.2011.591865. MR2876834

Department of Mathematics, Statistics and Computer Science, University of Illinois Chicago, Chicago, Illinois 60607
*Email address*: mdai@uic.edu

Department of Mathematics, Statistics and Computer Science, University of Illinois Chicago, Chicago, Illinois 60607
*Email address*: hliu94@uic.edu

Contemporary Mathematics
Volume **748**, 2020
https://doi.org/10.1090/conm/748/15056

# Two classical properties of the Bessel quotient $I_{\nu+1}/I_\nu$ and their implications in pde's

Nicola Garofalo

*In ricordo di mio padre*

ABSTRACT. Two elementary and classical results about the Bessel quotient $y_\nu = \frac{I_{\nu+1}}{I_\nu}$ state that on the half-line $(0,\infty)$ one has for $\nu \geq -1/2$:
(i) $0 < y_\nu < 1$;
(ii) $y_\nu$ is strictly increasing.
In this paper we show that (i) and (ii) have some nontrivial and interesting applications to pde's. As a consequence of them, we establish some sharp new results for a class of degenerate partial differential equations of parabolic type in $\mathbb{R}^{n+1}_+ \times (0,\infty)$ which arise in connection with the analysis of the fractional heat operator $(\partial_t - \Delta)^s$ in $\mathbb{R}^n \times (0,\infty)$, see Theorems 1.2, 1.4, 1.5 and 1.7 below.

## 1. Introduction

A random variable $Y$ taking values from the nonnegative integers is called a Bessel random variable, or Bessel distribution, with parameters $\nu > -1$ and $z > 0$ if

$$\Pr(Y = k) = \frac{1}{I_\nu(z)\Gamma(k)\Gamma(k+\nu+1)}\left(\frac{z}{2}\right)^{2k+\nu}, \qquad k \in \mathbb{N}\cup\{0\},$$

where $I_\nu$ is the modified Bessel function of the first kind, see (5.6) in [**49**] and also (1.1) in [**61**]. The function $I_\nu$ is not very stable and it is often useful to consider the much more stable Bessel quotient $y_\nu = \frac{I_{\nu+1}}{I_\nu}$. On the real half-line $z \geq 0$ these two functions are connected by the equation

$$I_\nu(z) = \frac{1}{\Gamma(\nu+1)}\left(\frac{z}{2}\right)^\nu \exp\left(\int_0^z y_\nu(t)dt\right),$$

see Proposition 8.2 below.

The Bessel quotient $y_\nu$ plays an important role in a variety of problems from the applied sciences. For instance, it enters in the von Mises-Fisher distribution as the logarithmic derivative of the reciprocal of the norming constant. For $n \geq 2$

2010 *Mathematics Subject Classification.* 33A40, 35J70, 35H20.
*Key words and phrases.* Modified Bessel functions, curvature-dimension inequalities, Harnack inequalities, monotonicity formulas.
The author was supported in part by a Progetto SID (Investimento Strategico di Dipartimento) "Non-local operators in geometry and in free boundary problems, and their connection with the applied sciences", University of Padova, 2017.

let $\mathbb{S}^{n-1}$ be the unit sphere in $\mathbb{R}^n$ and indicate with $\sigma_{n-1} = 2\pi^{n/2}/\Gamma(n/2)$ its $(n-1)$-dimensional measure. Let $d\sigma$ indicate the normalized measure on $\mathbb{S}^{n-1}$, so that $\int_{\mathbb{S}^{n-1}} d\sigma(x) = 1$. A random vector $x \in \mathbb{S}^{n-1}$ has the $(n-1)$-dimensional *von Mises-Fisher* distribution $M_n(\omega, z)$ if its probability density function with respect to the uniform distribution is

$$f(x;\omega,z) = \left(\frac{z}{2}\right)^{\frac{n}{2}-1} \frac{1}{\Gamma(n/2) I_{\frac{n}{2}-1}(z)} \exp\{z < \omega, x >\}.$$

The parameters $\omega \in \mathbb{S}^{n-1}$ and $z \geq 0$ are respectively called the *mean direction* and the *concentration parameter* of the distribution. For $n = 2$, $M_2(\omega, z)$ is the distribution on the circle introduced in 1918 by R. von Mises in [**60**] to study the deviations of measured atomic weights from integral values. When $n = 3$, $M_3(\omega, z)$ are called the Fisher distributions since they were systematically studied in 1953 by R. Fisher, who used them to investigate statistical problems in paleomagnetism, see [**28**]. But Fisher distributions first appeared in Physics in the 1905 work of P. Langevin [**38**], where he showed that in a collection of weakly interacting dipoles of moments $\mathbf{m}$ subject to an external electric field, the directions of the dipoles $\frac{\mathbf{m}}{|\mathbf{m}|}$ have a Fisher distribution. For an interesting account on the von Mises-Fisher distributions we refer the reader to the book of Mardia and Jupp [**41**]. Also, the paper by Schou [**53**] contains various interesting statistical results about the concentration parameter of the distribution.

Using Cavalieri's Principle and the Poisson representation in (8.6) below, it is easy to recognize that for $z > 0$

$$\int_{\mathbb{S}^{n-1}} \exp\{z < \omega, x >\} d\sigma(x) = \Gamma(n/2) \left(\frac{z}{2}\right)^{1-\frac{n}{2}} I_{\frac{n}{2}-1}(z),$$

and thus it is clear that for any $\omega \in \mathbb{S}^{n-1}$ and $z \geq 0$

$$\int_{\mathbb{S}^{n-1}} f(x;\omega,z) d\sigma(x) = 1.$$

In view of this latter identity the function of $z$,

$$a_n(z) = \left(\frac{z}{2}\right)^{\frac{n}{2}-1} \frac{1}{\Gamma(n/2) I_{\frac{n}{2}-1}(z)},$$

is called the *norming constant* of the distribution. From formula (8.15) below we see that

$$\frac{d}{dz} \log a_n^{-1}(z) = \frac{d}{dz} \log \left(\frac{I_{\frac{n}{2}-1}(z)}{z^{\frac{n}{2}-1}}\right) = y_{\frac{n}{2}-1}(z).$$

This formula underscores the key role of the Bessel quotient $y_{n/2-1}$ in the von Mises-Fisher distribution $M_n(\omega, z)$.

In 1965 Raj Pal Soni established the following elementary, yet quite important, inequality concerning $y_\nu$, see [**56**] and also Proposition 8.6 below: for every $z > 0$ one has

$$y_\nu(z) < 1, \qquad \nu > -1/2. \tag{1.1}$$

Later, Nasell observed in [**45**] that (1.1) is also true when $\nu = -1/2$, see (8.23) below. The inequality (1.1) fails for large enough values of $z$ when $-1 < \nu < -1/2$, see the Appendix in this paper, and in particular Proposition 8.4 and Remark 8.5.

Another stronger property of the Bessel quotient is the following, see Proposition 8.8 below:

(i) when $\nu \geq -1/2$ the function $y_\nu$ strictly increases on $(0,\infty)$ from $y_\nu(0) = 0$ to its asymptotic value $y_\nu(\infty) = 1$;
(ii) if instead $-1 < \nu < -1/2$, then $y_\nu$ first increases to its absolute maximum $> 1$, and then it becomes strictly decreasing to its asymptotic value $y_\nu(\infty) = 1$.

It is clear that the strict monotonicity of $y_\nu$ in (i), combined with $y_\nu(\infty) = 1$, implies the inequality (1.1).

In this paper we show that such two elementary and classical results (1.1) and (i) about the Bessel quotient $y_\nu$ have some nontrivial and interesting applications to pde's. As a consequence of them, we establish some sharp new results for a class of degenerate partial differential equations of parabolic type in $\mathbb{R}^{n+1}_+ \times (0,\infty)$ which arise in connection with the analysis of the fractional heat operator $(\partial_t - \Delta)^s$ in $\mathbb{R}^n \times (0,\infty)$, see Theorems 1.2, 1.4, 1.5 and 1.7 below.

To set the stage, consider the Bessel process on the half-line $z > 0$,

$$\mathscr{B}_z^{(a)} u = z^{-a} \frac{d}{dz}\left(z^a \frac{d}{dz} u\right) = u_{zz} + \frac{a}{z} u_z. \tag{1.2}$$

This operator has a fractal dimension given by the number $a+1$. Since we are interested in a positive dimension, we assume henceforth that $a > -1$.

The Bessel process plays an ubiquitous role in many branches of pure and applied sciences. It is well known that if we consider the Laplace operator in $\mathbb{R}^n$, then $\Delta$ acts on functions $u(x) = f(z)$, which depend only on the distance to the origin $z = |x|$, as follows

$$\Delta u(x) = \mathscr{B}_z^{(n-1)} f(z).$$

More in general, if in $\mathbb{R}^n$ we consider a function with cylindrical symmetry $u(x_1, x_2, ..., x_n) = f(x_1, z)$, where $z = \sqrt{x_2^2 + ... + x_n^2}$, then letting $x_1 = x$, we have

$$\Delta u = \frac{\partial^2 f}{\partial x^2} + \frac{\partial^2 f}{\partial z^2} + \frac{n-2}{z}\frac{\partial f}{\partial z} = \frac{\partial^2 f}{\partial x^2} + \mathscr{B}_z^{(n-2)} f. \tag{1.3}$$

This observation was one of the motivating elements in the 1965 seminal paper of Muckenhoupt and Stein [**44**] (see equation (1.2) in their paper and the subsequent discussion). The operator in the right-hand side of (1.3) also arose in Molchanov's 1967 paper [**42**] on the Martin boundary for invariant processes on solvable groups. He focused on the group $\mathbb{G} = \begin{pmatrix} z & y \\ 0 & 1 \end{pmatrix}$ of affine transformations on the line, and on the following subclass

$$L_\nu = \partial_{yy} + \partial_{zz} + \frac{2\nu+1}{z}\partial_z = \partial_{yy} + \mathscr{B}_z^{(2\nu+1)} \tag{1.4}$$

of that of all left-invariant second-order elliptic operators with respect to such group action, see formulas (2) and (3) in [**42**]. In 1969 the work of Molchanov and Ostrovskii [**43**] introduced in probability the idea of traces of Bessel processes.

In 2007 Caffarelli and Silvestre's celebrated extension paper [**18**] gave a renewed prominence to the Bessel operator in pde's and free boundaries. Among other things, they showed that, if for a given $a \in (-1,1)$ and a $u \in \mathscr{S}(\mathbb{R}^n)$, one indicates with $U(X)$, with $X = (x,z) \in \mathbb{R}^{n+1}_+$, $x \in \mathbb{R}^n$, $z > 0$, the solution to the Dirichlet problem

$$\begin{cases} \mathscr{L}_a U = \operatorname{div}_X(z^a \nabla_X U) = 0 & \text{in } \mathbb{R}^{n+1}_+, \\ U(x,0) = u(x), \end{cases} \tag{1.5}$$

then with $s = \frac{1-a}{2} \in (0,1)$ the following Dirichlet-to-Neumann relation holds

$$(-\Delta)^s u(x) = - \frac{2^{-a}\Gamma\left(\frac{1-a}{2}\right)}{\Gamma\left(\frac{1+a}{2}\right)} \lim_{z\to 0^+} z^a \frac{\partial U}{\partial z}(X). \tag{1.6}$$

We note that when $s = 1/2$, then $a = 0$, and the operator in (1.5) is the standard Laplacean in $\mathbb{R}^{n+1}_+$.

The Caffarelli-Silvestre extension procedure (1.5) has played a pivotal role in the analysis of nonlocal operators such as $(-\Delta)^s$, since via (1.6) it allows to turn problems involving the latter into ones involving the differential (local) operator $\mathscr{L}_a$. The Bessel operator occupies a central position in such procedure since the extension operator $\mathscr{L}_a$ can be written as follows

$$\mathscr{L}_a = z^a(\Delta_x + \mathscr{B}_z^{(a)}). \tag{1.7}$$

The reader should note the similarity between (1.3), (1.4) and (1.7). In this perspective, one should think of the operator between parenthesis in the right-hand side of (1.7) as the standard Laplacean in the space $\mathbb{R}^{n+a+1}$ of fractal dimension $n+a+1$, with variables $(x,y)$, where $x \in \mathbb{R}^n$ and $y \in \mathbb{R}^{a+1}$, acting on a "cylindrical" function $u(x,y) = f(x,z)$, with $z = |y|$. In more recent years Stinga and Torrea have generalized the extension procedure to different classes of operators, including uniformly elliptic operators in divergence form $L = \operatorname{div}(A(x)\nabla)$, see **[57]**, or the heat operator $H = \frac{\partial}{\partial t} - \Delta_x$, see **[58]**. This latter result was also established simultaneously and independently by Nyström and Sande in **[46]**.

Motivated by such developments, and also by the new ones in **[9]**, **[8]**, **[5]**, **[30]**, **[10]** and **[31]**, in this paper we establish some properties of the Bessel semigroup, and provide some interesting applications of these results to the following degenerate parabolic operator in $\mathbb{R}^{n+1}_+ \times (0,\infty)$

$$\partial_t(z^a U) - \mathscr{L}_a U = \partial_t(z^a U) - \operatorname{div}_X(z^a \nabla_X U), \tag{1.8}$$

where $U = U(X,t)$ is a function defined in $\mathbb{R}^{n+1}_+ \times (0,\infty)$. Here, we have kept with the notations introduced before (1.5) above. We mention that, similarly to its elliptic predecessor (1.5), the operator (1.8) is the extension operator for the fractional powers $(\partial_t - \Delta)^s$, $0 < s < 1$, of the heat operator, see **[46]** and **[58]**.

A key remark concerning (1.8) is that it belongs to a general class of equations first introduced by Chiarenza and Serapioni in **[18]**. These authors considered degenerate parabolic equations in $\mathbb{R}^{n+1}$ of the type

$$\partial_t(\omega(x)u) - \operatorname{div}(A(x)\nabla u) = 0, \tag{1.9}$$

where $\omega$ is a $A_2$-weight of Muckenhoupt in $\mathbb{R}^n$, and $A(x)$ is a matrix-valued function with bounded measurable coefficients, for which $A(x) = A(x)^T$, and such that

$$< A(x)\xi, \xi > \cong \omega(x)|\xi|^2.$$

In their main result, they proved that nonnegative solutions of (1.9) satisfy a scale invariant Harnack inequality on the standard parabolic cylinders. Such result proved to be the appropriate parabolic counterpart of the elliptic one previously obtained by Fabes, Kenig and Serapioni in **[26]**. We note here that, since $\omega(X) = |z|^a$

belongs to $A_2(\mathbb{R}^{n+1})$ if and only if $|a| < 1$, the model equation (1.8) is a special case of (1.9).

The connection between (1.8) and the Bessel semigroup is in the fact that, similarly to (1.7), we can alternatively write the extension operator as

$$\partial_t(z^a U) - \mathscr{L}_a U = z^a(\partial_t - \Delta - \mathscr{B}_z^{(a)}).$$

The parabolic operator (1.8) has recently received increasing attention. In connection with the parabolic Signorini problem, which is intimately linked to the obstacle problem for the fractional power $(\partial_t - \Delta)^{1/2}$, the analysis of the case $a = 0$ in (1.8) was extensively developed in the monograph [**20**]. The general case $-1 < a < 1$ was studied in [**9**] in the problem of the unique continuation backward in time. In the paper [**4**] the authors established the optimal interior regularity of the solutions of the thin obstacle problem for (1.8). In [**5**] the authors obtained various interesting results on the nodal sets of the solutions of (1.8). The paper [**30**] studied the extension problem for hypoelliptic sub-Laplaceans of Hörmander type. Finally, in the forthcoming article [**8**] the authors develop the analysis of the singular part of the free boundary in the thin obstacle problem studied in [**4**]. Of course, this list of works is far from being exhaustive.

To introduce our results we recall that in their celebrated work [**40**] Li and Yau proved (among other things) that if $f > 0$ is a solution of the heat equation $\partial_t f - \Delta f = 0$ on a boundaryless, complete $n$-dimensional Riemannan manifold $\mathbb{M}$ having Ricci $\geq 0$, then its entropy $u = \log f$ satisfies the (deep) inequality on $\mathbb{M} \times (0, \infty)$,

$$|\nabla u|^2 - \partial_t u \leq \frac{n}{2t}. \tag{1.10}$$

We mention that (1.10) becomes an equality when $f$ is the heat kernel in flat $\mathbb{R}^n$, see (4.1) below. The importance of the inequality (1.10) is underscored by the fact that a direct remarkable consequence of it is the following sharp form of the Harnack inequality, valid for any $x, y \in \mathbb{M}$ and any $0 < s < t < \infty$,

$$f(x, s) \leq f(y, t) \left(\frac{t}{s}\right)^{\frac{n}{2}} \exp\left(\frac{d(x, y)^2}{4t}\right). \tag{1.11}$$

Such Harnack inequality is keen to that proved independently by Hadamard [**33**] and Pini [**47**] for the standard heat equation in the plane. One remarkable aspect of (1.11) is that the constant $\left(\frac{t}{s}\right)^{\frac{n}{2}} \exp\left(\frac{d(x,y)^2}{4t}\right)$ in its right-hand side is explicit and best possible.

In this note we start from a seemingly very simple problem. Namely, we consider the Cauchy problem for the Bessel operator $\mathscr{B}_z^{(a)}$ on the half-line $\{z > 0\}$, with Neumann boundary condition (or *Feller's zero-flux condition*, see Section 5 below),

$$\begin{cases} \partial_t u - \mathscr{B}_z^{(a)} u = 0, \\ u(z, 0) = \varphi(z), \\ \lim_{z \to 0^+} z^a \partial_z u(z, t) = 0. \end{cases} \tag{1.12}$$

This corresponds to Brownian motion on the half-line $(0, \infty)$ reflected at $z = 0$, as opposed to killed Brownian motion, when a Dirichlet condition is imposed. We denote by $\{P_t^{(a)}\}_{t\geq 0}$ the semigroup associated with (1.12) and given by the formula (2.3) below. For the definition of the space $C^1_{(a)}(0,\infty)$ see Section 2 below. Our first main result is the following.

THEOREM 1.1 (Li-Yau type inequality). *Let $a \geq 0$. Given a function $\varphi \geq 0$ such that $\varphi \in \mathscr{C}^1_{(a)}(0,\infty)$, we have for any $z>0$ and $t>0$,*

$$\left(\partial_z \log P_t^{(a)}\varphi(z)\right)^2 - \partial_t \log P_t^{(a)}\varphi(z) < \frac{a+1}{2t}. \tag{1.13}$$

*When $z=0$ the inequality (1.13) is true for every $a>-1$ and with $\leq$ instead of $<$.*

Using Theorem 1.1 we then prove the following sharp result.

THEOREM 1.2 (Harnack inequality). *Let $a \geq 0$. For every $\varphi \geq 0$ such that $\varphi \in \mathscr{C}^1_{(a)}(0,\infty)$, we have for $z,\zeta \in \mathbb{R}^+$ and $0<s<t<\infty$,*

$$P_s^{(a)}\varphi(z) < P_t^{(a)}\varphi(\zeta) \left(\frac{t}{s}\right)^{\frac{a+1}{2}} \exp\left(\frac{(z-\zeta)^2}{4(t-s)}\right). \tag{1.14}$$

The reader should notice the striking similarity between (1.10), (1.11) above, and (1.13), (1.14). We emphasize that the factor $\frac{(z-\zeta)^2}{4(t-s)}$ in the exponential in the right-hand side of (1.14) reflects the invariance of the heat operator $\partial_t - \mathscr{B}_z^{(a)}$ with respect to the dilations $(z,t)\to(\lambda z,\lambda^2 t)$, whereas the factor $\left(\frac{t}{s}\right)^{\frac{a+1}{2}}$ indicates that the number $Q = a+1$ plays the role of a fractal dimension for the semigroup $\{P_t^{(a)}\}_{t\geq 0}$.

We also stress another aspect of Theorem 1.2 that should not go unnoticed. Consider the quadrant $\mathscr{Q}^+ = \{(z,t)\in\mathbb{R}^2 \mid z>0,\ t>0\}$. In every (elliptic or parabolic) Harnack inequality one expects the constant which multiplies the term in the right-hand side to blow-up as one approaches the boundary of the relevant domain. This does not happen for (1.14), as the factor $\left(\frac{t}{s}\right)^{\frac{a+1}{2}} \exp\left(\frac{(z-\zeta)^2}{4(t-s)}\right)$ does not seem to *see* the vertical portion $\{(0,t)\in\mathbb{R}^2 \mid t>0\}$ of $\partial\mathscr{Q}^+$, exactly as for the global inequality (1.11) above, in which there is no boundary. For instance, we can let $\zeta\to 0^+$, or even let $z,\zeta\to 0^+$, without causing the factor $\left(\frac{t}{s}\right)^{\frac{a+1}{2}} \exp\left(\frac{(z-\zeta)^2}{4(t-s)}\right)$ to blow-up. In other words, Theorem 1.2 behaves like an *interior* Harnack inequality in the whole half-plane $\mathscr{Q} = \{(z,t)\in\mathbb{R}^2 \mid z\in\mathbb{R},\ t>0\}$. The explanation for this is in the vanishing Neumann condition in (1.12) above. Such condition implies that solutions to (1.12) be smooth in $z$ (in fact, real analytic) up to the vertical line $z=0$. Therefore, if one defines $U(z,t)=u(|z|,t)$, one obtains a global solution on $\mathbb{R}\times(0,\infty)$ of the pde.

What is remarkable about Theorem 1.1 is that it ultimately hinges on the above inequality (1.1) for the Bessel quotient $y_\nu$. The key connection between Li-Yau and (1.1) is the identity (4.2) in Proposition 4.1 below. Since the link between $\nu$ and the parameter $a$ in (1.2) is given by the equation $\nu = \frac{a-1}{2}$, from our work in Section 4 it will be evident that, in its sharp form (1.13) above, such inequality fails to hold when $-1<a<0$, see also Section 8.

Having said this, the question naturally arises of whether our approach can be pushed to establish a Harnack inequality for the semigroup $\{P_t^{(a)}\}_{t\geq 0}$ also in the range $-1<a<0$. We presently only have some inconclusive indication about the answer. Nonetheless, we emphasize that a Harnack inequality in the range

$a \in (-1, 0)$ is in fact known. For this one can invoke either Theorem 2.1 in [**18**], or Theorem 4.1 in the more recent work [**23**]. We have already discussed the Harnack inequality in [**18**]. In [**23**] the authors prove a Harnack inequality for a general class of degenerate parabolic operators on manifolds with corners which arise in population biology. The Kimura equations treated in [**23**] include as special case the model Cauchy problem (5.3), (5.4) below. As we show in Proposition 5.1, such model is equivalent to the problem (1.12), and thus a Harnack inequality for the latter can be obtained from the cited [**23**, Theorem 4.1].

We now discuss the second set of main results in this paper. Given a function $\varphi \in C_0^\infty(\mathbb{R}^{n+1}_+)$, consider the Cauchy problem with Neumann condition for the operator (1.8) above

$$
\begin{cases}
\partial_t(z^a U) - \mathscr{L}_a U = 0 & \text{in } \mathbb{R}^{n+1}_+ \times (0,\infty) \\
U(X,0) = \varphi(X), & X \in \mathbb{R}^{n+1}_+, \\
\lim_{z\to 0^+} z^a \partial_z U(x,z,t) = 0. &
\end{cases}
\tag{1.15}
$$

The solution to (1.15) is represented by the formula

$$
U(X,t) = \mathscr{P}_t^{(a)}\varphi(X) = \int_{\mathbb{R}^{n+1}_+} \varphi(Y)\mathscr{G}_a(X,Y,t)\zeta^a dY,
\tag{1.16}
$$

where $\mathscr{G}_a(X,Y,t)$ is given in (6.2) below. We note that $\{\mathscr{P}_t^{(a)}\}_{t\geq 0}$ defines a semigroup on $C_0^\infty(\mathbb{R}^{n+1}_+)$. Concerning such semigroup we have the following sharp Harnack inequality.

THEOREM 1.3. *Let $a \geq 0$. Let $\varphi \geq 0$ be a function for which $U$ given by* (1.16) *represents a classical solution to* (1.15). *Then, for every $X, Y \in \mathbb{R}^{n+1}_+$ and every $0 < s < t < \infty$, we have*

$$
U(X,s) \le U(Y,t) \left(\frac{t}{s}\right)^{\frac{n+a+1}{2}} \exp\left(\frac{|X-Y|^2}{4(t-s)}\right).
\tag{1.17}
$$

*When $X = (x,0), Y = (y,0)$ the inequality is valid in the full range $a > -1$, and becomes*

$$
U(x,0,s) \le U(y,0,t) \left(\frac{t}{s}\right)^{\frac{n+a+1}{2}} \exp\left(\frac{|x-y|^2}{4(t-s)}\right).
\tag{1.18}
$$

Once again, one should compare Theorem 1.3 to the Harnack inequality (1.11) of Li and Yau. Concerning Theorem 1.3 two comments are in order:

1) For any $a \in (-1,1)$ a Harnack inequality for positive solutions of (1.15) on the parabolic cylinders $B(r) \times (\alpha r^2, \beta r^2)$, where $B(r)$ is a Euclidean ball in $\mathbb{R}^{n+1}$, can be obtained from Theorem 2.1 in [**18**]. One must first prove that the Neumann condition (1.15) above implies that $U$ is smooth in $z$ up to the thin manifold $\mathbb{M} = (\mathbb{R}^n \times \{0\}) \times (0,\infty)$, and then show that the even reflection of $U$ in $z$ is a solution across such manifold. At that point, one can appeal to the above cited interior Harnack inequality in [**18**]. The novelty in Theorem 1.3 with respect to such approach is that it produces the explicit sharp constant $\left(\frac{t}{s}\right)^{\frac{n+a+1}{2}} \exp\left(\frac{|X-Y|^2}{4(t-s)}\right)$. Furthermore, our direct proof has already encoded the information of being an "interior" Harnack inequality, and we do not need to even reflect

the solution across the thin manifold $\mathbb{M}$. In fact, with $X = (x,z)$ and $Y = (y,\zeta)$, we can let either $z$ or $\zeta$, or both tend to zero, and yet the constant $\exp\left(\frac{|X-Y|^2}{4(t-s)}\right) = \exp\left(\frac{|x-y|^2+(z-\zeta)^2}{4(t-s)}\right)$ does not blow up, see (1.18) above.

2) Theorem 1.3 is valid in the whole range $a \geq 0$, whereas Theorem 2.1 in [**18**] is not applicable when $a \geq 1$, since in such range $\omega(X) = |z|^a$ is not even locally integrable.

The proof of Theorem 1.3 is based on the following inequality of Li-Yau type.

THEOREM 1.4. *Let $a \geq 0$ and $\varphi$ and $U$ be as in Theorem 1.3. Then, for any $X \in \mathbb{R}^{n+1}_+$ and $t > 0$ one has*

$$|\nabla_X \log U(X,t)|^2 - \partial_t \log U(X,t) < \frac{n+a+1}{2t}. \tag{1.19}$$

*If instead $X = (x,0), Y = (0,y) \in \mathbb{R}^n \times \{0\}$, then the inequality (1.19) is true for every $a > -1$, in the following form*

$$|\nabla_X \log U(x,0,t)|^2 - \partial_t \log U(x,0,t) \leq \frac{n+a+1}{2t}. \tag{1.20}$$

It is remarkable that, similarly to that of Theorem 1.1, also Theorem 1.4 ultimately rests on the elementary inequality (1.1) above.

The final set of results in this paper has to do with monotonicity formulas. In the paper [**9**] the authors have studied the problem of strong unique continuation backward in time for the nonlocal equation

$$(\partial_t - \Delta)^s u = V(x,t)u, \qquad 0 < s < 1.$$

Given $s$ in such range, let $a = 1 - 2s$. One of the central results in [**9**] was a monotonicity formula for solutions of the extension operator (1.8) above, with the Neumann condition

$$-\frac{2^{-a}\Gamma\left(\frac{1-a}{2}\right)}{\Gamma\left(\frac{1+a}{2}\right)} \lim_{z\to 0^+} z^a \partial_z U(x,z,t) = V(x,t)u(x,t).$$

Monotonicity formulas for the heat equation go back to the work of Struwe [**59**]. In [**48**] Poon first proved a monotonicity formula for the heat equation analogous to the celebrated one established by Almgren in [**2**] for multiple-valued harmonic functions. The work [**20**] contains, among other things, generalizations of Poon's result to the parabolic Signorini problem. This is the case $a = 0$ of the the extension operator (1.8) above. The forthcoming article [**8**] extends the results in [**20**] to the full range $a \in (-1,1)$.

We emphasize that the problems studied in [**20**] and [**8**] are thin obstacle problems for the operator (1.8) in which the free boundary lives in the thin manifold $\{z = 0\}$. Because of this reason, in all the results in these papers the various quantities at play are "centered" at one convenient point, the origin, of the thin manifold. By this we mean that the relevant Gaussian measures in the relevant monotonicity formulas are centered at the origin. Such choice is for all practical purposes immaterial if one deals with a problem in which the focus is the thin manifold. But it becomes relevant in situations when this is no longer the case.

Such considerations lead to the question of the stability of monotonicity properties in dependence of the center of the Gaussian measure. A natural testing ground in this direction is the prototypical non-translation invariant parabolic pde in (1.12) above. We have discovered that, interestingly, there is a discrepancy in the resulting monotonicity formulas, according to whether the "center" of the relevant Gaussian measure is located at a point $(\zeta, t)$ with $\zeta > 0$, or $\zeta = 0$. Such discrepancy is similar to that in Theorems 1.1 and 1.3, but it no longer rests on (1.1) above. Remarkably, the monotonicity of the relevant energy and frequency functions now ultimately depends on the monotonicity of the Bessel quotient $y_\nu$. For the proof of such property see Proposition 8.8 below.

For the definition of the scaled energy, with respect to the backward Gaussian-Bessel measure centered at $(z, T)$, $E^{(a)}_{z,T}(t)$, the reader should see (7.3) below. The following are our main results.

THEOREM 1.5 (Struwe type monotonicity formula). *Suppose that $u$ be a solution to (7.1) in $(0,\infty)\times(0,\infty)$ satisfying the condition (7.2) (and other reasonable growth assumptions). Then, for any fixed $z > 0$ the function $t \to E^{(a)}_{z,T}(t)$ is strictly decreasing on $(0, T)$ when $a \geq 0$. Precisely, we have*

$$\frac{dE^{(a)}_{z,T}}{dt}(t) = -(T-t)\int_0^\infty \left[u_t + u_\zeta \frac{p^{(a)}_\zeta}{p^{(a)}}\right]^2 p^{(a)}\zeta^a d\zeta + G^{(a)}_{z,T}(t), \tag{1.21}$$

*where*

$$G^{(a)}_{z,T}(t) = -(T-t)\int_0^\infty u_\zeta^2 \left[\frac{p^{(a)}_{\zeta\zeta}}{p^{(a)}} - \frac{(p^{(a)}_\zeta)^2}{(p^{(a)})^2} + \frac{1}{2(T-t)}\right] p^{(a)}\zeta^a d\zeta < 0. \tag{1.22}$$

*When $z = 0$, then for any $a > -1$ we have $G^{(a)}_T(t) = G^{(a)}_{0,T}(t) \equiv 0$, and the function $t \to E^{(a)}_T(t)$ is monotone decreasing on $(0, T)$.*

REMARK 1.6. We emphasize the remarkable discrepancy in Theorem 1.5 between the case $z > 0$, in which the (strict) monotonicity of $t \to E^{(a)}_{z,T}(t)$ holds only when $a \geq 0$, and that when $z = 0$, in which we have monotonicity in the full range $a > -1$.

The next theorem is the second main result about monotonicity formulas. For the meaning of the frequency centered at $z \geq 0$, $N^{(a)}_z(r)$, we refer the reader to Definition 7.3 below. When $z = 0$ we simply write $N^{(a)}(r)$.

THEOREM 1.7 (Poon type monotonicity formula). *Let $u$ be a solution to (7.1), satisfying (7.2). For any $z > 0$ the frequency $r \to N^{(a)}_z(r)$ is strictly increasing when $a \geq 0$. If instead $z = 0$, then the frequency is non-decreasing for any $a > -1$. Furthermore, in this second case we have $N^{(a)}(r) \equiv \kappa$ if and only if $u$ is homogeneous of degree $\kappa$ with respect to the parabolic dilations $(\zeta, t) \to (\lambda\zeta, \lambda^2 t)$.*

## 2. The Bessel semigroup

In this section we collect some known facts concerning the Cauchy problem for the Bessel operator with Neumann boundary condition

$$\begin{cases} \partial_t u - \mathscr{B}^{(a)}_z u = 0, \\ u(z,0) = \varphi(z), \\ \lim\limits_{z\to 0^+} z^a \partial_z u(z,t) = 0. \end{cases} \tag{2.1}$$

We begin by introducing the following classes of functions

$$\mathscr{C}_{(a)}(0,\infty)=\left\{\varphi\in C(0,\infty)\mid\int_0^R|\varphi(z)|z^a dz<\infty,\quad\int_R^\infty|\varphi(z)|z^{\frac{a}{2}}dz<\infty,\forall R>0\right\},$$

and

$$\mathscr{C}^1_{(a)}(0,\infty)=\left\{\varphi\in C^1(0,\infty)\mid\varphi,\frac{1}{z}\varphi'\in\mathscr{C}_{(a)}(0,\infty)\right\}.$$

As it was observed in (22.8) of [**29**], membership in $\mathscr{C}^1_{(a)}(0,\infty)$ imposes, in particular, the weak *Neumann condition*

$$\liminf_{z\to 0^+} z^a|\varphi'(z)|=0. \tag{2.2}$$

PROPOSITION 2.1. *Given* $\varphi\in\mathscr{C}^1_{(a)}(0,\infty)$, *the Cauchy problem* (2.1) *admits the following solution*

$$u(z,t)=P_t^{(a)}\varphi(z)\overset{def}{=}\int_0^\infty\varphi(\zeta)p^{(a)}(z,\zeta,t)\zeta^a d\zeta, \tag{2.3}$$

*where for* $z,\zeta,t>0$ *we have denoted by*

$$\begin{aligned}p^{(a)}(z,\zeta,t)&=(2t)^{-\frac{a+1}{2}}\left(\frac{z\zeta}{2t}\right)^{\frac{1-a}{2}}I_{\frac{a-1}{2}}\left(\frac{z\zeta}{2t}\right)e^{-\frac{z^2+\zeta^2}{4t}}\\&=\frac{1}{2t}(z\zeta)^{\frac{1-a}{2}}I_{\frac{a-1}{2}}\left(\frac{z\zeta}{2t}\right)e^{-\frac{z^2+\zeta^2}{4t}},\end{aligned} \tag{2.4}$$

*the heat kernel of* $\mathscr{B}_z^{(a)}$ *on* $(\mathbb{R}^+,z^a dz)$, *with Neumann boundary conditions. For* $t\le 0$ *we set* $p^{(a)}(z,\zeta,t)\equiv 0$.

In (2.4) we have denoted by $I_\nu(z)$ the modified Bessel function of the first kind and order $\nu\in\mathbb{C}$ defined by the series (8.1) below. Formulas (2.3), (2.4) are well-known to workers in probability (see for instance formula (6.14) on p. 238 in [**36**]), but not equally known to those in partial differential equations. For a direct proof based exclusively on analytic tools we refer the reader to Proposition 22.3 in [**29**]. Another analytical proof can be found in the paper [**21**], where Epstein and Mazzeo construct the fundamental solution (5.6) for the Cauchy problem (5.3) below. For this aspect we refer the reader to Section 5.

We next collect some important properties of the Bessel heat kernel $p^{(a)}(z,\zeta,t)$ in (2.4) above. Since we have not found in the literature a direct source which is suitable for workers in analysis, we provide details of their proofs. We begin by noting the following simple facts:

(i) $p^{(a)}(z,\zeta,t)>0$ for every $z,\zeta>0$ and $t>0$;
(ii) $p^{(a)}(z,\zeta,t)=p^{(a)}(\zeta,z,t)$;
(iii) $p^{(a)}(\lambda z,\lambda\zeta,\lambda^2 t)=\lambda^{-(a+1)}p^{(a)}(\zeta,z,t)$.

Property (i) follows from the fact that $I_\nu(z)>0$ for any $z>0$, and any $\nu\ge -1$, see (8.1) and the comments following it. Property (ii) is obvious from (2.4) and it is a reflection of the symmetry of the operator $\mathscr{B}_z^{(a)}$ on $(0,\infty)$ equipped with the measure $d\mu^{(a)}(z)=z^a dz$. Property (iii) reflects the invariance of the heat operator $\partial_t-\mathscr{B}_z^{(a)}$ with respect to the parabolic scalings $\lambda\to(\lambda z,\lambda^2 t)$. In particular, (iii) implies that

$$p^{(a)}(z,\zeta,t)=t^{-\frac{a+1}{2}}p^{(a)}(\frac{z}{\sqrt{t}},\frac{\zeta}{\sqrt{t}},1).$$

From (2.4) and (8.2) below we see that for any $\zeta > 0$ and $t > 0$,

$$p^{(a)}(0,\zeta,t) = \lim_{z\to 0^+} p^{(a)}(z,\zeta,t) = \frac{1}{2^a\Gamma(\frac{a+1}{2})} t^{-\frac{a+1}{2}} e^{-\frac{\zeta^2}{4t}}. \tag{2.5}$$

We also have for any $\zeta > 0$ and $t > 0$,

$$\lim_{z\to 0^+} z^a \partial_z p^{(a)}(z,\zeta,t) = 0. \tag{2.6}$$

The limit relation (2.6) can be proved using (4.5) and (8.10) below. Since $p^{(a)}(z,\zeta,t)$ is the Neumann fundamental solution for problem (2.1), the property (2.6) should come as no surprise.

REMARK 2.2. We note explicitly that although $\mathscr{B}_z^{(a)}$, originally defined on $C_0^\infty(0,\infty)$, is symmetric with respect to the measure $d\mu_{(a)} = z^a dz$ for every $a > -1$, it is essentially self-adjoint only when either $a < 0$, or $a > 2$. For this see [**7**, Proposition 2.4.1].

If we fix $z > 0, t > 0$, then the asymptotic behavior of $p^{(a)}(z,\zeta,t)\zeta^a$ as $\zeta \to 0^+$, or $\zeta \to \infty$, follows from that of the Bessel function $I_\nu$. Keeping in mind that (8.2) and (8.5) give

$$I_{\frac{a-1}{2}}(z) \cong \begin{cases} \dfrac{z^{\frac{a-1}{2}}}{2^{\frac{a-1}{2}}\Gamma(\frac{a+1}{2})}, & \text{as } z \to 0^+, \\ \frac{e^z}{(2\pi z)^{1/2}}\left(1 + O(|z|^{-1})\right), & \text{as } z \to \infty, \end{cases} \tag{2.7}$$

from (2.4) and (2.7) we see that for every fixed $(z,t) \in (0,\infty)\times(0,\infty)$

$$p^{(a)}(z,\zeta,t)\zeta^a = \begin{cases} O(\zeta^a), & \text{as } \zeta \to 0^+, \\ O(\zeta^{\frac{a}{2}} e^{-\frac{(\zeta-z)^2}{4t}}), & \text{as } \zeta \to \infty, \end{cases} \tag{2.8}$$

Since $a > -1$ we infer from (2.8) that for every fixed $(z,t) \in (0.\infty)\times(0,\infty)$,

$$\int_0^\infty p^{(a)}(z,\zeta,t)\zeta^a d\zeta < \infty.$$

Thus, it is possible to consider $P_t^{(a)}\varphi$ for every $\varphi \in L^\infty(0,\infty)$. In particular, it makes sense to consider $P_t^{(a)}1$. The next result provides an important information in this connection.

PROPOSITION 2.3 (Stochastic completeness). *Let $a > -1$. For every $(z,t) \in (0,\infty)\times(0,\infty)$ one has*

$$P_t^{(a)}1(z) = \int_0^\infty p^{(a)}(z,\zeta,t)\zeta^a d\zeta = 1.$$

PROOF. We have

$$\int_0^\infty p^{(a)}(z,\zeta,t)\zeta^a d\zeta = (2t)^{-\frac{a+1}{2}} e^{-\frac{z^2}{4t}} \int_0^\infty \left(\frac{z\zeta}{2t}\right)^{\frac{1-a}{2}} I_{\frac{a-1}{2}}\left(\frac{z\zeta}{2t}\right) e^{-\frac{\zeta^2}{4t}} \zeta^{a+1}\frac{d\zeta}{\zeta}$$

$$\left(\text{change of variable } y = \frac{z\zeta}{2t}, \text{ so that } \zeta = \frac{2t}{z}y\right)$$

$$= \frac{(2t)^{-\frac{a+1}{2}} e^{-\frac{z^2}{4t}} 2^{a+1} t^{a+1}}{z^{a+1}} \int_0^\infty y^{\frac{a+1}{2}} I_{\frac{a-1}{2}}(y) e^{-\frac{t}{z^2}y^2} dy$$

$$= \frac{e^{-\frac{z^2}{4t}} 2^{\frac{a+1}{2}} t^{\frac{a+1}{2}}}{z^{a+1}} \int_0^\infty y^{\frac{a+1}{2}} I_{\frac{a-1}{2}}(y) e^{-\frac{t}{z^2}y^2} dy$$

If we set

$$\nu = \frac{a-1}{2},$$

then $\nu > -1$ and $\nu + 1 = \frac{a+1}{2}$. Applying Lemma 8.1 below with such choice of $\nu$ and $\alpha = \frac{t}{z^2}$, we find

$$\int_0^\infty y^{\frac{a+1}{2}} I_{\frac{a-1}{2}}(y) e^{-\frac{t}{z^2}y^2} dy = \frac{e^{1/4\alpha}}{2^{\nu+1}\alpha^{\nu+1}} = \frac{e^{\frac{z^2}{4t}} z^{a+1}}{2^{\frac{a+1}{2}} t^{\frac{a+1}{2}}}.$$

This proves the proposition. □

We next prove that (2.3) defines a semigroup of operators.

PROPOSITION 2.4 (Chapman-Kolmogorov equation). *Let $a > -1$. For every $z, \eta > 0$ and every $0 < s, t < \infty$ one has*

$$p^{(a)}(z,\eta,t+s) = \int_0^\infty p^{(a)}(z,\zeta,t) p^{(a)}(\zeta,\eta,s)\zeta^a d\zeta.$$

PROOF. We begin with the right-hand side in the above equation

$$\int_0^\infty p^{(a)}(z,\zeta,t) p^{(a)}(\zeta,\eta,s)\zeta^a d\zeta$$

$$= (2t)^{-1}(2s)^{-1}(z\eta)^{\frac{1-a}{2}} e^{-\frac{z^2}{4t}} e^{-\frac{\eta^2}{4s}} \int_0^\infty \zeta^{1-a} I_{\frac{a-1}{2}}\left(\frac{z\zeta}{2t}\right) I_{\frac{a-1}{2}}\left(\frac{\eta\zeta}{2s}\right) e^{-\frac{\zeta^2}{4t}} e^{-\frac{\zeta^2}{4s}} \zeta^a d\zeta$$

$$= (2t)^{-1}(2s)^{-1}(z\eta)^{\frac{1-a}{2}} e^{-\frac{z^2}{4t}} e^{-\frac{\eta^2}{4s}} \int_0^\infty \zeta e^{-\left(\frac{1}{4t}+\frac{1}{4s}\right)\zeta^2} I_{\frac{a-1}{2}}\left(\frac{z\zeta}{2t}\right) I_{\frac{a-1}{2}}\left(\frac{\eta\zeta}{2s}\right) d\zeta$$

At this point we invoke the following formula, see e.g. no. 8 on p. 321 in [**50**],

$$\int_0^\infty \zeta e^{-p\zeta^2} I_\nu(b\zeta) I_\nu(c\zeta) d\zeta = \frac{1}{2p} e^{\frac{b^2+c^2}{4p}} I_\nu\left(\frac{bc}{2p}\right), \qquad \nu > -1,\ \Re p > 0. \tag{2.9}$$

Applying (2.9) with $\nu = \frac{a-1}{2}$ and

$$p = \frac{1}{4t} + \frac{1}{4s} = \frac{t+s}{4ts}, \qquad b = \frac{z}{2t}, \qquad c = \frac{\eta}{2s},$$

we find after some elementary reductions

$$\int_0^\infty \zeta e^{-\left(\frac{1}{4t}+\frac{1}{4s}\right)\zeta^2} I_{\frac{a-1}{2}}\left(\frac{z\zeta}{2t}\right) I_{\frac{a-1}{2}}\left(\frac{\eta\zeta}{2s}\right) d\zeta = \frac{2ts}{t+s} e^{\frac{s}{4t(t+s)}z^2} e^{\frac{t}{4s(t+s)}\eta^2} I_{\frac{a-1}{2}}\left(\frac{z\eta}{2(t+s)}\right).$$

Substituting this identity in the above equation and simplifying we obtain

$$\begin{aligned}
&\int_0^\infty p^{(a)}(z,\zeta,t)p^{(a)}(\zeta,\eta,s)\zeta^a d\zeta \\
&= (2t)^{-1}(2s)^{-1}(z\eta)^{\frac{1-a}{2}} e^{-\frac{z^2}{4t}} e^{-\frac{\eta^2}{4s}} \frac{2ts}{t+s} e^{\frac{s}{4t(t+s)}z^2} e^{\frac{t}{4s(t+s)}\eta^2} I_{\frac{a-1}{2}}\left(\frac{\zeta\eta}{2(t+s)}\right) \\
&= \frac{1}{2(t+s)}(z\eta)^{\frac{1-a}{2}} e^{-\frac{z^2}{4t}(1-\frac{s}{t+s})} e^{-\frac{\eta^2}{4s}(1-\frac{t}{t+s})} I_{\frac{a-1}{2}}\left(\frac{\zeta\eta}{2(t+s)}\right) \\
&= \frac{1}{2(t+s)}(z\eta)^{\frac{1-a}{2}} e^{-\frac{z^2+\eta^2}{4(t+s)}} I_{\frac{a-1}{2}}\left(\frac{\zeta\eta}{2(t+s)}\right) = p^{(a)}(z,\eta,t+s).
\end{aligned}$$

An immediate consequence of Proposition 2.4 is the following.

PROPOSITION 2.5 (Semigroup property). *For every $a > -1$ and every $t, s > 0$ one has*

$$P_t^{(a)} \circ P_s^{(a)} = P_{t+s}^{(a)}.$$

□

We close this section by analyzing explicitly the case in which $a = 0$ in (2.4). In such case we have

$$p^{(0)}(z,\zeta,t) = (2t)^{-\frac{1}{2}} \left(\frac{z\zeta}{2t}\right)^{\frac{1}{2}} I_{-1/2}\left(\frac{z\zeta}{2t}\right) e^{-\frac{z^2+\zeta^2}{4t}}, \tag{2.10}$$

We now note the following well-known formulas, see (5.8.5) on p. 112 in [**39**],

$$I_{-1/2}(z) = \left(\frac{2}{\pi z}\right)^{1/2} \cosh z, \qquad I_{1/2}(z) = \left(\frac{2}{\pi z}\right)^{1/2} \sinh z. \tag{2.11}$$

Using (2.11) in (2.10), we obtain

$$\begin{aligned}
p^{(0)}(z,\zeta,t) &= (2t)^{-1/2}\left(\frac{z\zeta}{2t}\right)^{\frac{1}{2}} \frac{\sqrt{4t}}{\sqrt{\pi}\sqrt{z\zeta}} \cosh\left(\frac{z\zeta}{2t}\right) e^{-\frac{z^2+\zeta^2}{4t}} \\
&= (\pi t)^{-1/2}\cosh\left(\frac{z\zeta}{2t}\right) e^{-\frac{z^2+\zeta^2}{4t}} \\
&= (4\pi t)^{-1/2}\left(e^{\frac{z\zeta}{2t}} + e^{-\frac{z\zeta}{2t}}\right) e^{-\frac{z^2+\zeta^2}{4t}} \\
&= (4\pi t)^{-1/2}\left(e^{-\frac{(z-\zeta)^2}{4t}} + e^{-\frac{(z+\zeta)^2}{4t}}\right),
\end{aligned} \tag{2.12}$$

where in the last equality we have used the identities

$$e^{-\frac{(z-\zeta)^2}{4t}} = e^{-\frac{z^2+\zeta^2}{4t}} e^{\frac{z\zeta}{2t}}, \qquad e^{-\frac{(z+\zeta)^2}{4t}} = e^{-\frac{z^2+\zeta^2}{4t}} e^{-\frac{z\zeta}{2t}}. \tag{2.13}$$

From (2.3) we have

$$\begin{aligned}
P_t^{(0)}\varphi(z) &= (4\pi t)^{-1/2}\int_0^\infty \varphi(\zeta)e^{-\frac{(z-\zeta)^2}{4t}} d\zeta + (4\pi t)^{-1/2}\int_0^\infty \varphi(\zeta)e^{-\frac{(z+\zeta)^2}{4t}} d\zeta \\
&= (4\pi t)^{-1/2}\int_0^\infty \varphi(\zeta)e^{-\frac{(z-\zeta)^2}{4t}} d\zeta + (4\pi t)^{-1/2}\int_{-\infty}^0 \varphi(-\zeta)e^{-\frac{(z-\zeta)^2}{4t}} d\zeta \\
&= (4\pi t)^{-1/2}\int_{-\infty}^\infty \Phi(\zeta)e^{-\frac{(z-\zeta)^2}{4t}} d\zeta,
\end{aligned} \tag{2.14}$$

where we have let

$$\Phi(\zeta) = \varphi(|\zeta|), \qquad \zeta \in \mathbb{R}. \tag{2.15}$$

Formula (2.14) proves the following result.

PROPOSITION 2.6. *When $a = 0$, at any $z, t > 0$ the solution $u(z,t)$ to the problem* (2.1) *coincides with the function $U(z,t)$ that solves the problem*

$$\begin{cases} \partial_t U - \partial_{zz} U = 0 & \text{in } \mathbb{R}, \\ U(z,0) = \Phi(z), & z \in \mathbb{R}, \end{cases} \tag{2.16}$$

*where $\Phi$ is the even extension* (2.15) *to the whole line $\mathbb{R}$ of the function $\varphi$ on $[0,\infty)$.*

In connection with the Dirichlet Bessel semigroup, we mention the papers [**16**], [**17**] in which the authors establish various sharp asymptotic bounds for the heat kernel.

## 3. A curvature-dimension inequality

In the famous paper [**6**] Bakry and Émery introduced their so-called $\Gamma$-calculus as a different way of approaching global results in geometry through analytical tools. At the roots of such calculus there is the notion of *curvature-dimension* inequality. A $n$-dimensional Riemannian manifold $\mathbb{M}$ with Laplacean $\Delta$ is said to satisfy the curvature-dimension inequality $\mathrm{CD}(\rho, n)$ for some $\rho \in \mathbb{R}$ if for all functions $f \in C^\infty(\mathbb{M})$ one has

$$\Gamma_2(f) \geq \frac{1}{n}(\Delta f)^2 + \rho\Gamma(f). \tag{3.1}$$

Here, $\Gamma$ and $\Gamma_2$ respectively denote the *carré du champ* and the *Hessian* canonically associated with $\Delta$, see [**6**], and also the book [**7**]. A remarkable aspect of (3.1) is that it is equivalent to the lower bound Ricci $\geq \rho$ on the Ricci tensor of $\mathbb{M}$. In the paper [**15**] it was shown that many global properties of the heat semigroup, in a setting which includes the Riemannian one, can be derived exclusively from a generalization of the curvature-dimension inequality (3.1). In this connection one should also see [**14**], [**11**] and [**12**].

In this section we observe that the Bessel semigroup on $(\mathbb{R}_+, d\mu^{(a)})$, where $d\mu^{(a)}(z) = z^a dz$, with generator $\mathscr{B}_z^{(a)}$, satisfies a property similar to (3.1) provided that $a \geq 0$, see Proposition 3.1 below. Although we do not use such fact in the present paper, we have decided to include it since, interestingly, it displays the same "best possible" nature of the Bessel process which permeates all our results.

We begin with defining for every $f, g \in C^\infty(\mathbb{R})$ the *carré du champ* associated with $\mathscr{B}_z^{(a)}$,

$$\Gamma^{(a)}(f,g) = \frac{1}{2}\left[\mathscr{B}_z^{(a)}(fg) - f\mathscr{B}_z^{(a)}g - g\mathscr{B}_z^{(a)}f\right]. \tag{3.2}$$

One easily verifies that

$$\Gamma^{(a)}(f,g) = f'g', \qquad \text{hence} \qquad \Gamma^{(a)}(f) = (f')^2. \tag{3.3}$$

Next, we consider the Hessian associated with $\mathscr{B}_z^{(a)}$

$$\Gamma_2^{(a)}(f,g) = \frac{1}{2}\left[\mathscr{B}_z^{(a)}\Gamma^{(a)}(f,g) - \Gamma^{(a)}(f, \mathscr{B}_z^{(a)}g) - \Gamma^{(a)}(g, \mathscr{B}_z^{(a)}f)\right]. \tag{3.4}$$

A simple calculation shows that

(3.5) $\Gamma_2^{(a)}(f,g) = f''g'' + \frac{a}{z^2}f'g',$ therefore $\Gamma_2^{(a)}(f) = (f'')^2 + \frac{a}{z^2}(f')^2.$

We can now establish the relevant curvature-dimension inequality for the Bessel semigroup. The reader should also see [**7**], especially (1.16.9) and the discussion in Section 2.4.2.

PROPOSITION 3.1. *For every function $f \in C^\infty(\mathbb{R}_+)$ one has*

$$\Gamma_2^{(a)}(f) \geq \frac{1}{a+1}(\mathscr{B}_z^{(a)} f)^2, \tag{3.6}$$

*if and only if $a \geq 0$. In other words, $\mathscr{B}_z^{(a)}$ satisfies the curvature-dimension inequality* $\mathrm{CD}(0, a+1)$ *on $\mathbb{R}_+$ if and only if $a \geq 0$.*

PROOF. If we start from assuming that $a > -1$, then $a+1>0$, and therefore in view of (3.5) the desired conclusion is equivalent to

$$\left(f'' + \frac{a}{z}f'\right)^2 \leq (a+1)\left((f'')^2 + \frac{a}{z^2}(f')^2\right).$$

In turn, this inequality is equivalent to

$$a\left(f'' - \frac{1}{z}f'\right)^2 \geq 0,$$

which is true for any $f \in C^\infty(\mathbb{R})$ if and only if $a \geq 0$. When $-1 < a < 0$ the inequality (3.6) gets reversed. □

We note in closing that, when $a > 0$, then the fractal "dimension" $Q = a+1$ in the inequality $\mathrm{CD}(0, a+1)$ in (3.6) above is strictly bigger than the topological dimension of the ambient manifold $M = \mathbb{R}^+$.

## 4. An inequality of Li-Yau type for the Bessel semigroup

In this section we prove Theorem 1.1 above. As we have mentioned, such result ultimately hinges on the global property (1.1) of the modified Bessel function $I_\nu$.

The following proposition represents the Bessel semigroup counterpart of the simple (but important) fact that for the standard heat kernel $p(x,y,t) = (4\pi t)^{-n/2}\exp(-\frac{|x-y|^2}{4t})$ in $\mathbb{R}^n$, we have

$$|\nabla_x \log p(x,y,t)|^2 - \partial_t \log p(x,y,t) = \frac{n}{2t}. \tag{4.1}$$

Except that, as (4.3) and (4.7) in Propositions 4.1 and 4.2 show, one should not expect the equality as in (4.1).

Hereafter, for $\nu > -1$ we indicate with $y_\nu(z) = I_{\nu+1}(z)/I_\nu(z)$ the Bessel quotient, see (8.9) below. For a detailed analysis of the function $y_\nu$ we refer the reader to Section 8. Here, we note that $y_\nu(0) = 0$, see (8.11) below.

PROPOSITION 4.1. *Let $a > -1$. For every $z, \zeta \in \mathbb{R}^+$ and $t > 0$ one has*

$$\left(\partial_z \log p^{(a)}(z,\zeta,t)\right)^2 - \partial_t \log p^{(a)}(z,\zeta,t) = \frac{a+1}{2t} + \frac{\zeta^2}{4t^2}\left(y_{\frac{a-1}{2}}\left(\frac{z\zeta}{2t}\right)^2 - 1\right). \tag{4.2}$$

*In particular, if we let $z \to 0^+$ in (4.2) we obtain for any $\zeta > 0$ and $t > 0$,*

$$\left(\partial_z \log p^{(a)}(0,\zeta,t)\right)^2 - \partial_t \log p^{(a)}(0,\zeta,t) < \frac{a+1}{2t}. \tag{4.3}$$

PROOF. We define

$$\Lambda_\nu(z) = z^{-\nu} I_\nu(z),$$

and recall, see (8.15) below, that

$$\frac{d}{dz} \log \Lambda_\nu(z) = y_\nu(z), \qquad z > 0. \tag{4.4}$$

Notice that since the right-hand side is strictly positive for $z > 0$, the equation (4.4) says in particular that $\Lambda_\nu$ is strictly increasing on $(0, \infty)$. If we rewrite (2.4) as

$$p^{(a)}(z, \zeta, t) = (2t)^{-\frac{a+1}{2}} \Lambda_{\frac{a-1}{2}}\left(\frac{z\zeta}{2t}\right) e^{-\frac{z^2+\zeta^2}{4t}},$$

then we have

$$\log p^{(a)}(z, \zeta, t) = -\frac{a+1}{2} \log(2t) + \log \Lambda_{\frac{a-1}{2}}\left(\frac{z\zeta}{2t}\right) - \frac{z^2 + \zeta^2}{4t}.$$

The chain rule and (4.4) give

$$\partial_z \log p^{(a)}(z, \zeta, t) = y_{\frac{a-1}{2}}\left(\frac{z\zeta}{2t}\right) \frac{\zeta}{2t} - \frac{z}{2t}. \tag{4.5}$$

Analogously, we find

$$\partial_t \log p^{(a)}(z, \zeta, t) = -\frac{a+1}{2t} - y_{\frac{a-1}{2}}\left(\frac{z\zeta}{2t}\right) \frac{z\zeta}{2t^2} + \frac{z^2 + \zeta^2}{4t^2}. \tag{4.6}$$

From (4.5) and (4.6) we conclude

$$\begin{aligned}
&\left(\partial_z \log p^{(a)}(z, \zeta, t)\right)^2 - \partial_t \log p^{(a)}(z, \zeta, t) \\
&= y_{\frac{a-1}{2}}\left(\frac{z\zeta}{2t}\right)^2 \frac{\zeta^2}{4t^2} + \frac{z^2}{4t^2} + \frac{a+1}{2t} - \frac{z^2 + \zeta^2}{4t^2} \\
&= \frac{a+1}{2t} + \frac{\zeta^2}{4t^2}\left(y_{\frac{a-1}{2}}\left(\frac{z\zeta}{2t}\right)^2 - 1\right).
\end{aligned}$$

We conclude that (4.2) is valid. To establish (4.3) it suffices to observe that in view of (8.10) we obtain for any fixed $\zeta > 0$ and $t > 0$, as $z \to 0^+$

$$y_{\frac{a-1}{2}}\left(\frac{z\zeta}{2t}\right) = O\left(\frac{z\zeta}{2t}\right) \longrightarrow 0.$$

Since by (2.5) $z \to p^{(a)}(z, \zeta, t)$ is continuous up to $z = 0$, this shows that

$$\left(\partial_z \log p^{(a)}(0, \zeta, t)\right)^2 - \partial_t \log p^{(a)}(0, \zeta, t) = \frac{a+1}{2t} - \frac{\zeta^2}{4t^2} < \frac{a+1}{2t}.$$

$\square$

We show next that, if we restrict the range of $a$, then a global Li-Yau inequality similar to (4.3) above holds. By this we mean that $z = 0$ can be replaced by *any* $z > 0$.

PROPOSITION 4.2 (Inequality of Li-Yau type for $p^{(a)}(z, \zeta, t)$). *Let $a \geq 0$. Then, for every $z, \zeta \in \mathbb{R}^+$ and $t > 0$ one has*

$$\left(\partial_z \log p^{(a)}(z, \zeta, t)\right)^2 - \partial_t \log p^{(a)}(z, \zeta, t) < \frac{a+1}{2t}. \tag{4.7}$$

PROOF. Since for $\nu \geq -1$ both $I_\nu(z)$ and $I_{\nu+1}(z)$ are positive for $z > 0$, if we let $\nu = \frac{a-1}{2}$, then since $a \geq 0$ we have $\nu \geq -1/2$, and in view of Proposition 4.1, proving (4.7) is equivalent to showing

$$y_\nu(z) < 1,$$

for every $\nu \geq -\frac{1}{2}$, and for every $z > 0$. But this follows from Proposition 8.6 below. □

REMARK 4.3. One should note the strict inequality in (4.7) which follows from that in (8.20). It would be interesting to know whether a sharper Li-Yau inequality can be derived by using a sharper upper bound on the function $y_\nu(z)$. In this connection, one should see the papers [**3**], [**55**], [**54**], [**34**], [**35**] and [**52**].

With Proposition 4.1 we now return to formula (2.3) and establish the main result of this section.

THEOREM 4.4. *Let $a > -1$. Let $\varphi \geq 0$ be such that $\varphi \in C^1_{(a)}(0,\infty)$. For every $z > 0$ and $t > 0$ we have the following* adjusted Li-Yau inequality *for the function $P_t^{(a)}\varphi(z)$ defined by* (2.3)

$$\left(\partial_z \log P_t^{(a)}\varphi(z)\right)^2 - \partial_t \log P_t^{(a)}\varphi(z) \leq \frac{a+1}{2t} + \frac{1}{P_t^{(a)}\varphi(z)} \int_0^\infty \varphi(\zeta)\frac{\zeta^2}{4t^2}\left(y_{\frac{a-1}{2}}\left(\frac{z\zeta}{2t}\right)^2 - 1\right) p^{(a)}(z,\zeta,t)\zeta^a d\zeta. \tag{4.8}$$

*When $z = 0$, we have for every $t > 0$*

$$\left(\partial_z \log P_t^{(a)}\varphi(0)\right)^2 - \partial_t \log P_t^{(a)}\varphi(0) < \frac{a+1}{2t} \tag{4.9}$$

*When $a \geq 0$ an inequality similar to* (4.9) *continues to be valid globally, i.e., for every $z > 0$ and $t > 0$ one has*

$$\left(\partial_z \log P_t^{(a)}\varphi(z)\right)^2 - \partial_t \log P_t^{(a)}\varphi(z) < \frac{a+1}{2t} \tag{4.10}$$

PROOF. In what follows we denote for simplicity $u(z,t) = P_t^{(a)}\varphi(z)$. Differentiating (2.3) with respect to $z$ gives

$$\begin{aligned} \partial_z u(z,t)^2 &= \left(\int_0^\infty \varphi(\zeta)\partial_z p^{(a)}(z,\zeta,t)\zeta^a d\zeta\right)^2 \\ &= \left(\int_0^\infty \varphi(\zeta)\frac{\partial_z p^{(a)}(z,\zeta,t)}{p^{(a)}(z,\zeta,t)^{1/2}} p^{(a)}(z,\zeta,t)^{1/2}\zeta^a d\zeta\right)^2 \\ &\leq \int_0^\infty \varphi(\zeta)\frac{\partial_z p^{(a)}(z,\zeta,t)^2}{p^{(a)}(z,\zeta,t)}\zeta^a d\zeta \int_0^\infty \varphi(\zeta) p^{(a)}(z,\zeta,t)\zeta^a d\zeta \\ &\leq u(z,t)\int_0^\infty \varphi(\zeta)\frac{\partial_z p^{(a)}(z,\zeta,t)^2}{p^{(a)}(z,\zeta,t)}\zeta^a d\zeta, \end{aligned} \tag{4.11}$$

where in the second to the last inequality we have applied Cauchy-Schwarz. In the latter inequality we now substitute (4.2) from Proposition 4.1 which we rewrite as follows

$$\frac{\partial_z p^{(a)}(z,\zeta,t)^2}{p^{(a)}(z,\zeta,t)} = \partial_t p^{(a)}(z,\zeta,t) + \frac{a+1}{2t} p^{(a)}(z,\zeta,t)$$
$$+ \frac{\zeta^2}{4t^2}\left(y_{\frac{a-1}{2}}(\frac{z\zeta}{2t})^2 - 1\right) p^{(a)}(z,\zeta,t).$$

We find

$$\int_0^\infty \varphi(\zeta)\frac{\partial_z p^{(a)}(z,\zeta,t)^2}{p^{(a)}(z,\zeta,t)}\zeta^a d\zeta = \int_0^\infty \varphi(\zeta)\partial_t p^{(a)}(z,\zeta,t)\zeta^a d\zeta$$
$$+ \frac{a+1}{2t}\int_0^\infty \varphi(\zeta) p^{(a)}(z,\zeta,t)\zeta^a d\zeta$$
$$+ \int_0^\infty \varphi(\zeta)\frac{\zeta^2}{4t^2}\left(y_{\frac{a-1}{2}}(\frac{z\zeta}{2t})^2 - 1\right) p^{(a)}(z,\zeta,t)\zeta^a d\zeta$$
$$= \partial_t u(z,t) + \frac{a+1}{2t}u(z,t) + \int_0^\infty \varphi(\zeta)\frac{\zeta^2}{4t^2}\left(y_{\frac{a-1}{2}}(\frac{z\zeta}{2t})^2 - 1\right) p^{(a)}(z,\zeta,t)\zeta^a d\zeta.$$

Substituting in (4.11) and dividing by $u(z,t)^2$ in the resulting inequality we conclude that (4.8) does hold. If we argue similarly, but use (4.3) instead of (4.2), we obtain (4.9).

Finally, to establish (4.10) we use Corollary 4.2 instead of Proposition 4.1, or simply observe that Proposition 8.6 guarantees that

$$\int_0^\infty \varphi(\zeta)\frac{\zeta^2}{4t^2}\left(y_{\frac{a-1}{2}}(\frac{z\zeta}{2t})^2 - 1\right) p^{(a)}(z,\zeta,t)\zeta^a d\zeta < 0.$$

□

REMARK 4.5. It is natural to wonder whether, in the range $-1 < a < 0$, there is a "good" Li-Yau inequality that can be derived from (4.8), similarly to what happens for the case of negative Ricci lower bounds in [**40**]. It should be clear to the reader that the main obstruction to answering this question in the affirmative is represented by the term

$$w(z,t) \overset{def}{=} \frac{1}{4t^2}\int_0^\infty \varphi(\zeta)\zeta^2\left(y_{\frac{a-1}{2}}(\frac{z\zeta}{2t})^2 - 1\right) p^{(a)}(z,\zeta,t)\zeta^a d\zeta.$$

The difficulty here is created by the presence of the factor $\zeta^2$ in the integral in the right-hand side. If we had $\zeta$ instead, then we could use Proposition 8.4 to control $w(z,t)$ in terms of $u(z,t)$.

With Theorem 4.4 in hands, we can now establish the scale invariant Harnack inequality for $P_t^{(a)}$ in Theorem 1.2 above. Since the proof is analogous to that of Theorem 1.3 in Section 6 below, we omit it and refer the reader to that source.

## 5. A comparison with the results of Chiarenza-Serapioni and of Epstein-Mazzeo

Because of its relevance in extension problems, it is interesting to understand what happens with Theorem 1.2 in the remaining range $-1 < a < 0$. Remarkably, in such range a Harnack inequality continues to hold.

One way to see this is to observe that a solution to the equation $\partial_t u - \mathscr{B}_z^{(a)} u = 0$, which in addition satisfies the condition $\lim_{z\to 0^+} z^a \partial_z u(z,t) = 0$ for every $t > 0$, must be smooth in $z$ (and in fact, real analytic) up to the vertical line $z = 0$. Therefore, if we set $U(z,t) = u(|z|,t)$ we see that $U$ is a nonnegative solution in the whole half-plane $\mathbb{R} \times (0,\infty)$ to

$$\partial_t(|z|^a U) - \partial_z(|z|^a \partial_z U) = 0,$$

which is a special case of (1.9) above, with $\omega(z) = |z|^a$. Since $\omega \in A_2(\mathbb{R})$ if and only if $|a| < 1$, by Theorem 2.1 [**18**] we conclude that a parabolic Harnack inequality holds for $U$ on $\mathbb{R} \times (0,\infty)$. From this, we immediately obtain a Harnack inequality for $u$ up to the vertical line $z = 0$ in the range $-1 < a < 1$. Two comments are in order though:

(i) while in Theorem 1.2 we obtain the sharp constant

$$\left(\frac{t}{s}\right)^{\frac{a+1}{2}} \exp\left(\frac{(z-\zeta)^2}{4(t-s)}\right), \tag{5.1}$$

that in the Harnack inequality (2.4) in Theorem 2.1 in [**18**] is not explicitly known;

(ii) Theorem 1.2 holds for any $a \geq 0$, whereas when $a > 1$ the results in [**18**] are no longer available as the weight $\omega(z) = |z|^a$ is not a $A_2$ weight of Muckenhoupt.

Another way to see that the Harnack inequality is true also in the range $-1 < a < 0$ is as follows. In their work [**21**] Epstein and Mazzeo studied the diffusion process associated with a class of degenerate parabolic equations from population biology. One the central models of interest for them was the Wright-Fisher operator

$$\mathscr{L}_{WF} = x(1-x)\frac{d^2}{dx^2},$$

which represents the diffusion limit of a Markov chain modeling the frequency of a gene with 2 alleles, without mutation or selection. One should also see the seminal paper by Feller [**27**], the papers [**37**] and [**49**], as well as the paper [**18**], which appeared in the same issue as [**21**], and the more recent works [**22**], [**23**] and [**24**].

In [**21**] and [**18**] the authors independently, and with different approaches, construct a parametrix for the Wright-Fisher operator $\mathscr{L}_{WF}$ by first localizing the analysis to a neighborhood of the boundary points $x = 0$ and $x = 1$. At this point, the approach in [**21**] is purely analytical, whereas that in [**18**] is more probabilistic. In [**21**] the authors by a suitable change of variable are thus led to consider the model operator on $(0,\infty)$

$$\mathscr{L}_b = x\frac{\partial^2}{\partial x^2} + b\frac{\partial}{\partial x}, \tag{5.2}$$

where $b > 0$ is a given number, and they construct the fundamental solution for the Cauchy problem (5.3) below. For a given $T > 0$ they consider the domain $D_T = [0,\infty) \times [0,T]$, and study the problem

$$\begin{cases} \frac{\partial v}{\partial t} - \mathscr{L}_b v = 0 & \text{in } D_T, \\ v(x,0) = f(x), & x > 0, \end{cases} \tag{5.3}$$

under Feller's zero flux condition

$$\lim_{x\to 0^+} x^b \frac{\partial v}{\partial x}(x,t) = 0. \tag{5.4}$$

Since solutions to (5.3) may not be smooth up to $x = 0$, even if the initial datum $f$ is smooth, the condition (5.4) is needed to single out smooth solutions. Following a suggestion by C. Fefferman, in (6.13) of [**21**] the authors find the following representation for the solution of (5.3)

$$v(x,t) = \int_0^\infty k_b(x,y,t) f(y) dy, \tag{5.5}$$

where (the following is formula (6.14) in [**21**])

$$k_b(x,y,t) = \frac{1}{t}\left(\frac{x}{y}\right)^{\frac{1-b}{2}} e^{-\frac{x+y}{t}} I_{b-1}\left(\frac{2\sqrt{xy}}{t}\right). \tag{5.6}$$

We recognize next that, via a simple change of variable the Cauchy problem (5.3), (5.4) is the same as (2.1) above.

PROPOSITION 5.1. *The transformation*

$$x = \frac{z^2}{4}, \qquad\qquad b = \frac{a+1}{2}, \tag{5.7}$$

*sends in a one-to-one, onto fashion solutions of* (2.1) *with* $a > -1$ *into solutions of* (5.3) *with* $b > 0$ *and with the Neumann condition* (5.4). *As a consequence, the problem* (5.3) *with the boundary condition* (5.4) *is equivalent to the problem* (2.1) *for the Bessel operator* $\mathscr{B}_z^{(a)}$.

PROOF. To see this, consider a function $v(x,t)$ and define

$$u(z,t) = v(z^2/4, t).$$

Then, the chain rule gives

$$u_z = v_x\ x_z, \quad u_{zz} = v_{xx}\ (x_z)^2 + v_x\ x_{zz},$$

and thus

$$u_t - \mathscr{B}_z^{(a)} u = v_t - v_{xx}\ (x_z)^2 - v_x\left(x_{zz} + \frac{a}{z} x_z\right).$$

Now, we have

$$x_{zz} + \frac{a}{z} x_z = \frac{a+1}{2} \overset{def}{=} b, \qquad (x_z)^2 = \frac{z^2}{4} = x.$$

We conclude that

$$u_t(z,t) - \mathscr{B}_z^{(a)} u(z,t) = v_t(x,t) - \mathscr{L}_b v(x,t). \tag{5.8}$$

Furthermore, one easily verifies that

$$z^a u_z(z,t) = 2^{2b-1} x^b v_x(x,t). \tag{5.9}$$

The equations (5.8), (5.9) prove the proposition. □

As a consequence of Proposition 5.1, letting $x = z^2/4$ in the left-hand side of (5.5), and making the change of variable $y = \zeta^2/4$ in the integral in the right-hand side, we expect the representation formula (5.5) to become exactly the formula (2.3) above. This is precisely the case, as the following simple verification shows:

$$\begin{aligned} u(z,t) = v(z^2/4,t) &= \int_0^\infty y k_b(z^2/4,y,t) f(y) \frac{dy}{y} \\ &= \frac{1}{2}\int_0^\infty \zeta k_b(z^2/4,\zeta^2/4,t) f(\zeta^2/4) d\zeta \\ &= \frac{1}{2t}\int_0^\infty \zeta k_{\frac{a+1}{2}}(z^2/4,\zeta^2/4,t) f(\zeta^2/4) d\zeta \\ &= \frac{1}{2t}\int_0^\infty (z\zeta)^{\frac{1-a}{2}} e^{-\frac{z^2+\zeta^2}{4t}} I_{\frac{a-1}{2}}\left(\frac{z\zeta}{2t}\right) \varphi(\zeta)\zeta^a d\zeta \\ &= \int_0^\infty \varphi(\zeta) p^{(a)}(z,\zeta,t)\zeta^a d\zeta, \end{aligned}$$

where we have let $\varphi(\zeta) = f(\zeta^2/4)$.

In Theorem 4.1 in their paper [**23**] Epstein and Mazzeo by a remarkable adaptation of the method of De Giorgi-Nash-Moser, and subsequent contributions of Saloff-Coste and Grigor'yan, establish a scale invariant Harnack inequality for a large class of degenerate parabolic equations defined on manifolds with corners. The relevant partial differential operators, known as generalized Kimura operators, arise in population biology and they contain as a special case the model (5.3) for the full range $b > 0$. As a consequence of the results in [**23**], and of Proposition 5.1 above, one obtains a Harnack inequality for positive solutions of (2.1) also in the range $-1 < a < 0$ which is not covered by our Theorem 1.2. However, it is not clear to this author that the Harnack inequality in (143) in [**23**] is capable of producing the sharp constant (5.1) in the right-hand side of (1.14) above. We also mention the paper [**24**] that contains a very different approach to the Harnack inequality, based on a stochastic representation of the solutions, for more more general classes of Kimura operators.

## 6. A sharp Harnack inequality for the parabolic extension problem

In this section we consider in $\mathbb{R}^{n+1}_+ \times (0,\infty)$ the so-called extension operator for the fractional powers $(\partial_t - \Delta)^s$, $0 < s < 1$, of the heat operator. Hereafter, for $x \in \mathbb{R}^n$ and $z > 0$ we denote by $X = (x,z) \in \mathbb{R}^{n+1}_+$, and by $(X,t)$ the generic point in $\mathbb{R}^{n+1}_+ \times (0,\infty)$. We also indicate $Y = (y,\zeta) \in \mathbb{R}^{n+1}_+$ and $(Y,\tau)$. Given a number $a \in (-1,1)$, the extension operator is the degenerate parabolic operator defined by

$$\mathscr{L}_a u = \partial_t(z^a u) - \operatorname{div}_X(z^a \nabla_X u). \tag{6.1}$$

It was recently introduced independently by Nyström-Sande in [**46**], and Stinga-Torrea in [**57**]. These authors proved that if for a given $\varphi \in \mathscr{S}(\mathbb{R}^{n+1})$, the function $u$ solves the problem

$$\begin{cases} \mathscr{L}_a u = 0 & \text{in } \mathbb{R}^{n+1}_+ \times (0,\infty), \\ u(x,0,t) = \varphi(x,t), \end{cases}$$

then, with $s \in (0,1)$ determined by the equation $a = 1 - 2s$, one has

$$-\frac{2^{-a}\Gamma\left(\frac{1-a}{2}\right)}{\Gamma\left(\frac{1+a}{2}\right)} \lim_{z\to 0^+} z^a \frac{\partial u}{\partial z}(x,z,t) = (\partial_t - \Delta)^s \varphi(x,t).$$

The reader should compare the latter equation with (1.6) above.

In what follows we establish a remarkable sharp Harnack inequality for the semigroup associated with the operator $\mathscr{L}_a$. The first (important) observation is that the Neumann fundamental solution for $\mathscr{L}_a$, with singularity at $(Y,0) = (y,\zeta,0)$, is given by

$$\mathscr{G}_a(X,Y,t) = p(x,y,t)p^{(a)}(z,\zeta,t), \tag{6.2}$$

where $p(x,y,t) = (4\pi t)^{-n/2}\exp(-\frac{|x-y|^2}{4t})$ is the standard heat kernel in $\mathbb{R}^n \times (0,\infty)$ and $p^{(a)}(z,\zeta,t)$ is given by (2.4) above.

Using Proposition 2.3, and the well-known fact that $\int_{\mathbb{R}^n} p(x,y,t)dy = 1$ for every $x \in \mathbb{R}^n$ and $t > 0$, it is a trivial exercise to verify that for every $X \in \mathbb{R}^{n+1}_+$ and $t > 0$ one has

$$\int_{\mathbb{R}^{n+1}_+} \mathscr{G}_a(X,Y,t)\zeta^a dY = 1. \tag{6.3}$$

Given a function $\varphi \in C_0^\infty(\mathbb{R}^{n+1}_+)$, consider the Cauchy problem with Neumann condition

$$\begin{cases} \mathscr{L}_a u = 0 & \text{in } \mathbb{R}^{n+1}_+ \times (0,\infty) \\ u(X,0) = \varphi(X), & X \in \mathbb{R}^{n+1}_+, \\ \lim_{z\to 0+} z^a \partial_z u(x,z,t) = 0. \end{cases} \tag{6.4}$$

The solution to (6.4) is represented by the formula

$$u(X,t) = \int_{\mathbb{R}^{n+1}_+} \varphi(Y)\mathscr{G}_a(X,Y,t)\zeta^a dY. \tag{6.5}$$

The following is the main result of this section.

THEOREM 6.1. *Let $a \geq 0$. Let $\varphi \geq 0$ be a function for which $u$ given by (6.5) represents a classical solution to (6.4). Then, for every $X, Y \in \mathbb{R}^{n+1}_+$ and every $0 < s < t < \infty$, we have*

$$u(X,s) < u(Y,t)\left(\frac{t}{s}\right)^{\frac{n+a+1}{2}} \exp\left(-\frac{|X-Y|^2}{4(t-s)}\right).$$

The proof of Theorem 6.1 is based on the following inequality of Li-Yau type for $u$.

THEOREM 6.2. *Let $a \geq 0$ and $\varphi$ and $u$ be as in Theorem 6.1. Then, for any $X \in \mathbb{R}^{n+1}_+$ and $t > 0$ one has*

$$|\nabla_X \log u(X,t)|^2 - \partial_t \log u(X,t) < \frac{n+a+1}{2t}.$$

PROOF. Differentiating under the integral sign in (6.5) and applying the Cauchy-Schwarz inequality similarly to the proof of (4.11), we find

(6.6)

$$|\nabla_X u(X,t)|^2 \leq \int_{\mathbb{R}^{n+1}_+} \varphi(Y)\frac{|\nabla_X \mathscr{G}_a(X,Y,t)|^2}{\mathscr{G}_a(X,Y,t)}\zeta^a dY \int_{\mathbb{R}^{n+1}_+} \varphi(Y)\mathscr{G}_a(X,Y,t)\zeta^a dY$$
$$= u(X,t)\int_{\mathbb{R}^{n+1}_+} \varphi(Y)\frac{|\nabla_X \mathscr{G}_a(X,Y,t)|^2}{\mathscr{G}_a(X,Y,t)}\zeta^a dY.$$

Next, we prove the following crucial result: for every $X, Y \in \mathbb{R}^{n+1}_+$ and every $t > 0$ one has

$$|\nabla_X \log \mathscr{G}_a(X,Y,t)|^2 - \partial_t \log \mathscr{G}_a(X,Y,t) < \frac{n+a-1}{2t}. \tag{6.7}$$

To establish (6.7) we note that from the equation (6.2) we obtain

$$|\nabla_X \log \mathscr{G}_a(X,Y,t)|^2 - \partial_t \log \mathscr{G}_a(X,Y,t) = |\nabla_X \log p(x,y,t)|^2 - \partial_t \log p(x,y,t)$$
$$+ (\partial_z \log p^{(a)}(z,\zeta,t))^2 - \partial_t \log p^{(a)}(z,\zeta,t).$$

The claim (6.7) now follows from (4.1) and from Proposition 4.2 above.

With (6.7) in hands, we return to the integral in the right-hand side of (6.6) and proceed as follows

$$\int_{\mathbb{R}^{n+1}_+} \varphi(Y) \frac{|\nabla_X \mathscr{G}_a(X,Y,t)|^2}{\mathscr{G}_a(X,Y,t)} \zeta^a dY < \int_{\mathbb{R}^{n+1}_+} \varphi(Y) \mathscr{G}_a(X,Y,t) \partial_t \log \mathscr{G}_a(X,Y,t) \zeta^a dY$$
$$+ \frac{n+a+1}{2t} \int_{\mathbb{R}^{n+1}_-} \varphi(Y) \mathscr{G}_a(X,Y,t) \zeta^a dY$$
$$= \partial_t \int_{\mathbb{R}^{n+1}_+} \varphi(Y) \mathscr{G}_a(X,Y,t) \zeta^a dY + \frac{n+a+1}{2t} u(X,t)$$
$$= \partial_t u(X,t) + \frac{n+a+1}{2t} u(X,t).$$

We conclude that

$$|\nabla_X u(X,t)|^2 < u(X,t) \partial_t u(X,t) + \frac{n+a+1}{2t} u(X,t)^2.$$

Dividing by $u(X,t)^2$ we reach the desired conclusion. □

Having established Theorem 1.4 we finally turn to the

PROOF OF THEOREM 1.3. The argument is the same as that in [**40**]. We repeat it here for the sake of completeness. Fix two points $(X,s), (Y,t) \in \mathbb{R}^{n+1}_+ \times (0,\infty)$, with $0 < s < t < \infty$, and consider the straight-line segment (a geodesic line) which starts from $(Y,t)$ and ends in $(X,s)$. We parametrize it by

$$\alpha(\tau) = (Y + \tau(X-Y), t - \tau(t-s)), \qquad 0 \le \tau \le 1.$$

Clearly,

$$\alpha'(\tau) = (X - Y, -(t-s)), \qquad 0 \le \tau \le 1.$$

We now consider the function

$$h(\tau) = \log U(\alpha(\tau)).$$

We have

$$\log \frac{U(X,s)}{U(Y,t)} = h(1) - h(0) = \int_0^1 h'(\tau) d\tau$$
$$= \int_0^1 < \nabla_X \log U(\alpha(\tau)), X - Y > d\tau - (t-s) \int_0^1 \partial_t \log U(\alpha(\tau)) d\tau$$
$$\le |X - Y| \int_0^1 |\nabla_X \log U(\alpha(\tau))| d\tau - (t-s) \int_0^1 \partial_t \log U(\alpha(\tau)) d\tau$$
$$\le \frac{|X-Y|^2}{2\varepsilon} + \frac{\varepsilon}{2} \int_0^1 |\nabla_X \log U(\alpha(\tau))|^2 d\tau - (t-s) \int_0^1 \partial_t \log U(\alpha(\tau)) d\tau.$$

At this point we observe that Theorem 1.4 implies for every $0 \le \tau \le 1$

$$-\partial_t \log u(\alpha(\tau)) < \frac{n+a+1}{2(t-\tau(t-s))} - |\nabla_X \log U(\alpha(\tau))|^2.$$

Replacing this information in the above inequality, we find

$$\log \frac{U(X,s)}{U(Y,t)} \le \frac{|X-Y|^2}{2\varepsilon} + \frac{\varepsilon}{2}\int_0^1 |\nabla_X \log U(\alpha(\tau))|^2 d\tau$$
$$+ \log\left(\frac{t}{s}\right)^{\frac{n+a+1}{2}} - (t-s)\int_0^1 |\nabla_X \log U(\alpha(\tau))|^2 d\tau.$$

If we now choose $\varepsilon > 0$ such that $\varepsilon = 2(t-s)$, we finally obtain

$$\log \frac{U(X,s)}{U(Y,t)} \le \frac{|X-Y|^2}{4(t-s)} + \log\left(\frac{t}{s}\right)^{\frac{n+a+1}{2}}.$$

Exponentiating both sides of this inequality we reach the desired conclusion. □

## 7. Monotonicity formulas of Struwe and Almgren-Poon type for the Bessel semigroup

In this section we prove Theorems 1.5 and 1.7. As we have pointed out, these results provide one more interesting instance of the underlying theme of this paper. As it will be apparent from the proofs, remarkably this time is not the inequality (1.1) that lurks in the shadows, but rather the stronger monotonicity property of the Bessel quotient $y_\nu = I_{\nu+1}/I_\nu$ in Proposition 8.8 below.

In $\mathscr{Q} = (0,\infty)\times(0,\infty)$ we consider a solution $u$ of the heat equation

$$\partial_t u - \mathscr{B}_\zeta^{(a)} u = 0, \tag{7.1}$$

which for every $t > 0$ satisfies the Neumann condition

$$\lim_{\zeta\to 0^+} \zeta^a \partial_\zeta u(\zeta,t) = 0. \tag{7.2}$$

For a given $z > 0$ and $T > 0$ we introduce the following scaled energy, *centered at* $(z,T)$, with respect to the backward Gaussian-Bessel measure

$$E_{z,T}^{(a)}(t) \overset{def}{=} \frac{T-t}{2}\int_0^\infty (\partial_\zeta u(\zeta,t))^2 p^{(a)}(z,\zeta,T-t)\zeta^a d\zeta. \tag{7.3}$$

It is obvious that without further assumptions it is not guaranteed that the integral (7.3), as well as those that will appear in the proofs of this section, be convergent. This difficulty is serious and in order to circumvent it one needs to:

(a) multiply the function $u$ by a suitable cutoff function as it was done in the monograph [**20**] in the study of closely related monotonicity properties. Since doing this changes the equation satisfied by $u$, the analysis becomes considerably more complicated;

(b) develop the regularity theory which is necessary to rigorously justify all integration by parts that occur when differentiating (7.3).

Since our intent is to point to a new phenomenon, we will "wave our hands" on these important aspects, and refer the reader to [**20**], [**9**] and [**8**] for a rigorous treatment. As a consequence we will from now on assume that *all* integrations by parts are justifed, and *all* boundary terms vanish.

We note explicitly that, in the $\Gamma$-language of Section 3, the above energy can be written (see (3.4) above)

$$E_{z,T}^{(a)}(t) = \frac{T-t}{2} P_{T-t}^{(a)}(\Gamma^{(a)}(u))(z).$$

In the proof of Theorem 1.5 below, we will also distinguish between the cases $z > 0$ and $z = 0$. When $z = 0$ in the integral in (7.3) above we integrate against $p^{(a)}(0, \zeta, T-t)$ (for the value of this function see (2.5) above). For simplicity, in this case we will write $E_T(t)$ instead of $E_{0,T}(t)$.

For later use, we also observe that the function $g^{(a)}(\zeta, t) = p^{(a)}(z, \zeta, T-t)$ satisfies the backward heat equation

$$\partial_t g^{(a)}(\zeta, t) = -\mathscr{B}_\zeta^{(a)} g^{(a)}(\zeta, t). \tag{7.4}$$

Henceforth, to simplify the notation we will indicate partial derivatives with $u_\zeta, u_{\zeta\zeta}, u_t$, etc. We will also routinely drop the arguments of all functions appearing in the integral in (7.3), and write $u_\zeta, p^{(a)}, p_\zeta^{(a)}$, etc., instead of $u_\zeta(\zeta, t), p^{(a)}(z, \zeta, T-t), p_\zeta^{(a)}(z, \zeta, T-t)$, etc.

Proof of Theorem 1.5. Differentiating (7.3) we find

$$\frac{dE_{z,T}^{(a)}}{dt}(t) = \frac{T-t}{2} \int_0^\infty -\frac{1}{2(T-t)} u_\zeta^2 p^{(a)} \zeta^a d\zeta + (T-t) \int_0^\infty u_\zeta u_{t\zeta} p^{(a)} \zeta^a d\zeta$$
$$- \frac{T-t}{2} \int_0^\infty u_\zeta^2 \mathscr{B}_\zeta^{(a)} p^{(a)} \zeta^a d\zeta,$$

where in the last term we have used (7.4). We now integrate by parts in the second integral in the right-hand side, obtaining

$$\int_0^\infty u_\zeta u_{t\zeta} p^{(a)} \zeta^a d\zeta = \zeta^a u_\zeta u_t p^{(a)} \Big|_{\zeta=0}^\infty - \int_0^\infty u_{\zeta\zeta} u_t p^{(a)} \zeta^a d\zeta - \int_0^\infty \frac{a}{\zeta} u_\zeta u_t p^{(a)} \zeta^a d\zeta$$
$$- \int_0^\infty u_\zeta u_t p_\zeta^{(a)} \zeta^a d\zeta = - \int_0^\infty \mathscr{B}_\zeta^{(a)} u \; u_t p^{(a)} \zeta^a d\zeta - \int_0^\infty u_\zeta u_t p_\zeta^{(a)} \zeta^a d\zeta$$
$$= - \int_0^\infty u_t^2 p^{(a)} \zeta^a d\zeta - \int_0^\infty u_t u_\zeta p_\zeta^{(a)} \zeta^a d\zeta,$$

where in the last equality we have used (7.1). Substituting in the above identity, we find

$$\frac{dE_{z,T}^{(a)}}{dt}(t) = \frac{T-t}{2} \int_0^\infty -\frac{1}{2(T-t)} u_\zeta^2 p^{(a)} \zeta^a d\zeta - \frac{T-t}{2} \int_0^\infty u_\zeta^2 \mathscr{B}_\zeta^{(a)} p^{(a)} \zeta^a d\zeta$$
$$- (T-t) \int_0^\infty u_t^2 p^{(a)} \zeta^a d\zeta - (T-t) \int_0^\infty u_t u_\zeta p_\zeta^{(a)} \zeta^a d\zeta$$
$$= -(T-t) \int_0^\infty \left[ u_t + u_\zeta \frac{p_\zeta^{(a)}}{p^{(a)}} \right]^2 p^{(a)} \zeta^a d\zeta + (T-t) \int_0^\infty u_t u_\zeta p_\zeta^{(a)} \zeta^a d\zeta$$
$$+ (T-t) \int_0^\infty u_\zeta^2 \frac{(p_\zeta^{(a)})^2}{(p^{(a)})^2} p^{(a)} \zeta^a d\zeta + \frac{T-t}{2} \int_0^\infty -\frac{1}{2(T-t)} u_\zeta^2 p^{(a)} \zeta^a d\zeta$$
$$- \frac{T-t}{2} \int_0^\infty u_\zeta^2 \mathscr{B}_\zeta^{(a)} p^{(a)} \zeta^a d\zeta.$$

We finally write

$$\frac{dE_{z,T}^{(a)}}{dt}(t) = -(T-t)\int_0^\infty \left[u_t + u_\zeta \frac{p_\zeta^{(a)}}{p^{(a)}}\right]^2 p^{(a)}\zeta^a d\zeta + G_{z,T}^{(a)}(t), \tag{7.5}$$

where we have set

$$\begin{aligned} G_{z,T}^{(a)}(t) = (T-t)\int_0^\infty u_\zeta^2 \left[\frac{(p_\zeta^{(a)})^2}{(p^{(a)})^2} - \frac{1}{2(T-t)} - \frac{\mathscr{B}_\zeta^{(a)} p^{(a)}}{2p^{(a)}}\right] p^{(a)}\zeta^a d\zeta \\ + (T-t)\int_0^\infty u_t u_\zeta p_\zeta^{(a)} \zeta^a d\zeta. \end{aligned} \tag{7.6}$$

Now, we integrate by parts in the second integral in the right-hand side of (7.6), obtaining

$$\begin{aligned} &\int_0^\infty u_t u_\zeta p_\zeta^{(a)}\zeta^a d\zeta = \int_0^\infty \mathscr{B}_\zeta^{(a)} u \, u_\zeta p_\zeta^{(a)}\zeta^a d\zeta = \int_0^\infty u_{\zeta\zeta} u_\zeta p_\zeta^{(a)}\zeta^a d\zeta \\ &+ \int_0^\infty \frac{a}{\zeta} u_\zeta u_\zeta p_\zeta^{(a)}\zeta^a d\zeta = \int_0^\infty \partial_\zeta\left(\frac{u_\zeta^2}{2}\right) p_\zeta^{(a)}\zeta^a d\zeta + \int_0^\infty \frac{a}{\zeta} u_\zeta^2 p_\zeta^{(a)}\zeta^a d\zeta \\ &= \zeta^a p_\zeta^{(a)}\left(\frac{u_\zeta^2}{2}\right)\Big|_{\zeta=0}^\infty - \int_0^\infty \frac{u_\zeta^2}{2} p_{\zeta\zeta}^{(a)}\zeta^a d\zeta - \int_0^\infty \frac{a}{\zeta}\frac{u_\zeta^2}{2} p_\zeta^{(a)}\zeta^a d\zeta \\ &+ \int_0^\infty \frac{a}{\zeta} u_\zeta^2 p_\zeta^{(a)}\zeta^a d\zeta = \int_0^\infty \frac{u_\zeta^2}{2}\frac{a}{\zeta} p_\zeta^{(a)}\zeta^a d\zeta - \int_0^\infty \frac{u_\zeta^2}{2} p_{\zeta\zeta}^{(a)}\zeta^a d\zeta. \end{aligned}$$

Substituting this result in (7.6), we find

$$\begin{aligned} G_{z,T}^{(a)}(t) &= (T-t)\int_0^\infty u_\zeta^2 \left[\frac{(p_\zeta^{(a)})^2}{(p^{(a)})^2} - \frac{1}{2(T-t)} - \frac{\mathscr{B}_\zeta^{(a)} p^{(a)}}{2p^{(a)}}\right] p^{(a)}\zeta^a d\zeta \\ &+ (T-t)\int_0^\infty \frac{u_\zeta^2}{2}\frac{a}{\zeta} p_\zeta^{(a)}\zeta^a d\zeta - (T-t)\int_0^\infty \frac{u_\zeta^2}{2} p_{\zeta\zeta}^{(a)}\zeta^a d\zeta \\ &= -(T-t)\int_0^\infty u_\zeta^2 \left[\frac{p_{\zeta\zeta}^{(a)}}{p^{(a)}} - \frac{(p_\zeta^{(a)})^2}{(p^{(a)})^2} + \frac{1}{2(T-t)}\right] p^{(a)}\zeta^a d\zeta. \end{aligned} \tag{7.7}$$

From (7.5) we see that the proof of the theorem will be completed if we establish the following

CLAIM. For every $z > 0$ and $0 < t < T$, we have when $a \geq 0$

$$G_{z,T}^{(a)}(t) < 0. \tag{7.8}$$

To prove (7.8) it suffices to show that for every $z, \zeta > 0$ and $0 < t < T$, one has

$$\frac{p_{\zeta\zeta}^{(a)}}{p^{(a)}} - \frac{(p_\zeta^{(a)})^2}{(p^{(a)})^2} + \frac{1}{2(T-t)} > 0, \tag{7.9}$$

and we thus turn to proving (7.9). Now, (4.5) gives

$$\frac{(p_\zeta^{(a)})^2}{(p^{(a)})^2} = \left(y_{\frac{a-1}{2}}\left(\frac{z\zeta}{2(T-t)}\right)\frac{z}{2(T-t)} - \frac{\zeta}{2(T-t)}\right)^2. \tag{7.10}$$

Similarly, we obtain

$$p^{(a)}_{\zeta\zeta} = \left(y'_{\frac{a-1}{2}}\left(\frac{z\zeta}{2(T-t)}\right)\frac{z^2}{4(T-t)^2} - \frac{1}{2(T-t)}\right)p^{(a)} + \left(y_{\frac{a-1}{2}}\left(\frac{z\zeta}{2(T-t)}\right)\frac{z}{2(T-t)} - \frac{\zeta}{2(T-t)}\right)^2 p^{(a)}.$$

From this formula and (7.10), we find

$$\frac{p^{(a)}_{\zeta\zeta}}{2p^{(a)}} - \frac{(p^{(a)}_\zeta)^2}{(p^{(a)})^2} + \frac{1}{2(T-t)} = y'_{\frac{a-1}{2}}\left(\frac{z\zeta}{2(T-t)}\right)\frac{z^2}{4(T-t)^2}.$$

It is at this point that the monotonicity of the Bessel quotient enters the stage. Invoking the crucial Proposition 8.8 we now see that, when $a \geq 0$ (or, equivalently, $\nu \geq -1/2$), we have $y'_{\frac{a-1}{2}}\left(\frac{z\zeta}{2(T-t)}\right) > 0$ for any $z, \zeta > 0$ and every $0 < t < T$. This proves (7.9), and therefore (7.8), thus completing the proof of the first part of Theorem 1.5.

As for the second part, suppose that $z = 0$. Then, for any $a > -1$ we have from (2.5) above

$$p^{(a)}(0, \zeta, T-t) = \frac{1}{2^a\Gamma(\frac{a+1}{2})}(T-t)^{-\frac{a+1}{2}}e^{-\frac{\zeta^2}{4(T-t)}}.$$

This gives

$$p^{(a)}_\zeta = -\frac{\zeta}{2(T-t)}p^{(a)}, \quad p^{(a)}_{\zeta\zeta} = \left(-\frac{1}{2(T-t)} + \frac{\zeta^2}{4(T-t)^2}\right)p^{(a)}. \tag{7.11}$$

From these formulas we immediately obtain

$$\frac{p^{(a)}_{\zeta\zeta}}{p^{(a)}} - \frac{(p^{(a)}_\zeta)^2}{(p^{(a)})^2} + \frac{1}{2(T-t)} \equiv 0.$$

In view of (7.7) we conclude that $G^{(a)}_T \equiv 0$. Substitution in (7.5) finally gives

$$\frac{dE^{(a)}_{z,T}}{dt}(t) = -(T-t)\int_0^\infty \left[u_t + u_\zeta \frac{p^{(a)}_\zeta}{p^{(a)}}\right]^2 p^{(a)}\zeta^a d\zeta \leq 0.$$

□

Next, we show that the Bessel semigroup satisfies a monotonicity property analogous to that proved by Poon for the standard heat equation in [**48**]. On a solution $u$ of (7.1) satisfying (7.2), we introduce the quantity

$$L^{(a)}_{z,T}(t) = \frac{1}{2}\int_0^\infty u^2(\zeta, t)p^{(a)}(z, \zeta, T-t)\zeta^a d\zeta, \tag{7.12}$$

where $z > 0$ is fixed. We have the following result.

PROPOSITION 7.1. *For every $0 < t < T$ one has*

$$\frac{dL^{(a)}_{z,T}}{dt}(t) = \int_0^\infty u\left[u_t + u_\zeta \frac{p^{(a)}_\zeta}{p^{(a)}}\right]p^{(a)}\zeta^a d\zeta.$$

PROOF. Differentiating (7.12) and using (7.4), we find

$$\begin{aligned}\frac{dL_{z,T}^{(a)}}{dt}(t) &= \int_0^\infty uu_t p^{(a)}\zeta^a d\zeta - \frac{1}{2}\int_0^\infty u^2 \mathscr{B}_\zeta^{(a)} p^{(a)}\zeta^a d\zeta \\ &= \int_0^\infty uu_t p^{(a)}\zeta^a d\zeta - \frac{1}{2}\int_0^\infty u^2 p_{\zeta\zeta}^{(a)}\zeta^a d\zeta - \frac{1}{2}\int_0^\infty u^2\frac{a}{\zeta}p_\zeta^{(a)}\zeta^a d\zeta.\end{aligned}$$

Next, we integrate by parts in the second integral in the right-hand side of the latter identity, obtaining

$$-\frac{1}{2}\int_0^\infty u^2 p_{\zeta\zeta}^{(a)}\zeta^a d\zeta = -\frac{1}{2}u^2\zeta^a p_\zeta^{(a)}\Big|_0^\infty + \int_0^\infty uu_\zeta p_\zeta^{(a)}\zeta^a d\zeta + \frac{1}{2}\int_0^\infty u^2\frac{a}{\zeta}p_\zeta^{(a)}\zeta^a d\zeta$$

Substituting in the above equation we find

$$\frac{dL_{z,T}^{(a)}}{dt}(t) = \int_0^\infty uu_t p^{(a)}\zeta^a d\zeta + \int_0^\infty uu_\zeta p_\zeta^{(a)}\zeta^a d\zeta = \int_0^\infty u\left[u_t + u_\zeta\frac{p_\zeta^{(a)}}{p^{(a)}}\right]p^{(a)}\zeta^a d\zeta,$$

which is the sought for conclusion. □

We next establish an alternative way of interpreting the function $E_{z,T}^{(a)}(t)$ in (7.3) above.

LEMMA 7.2. *For every $t \in (0,T)$ one has*

$$E_{z,T}^{(a)}(t) = -\frac{T-t}{2}\int_0^\infty u\left[u_t + u_\zeta\frac{p_\zeta^{(a)}}{p^{(a)}}\right]p^{(a)}\zeta^a d\zeta.$$

PROOF. We have from (7.3)

$$\begin{aligned}E_{z,T}^{(a)}(t) &= \frac{T-t}{2}\int_0^\infty u_\zeta u_\zeta p^{(a)}\zeta^a d\zeta = \frac{T-t}{2}\zeta^a u_\zeta u p^{(a)}\Big|_{\zeta=0}^\infty - \frac{T-t}{2}\int_0^\infty u_{\zeta\zeta} u p^{(a)}\zeta^a d\zeta \\ &\quad - \frac{T-t}{2}\int_0^\infty u_\zeta u p_\zeta^{(a)}\zeta^a d\zeta - \frac{T-t}{2}\int_0^\infty \frac{a}{\zeta}u_\zeta u p^{(a)}\zeta^a d\zeta \\ &= -\frac{T-t}{2}\int_0^\infty u\mathscr{B}_\zeta^{(a)}u\, p^{(a)}\zeta^a d\zeta - \frac{T-t}{2}\int_0^\infty uu_\zeta p_\zeta^{(a)}\zeta^a d\zeta \\ &= -\frac{T-t}{2}\int_0^\infty u\left[u_t + u_\zeta\frac{p_\zeta^{(a)}}{p^{(a)}}\right]p^{(a)}\zeta^a d\zeta,\end{aligned}$$

where in the last equality we have used (7.1). This proves the lemma. □

Now, we let $T = 0$, and for fixed $z > 0$ and any $r > 0$, we consider the two quantities

(7.13)

$$H_z^{(a)}(r) \overset{def}{=} L_{z,0}^{(a)}(-r^2) = L_z^{(a)}(-r^2), \qquad I_z^{(a)}(r) \overset{def}{=} E_{z,0}^{(a)}(-r^2) = E_z^{(a)}(-r^2).$$

We note that from (7.12) and (7.3) we have

$$H_z^{(a)}(r) = \frac{1}{2}\int_0^\infty u^2(\zeta,t)p^{(a)}(z,\zeta,r^2)\zeta^a d\zeta,$$

and

$$I_z^{(a)}(r) = \frac{r^2}{2}\int_0^\infty u_\zeta^2(\zeta,t)p^{(a)}(z,\zeta,r^2)\zeta^a d\zeta.$$

DEFINITION 7.3. *We define the* frequency centered at $z$ *of a solution* $u$ *to* (7.1), *satisfying* (7.2), *by the following equation:*

$$N_z^{(a)}(r) = \frac{I_z^{(a)}(r)}{H_z^{(a)}(r)}, \qquad 0 < r < \sqrt{T}.$$

*When* $z = 0$, *we simply write* $N^{(a)}(r)$, *instead of* $N_0^{(a)}(r)$.

We next establish the second main result of this section.

PROOF OF THEOREM 1.7. Suppose first that $z > 0$. Using Definition 7.3 we see that the logarithmic derivative of $N_z^a$ is given by

$$\frac{d}{dr}\log N_z^{(a)}(r) = \frac{\frac{dI_z^{(a)}}{dr}(r)}{I_z^{(a)}(r)} - \frac{\frac{dH_z^{(a)}}{dr}(r)}{H_z^{(a)}(r)}. \tag{7.14}$$

By (7.13) we have

$$\frac{dH_z^{(a)}}{dr}(r) = -2r\frac{dL_z^{(a)}}{dt}(-r^2), \qquad \frac{dI_z^{(a)}}{dr}(r) = -2r\frac{dE_z^{(a)}}{dt}(-r^2). \tag{7.15}$$

Combining the first identity in (7.15) with Proposition 7.1, we have

$$\frac{\frac{dH_z^{(a)}}{dr}(r)}{H_z^{(a)}(r)} = \frac{-2r\int_0^\infty u\left[u_t + u_\zeta \frac{p_\zeta^{(a)}}{p^{(a)}}\right]p^{(a)}\zeta^a d\zeta}{\frac{1}{2}\int_0^\infty u^2 p^{(a)}\zeta^a d\zeta}. \tag{7.16}$$

From the second identity in (7.15) and (7.5), we now find

$$\frac{dI_z^{(a)}}{dr}(r) = 2r^3\int_0^\infty \left[u_t + u_\zeta\frac{p_\zeta^{(a)}}{p^{(a)}}\right]^2 p^{(a)}\zeta^a d\zeta - 2rG_z^{(a)}(-r^2),$$

where $G_z^{(a)} = G_{z,0}^{(a)}$ is given by (7.7). This equation and Lemma 7.2 give

$$\frac{\frac{dI_z^{(a)}}{dr}(r)}{I_z^{(a)}(r)} = \frac{2r^3\int_0^\infty \left[u_t + u_\zeta\frac{p_\zeta^{(a)}}{p^{(a)}}\right]^2 p^{(a)}\zeta^a d\zeta}{-\frac{r^2}{2}\int_0^\infty u\left[u_t + u_\zeta\frac{p_\zeta^{(a)}}{p^{(a)}}\right]p^{(a)}\zeta^a d\zeta} - \frac{2rG_z^{(a)}(-r^2)}{\frac{r^2}{2}\int_0^\infty u_\zeta^2(\zeta,t)p^{(a)}(z,\zeta,r^2)\zeta^a d\zeta}.$$

We emphasize in the right-hand side of the latter equation we have used Lemma 7.2 to express $E_z^{(a)}$ in the denominator of the first quotient, whereas we have used the definition (7.3) to write the same quantity in the denominator of the second quotient. It is crucial that there is a minus sign in front of the second quotient. Since the denominator is positive, in view of (7.8) above we can thus conclude that, given $z > 0$, then for every $r > 0$ we have

$$\frac{\frac{dI_z^{(a)}}{dr}(r)}{I_z^{(a)}(r)} > \frac{4r\int_0^\infty \left[u_t + u_\zeta\frac{p_\zeta^{(a)}}{p^{(a)}}\right]^2 p^{(a)}\zeta^a d\zeta}{-\int_0^\infty u\left[u_t + u_\zeta\frac{p_\zeta^{(a)}}{p^{(a)}}\right]p^{(a)}\zeta^a d\zeta}, \qquad a \geq 0. \tag{7.17}$$

Moreover, since in the end of the proof of Theorem 1.5 we showed that, when $z = 0$, then $G^{(a)} \equiv 0$ for every $a > -1$, we also infer that

$$\text{(7.18)} \qquad \frac{\frac{dI^{(a)}}{dr}(r)}{I^{(a)}(r)} = \frac{4r\int_0^\infty \left[u_t + u_\zeta \frac{p_\zeta^{(a)}}{p^{(a)}}\right]^2 p^{(a)}\zeta^a d\zeta}{-\int_0^\infty u\left[u_t + u_\zeta \frac{p_\zeta^{(a)}}{p^{(a)}}\right]p^{(a)}\zeta^a d\zeta}, \qquad a > -1.$$

Combining (7.14), (7.16), (7.17) and (7.18), we finally conclude that for every $z > 0$ and $r > 0$, we have for every $a \geq 0$,

$$\text{(7.19)} \qquad \frac{d}{dr}\log N_z^{(a)}(r) > \frac{4r\int_0^\infty \left[u_t + u_\zeta \frac{p_\zeta^{(a)}}{p^{(a)}}\right]^2 p^{(a)}\zeta^a d\zeta}{-\int_0^\infty u\left[u_t + u_\zeta \frac{p_\zeta^{(a)}}{p^{(a)}}\right]p^{(a)}\zeta^a d\zeta} - \frac{-4r\int_0^\infty u\left[u_t + u_\zeta \frac{p_\zeta^{(a)}}{p^{(a)}}\right]p^{(a)}\zeta^a d\zeta}{\int_0^\infty u^2 p^{(a)}\zeta^a d\zeta}.$$

If instead $z = 0$, then we obtain for every $a > -1$

$$\text{(7.20)} \qquad \frac{d}{dr}\log N^{(a)}(r) = \frac{4r\int_0^\infty \left[u_t + u_\zeta \frac{p_\zeta^{(a)}}{p^{(a)}}\right]^2 p^{(a)}\zeta^a d\zeta}{-\int_0^\infty u\left[u_t + u_\zeta \frac{p_\zeta^{(a)}}{p^{(a)}}\right]p^{(a)}\zeta^a d\zeta} - \frac{-4r\int_0^\infty u\left[u_t + u_\zeta \frac{p_\zeta^{(a)}}{p^{(a)}}\right]p^{(a)}\zeta^a d\zeta}{\int_0^\infty u^2 p^{(a)}\zeta^a d\zeta}.$$

Since by the inequality of Cauchy-Schwarz the right-hand side in (7.19), (7.20) is always $\geq 0$, we finally obtain

$$\text{(7.21)} \qquad \frac{d}{dr}\log N^{(a)}(r)\begin{cases} > 0, & \text{if } z > 0, \quad a \geq 0, \\ \geq 0, & \text{if } z = 0, \quad a > -1. \end{cases}$$

This almost concludes the proof of the theorem. We are left with showing the last part, i.e., the characterization of the case of constant frequency when $z = 0$. If for $\kappa > 0$ we have $u(\lambda\zeta, \lambda^2 t) = \lambda^\kappa u(\zeta, t)$ for every $\lambda > 0$, differentiating with respect to $\lambda$ and setting $\lambda = 1$, we find $\zeta u_\zeta + 2tu_t = \kappa u$. If we denote by $Z = Z(\zeta, t) = \zeta\partial_\zeta + 2t\partial_t$ the generator of the parabolic dilations, then we can write the latter equation as $Zu = \kappa u$. Now,

$$Zu(\zeta, -r^2) = \zeta u_\zeta - 2r^2 u_t.$$

On the other hand, with $p^{(a)} = p^{(a)}(0, \zeta, r^2)$, we find from (7.11)

$$p_\zeta^{(a)} = -\frac{\zeta}{2r^2}p^{(a)}.$$

Therefore,

$$\text{(7.22)} \qquad u_t + u_\zeta \frac{p_\zeta^{(a)}}{p^{(a)}} = u_t - \frac{\zeta}{2r^2}u_\zeta = \frac{Zu(\zeta, -r^2)}{2(-r^2)}$$

Since by the hypothesis we have $Zu(\zeta,-r^2)=\kappa u(\zeta,-r^2)$, we find that

$$I^{(a)}(r)=-\frac{r^2}{2}\int_0^\infty \frac{u(\zeta,-r^2)Zu(\zeta,-r^2)}{-2r^2}p^{(a)}\zeta^a d\zeta$$
$$=\frac{\kappa}{4}\int_0^\infty u^2p^{(a)}\zeta^a d\zeta=\frac{\kappa}{2}H^{(a)}(r).$$

In conclusion, if $u$ is homogeneous of degree $\kappa$, then $N^{(a)}(r)\equiv\frac{\kappa}{2}$.

Vice-versa, suppose that $N^{(a)}(r)\equiv\frac{\kappa}{2}$. We want to show that $u$ must be parabolically homogeneous of degree $\kappa$, i.e., $Zu=\kappa u$. From our assumption, we have $\frac{d}{dr}\log N^{(a)}(r)\equiv 0$. If $z=0$, then by the second option in (7.21) we obtain that for any $a>-1$ there must be equality in the Cauchy-Schwarz inequality in (7.20). This means that for every $r>0$ there exists $\alpha(r)\in\mathbb{R}$ such that for every $\zeta>0$

$$\left[u_t+u_\zeta\frac{p_\zeta^{(a)}}{p^{(a)}}\right](\zeta,-r^2)=\alpha(r)u(\zeta,-r^2).$$

From (7.22) we can rewrite this equation as

$$\frac{Zu(\zeta,-r^2)}{2(-r^2)}=\alpha(r)u(\zeta,-r^2). \tag{7.23}$$

This information implies that

$$\frac{I^{(a)}(r)}{H^{(a)}(r)}=-r^2\alpha(r).$$

We conclude that $-r^2\alpha(r)\equiv\kappa/2$, hence

$$\alpha(r)\equiv-\frac{\kappa}{2r^2}.$$

From (7.23) we finally have that for every $\zeta,r>0$ it must be $Zu(\zeta,-r^2)=\kappa u(\zeta,-r^2)$, which is the sought for conclusion. ☐

In closing we note that the frequency centered at any point $z>0$ does not detect homogeneity. One reason for this is that the critical information (7.22) above is no longer available when the pole of the Bessel-Gaussian $p^{(a)}(z,\zeta,r^2)$ is at a point $z>0$. A second (related) obstruction is in the fact that from the first option in (7.21) we obtain for $r>0$ the strict inequality

$$\frac{d}{dr}\log N^{(a)}(r)>0.$$

## 8. Appendix: The modified Bessel function $I_\nu(z)$

In this section we collect some facts concerning the modified Bessel function $I_\nu$ which are probably well-known to the special functions community, but not so to people working in pde's. For the reader's convenience we also provide a proof of Propositions 8.6 and 8.8, which are crucial to our work.

We recall that the modified Bessel function of the first kind and order $\nu\in\mathbb{C}$ is defined by the series

$$I_\nu(z)=\sum_{k=0}^\infty\frac{(z/2)^{\nu+2k}}{\Gamma(k+1)\Gamma(k+\nu+1)},\qquad |z|<\infty,\ |\arg z|<\pi, \tag{8.1}$$

in the complex plane cut along the negative real axis. From (8.1) we see that when $\nu > -1$, then $\Gamma(k+\nu+1) > 0$ for any $k \in \mathbb{N} \cup \{0\}$, and therefore $I_\nu(z) > 0$ when $z > 0$. The same is true also when $\nu = -1$.

From (8.1) the behavior of $I_\nu(z)$ for small values of $z > 0$ easily follows, and we have

$$\lim_{z \to 0^+} 2^\nu \Gamma(\nu+1) z^{-\nu} I_\nu(z) = 1. \tag{8.2}$$

The behavior of $I_\nu(z)$ for large values of $|z|$ is given by the following asymptotic series first found by Hankel

$$I_\nu(z) = \frac{e^z}{(2\pi z)^{1/2}} \left[ \sum_{k=0}^{n} \frac{(-1)^k (\nu,k)}{(2z)^k} + O(|z|^{-n-1}) \right], \qquad |\arg z| \le \frac{\pi}{2} - \delta, \tag{8.3}$$

where we have denoted by $(\nu, k)$ the Hankel coefficients

$$(\nu,k) = \frac{(-1)^k \Gamma(1/2-\nu+k)\Gamma(1/2+\nu+k)}{\Gamma(k+1)\Gamma(1/2-\nu)\Gamma(1/2+\nu)}, \qquad \text{so that } (\nu,0) = 1, \tag{8.4}$$

see (5.11.10) on p. 123 in [**39**], or 9.7.1 on p. 377 in [**1**]. Note that (8.3) implies in particular that for large $z > 0$

$$I_\nu(z) = \frac{e^z}{(2\pi z)^{1/2}} \left(1 + O(|z|^{-1})\right). \tag{8.5}$$

We also note the following Poisson representation of $I_\nu(z)$

$$I_\nu(z) = \frac{(z/2)^\nu}{\sqrt{\pi}\,\Gamma(\nu+1/2)} \int_{-1}^{1} e^{zt} (1-t^2)^{\nu-1/2} dt, \qquad \Re \nu > -\frac{1}{2}, \tag{8.6}$$

see (10) on p. 81 in [**25**].

For later use, we recall the recurrence relations satisfied by $I_\nu(z)$, see e.g. (5.7.9) on p. 110 in [**39**]

$$\frac{d}{dz}[z^{\nu+1} I_{\nu+1}(z)] = z^{\nu+1} I_\nu(z), \qquad \frac{d}{dz}[z^{-\nu} I_\nu(z)] = z^{-\nu} I_{\nu+1}(z). \tag{8.7}$$

For $z > 0$ we have $I_\nu(z) > 0$ for $\nu > -1$, and therefore we can rewrite (8.7) as follows

$$\frac{I'_{\nu+1}(z)}{I_\nu(z)} = 1 - \frac{\nu+1}{z} \frac{I_{\nu+1}(z)}{I_\nu(z)}, \qquad \frac{I'_\nu(z)}{I_\nu(z)} = \frac{I_{\nu+1}(z)}{I_\nu(z)} + \frac{\nu}{z}, \tag{8.8}$$

We next establish for the function $I_\nu$ a generalization of Weber's integral for the function $J_\nu$ (for the latter see (5.15.2) on p. 132 in [**39**], or also formula 4. on p. 717 in [**32**], or formula (10) on p. 29 in [**25**]). Such result is needed in the proof of Proposition 2.3 above.

LEMMA 8.1. *Let $\alpha > 0$, $\Re\nu > -1$, then*

$$\int_0^\infty x^{\nu+1} I_\nu(x) e^{-\alpha x^2} dx = 2^{-\nu-1} \alpha^{-\nu-1} e^{1/4\alpha}.$$

PROOF. We begin by observing that in view of (8.2) we have near $x = 0$

$$|x^{\nu+1} I_\nu(x)| = O(|x|^{2\nu+1}),$$

which is integrable since $\Re\nu > -1$. Furthermore, from (8.5) and $\alpha > 0$ we see that the integrand is absolutely convergent near $\infty$. Thus the integrand belongs to

$L^1(\mathbb{R}^+, dx)$. To compute the integral we apply (8.1) and integrate term by term obtaining

$$\int_0^\infty x^{2\nu+1} I_\nu(x) e^{-\alpha x^2} dx = \sum_{k=0}^\infty \frac{1}{2^{\nu+2k}\Gamma(k+1)\Gamma(k+\nu+1)} \int_0^\infty x^{2\nu+2k+2} e^{-\alpha x^2} \frac{dx}{x}$$

(change of variable $y = \alpha x^2$)

$$= \sum_{k=0}^\infty \frac{1}{2^{\nu+2k+1}\Gamma(k+1)\Gamma(k+\nu+1)} \int_0^\infty \left(\frac{y}{\alpha}\right)^{\nu+k+1} e^{-y} \frac{dy}{y}$$

$$= \sum_{k=0}^\infty \frac{1}{2^{\nu+2k+1}\Gamma(k+1)\alpha^{k+\nu+1}} = \frac{1}{2^{\nu+1}\alpha^{\nu+1}} \sum_{k=0}^\infty \frac{1}{2^{2k}\Gamma(k+1)\alpha^k}$$

$$= \frac{e^{1/4\alpha}}{2^{\nu+1}\alpha^{\nu+1}}.$$

□

For every $\nu > -1$ we next introduce the *Bessel quotient*

$$y_\nu(z) = \frac{I_{\nu+1}(z)}{I_\nu(z)}, \qquad z > 0. \tag{8.9}$$

From (8.2) we easily see that

$$y_\nu(z) \cong \frac{z}{2(\nu+1)}, \qquad \text{as } z \to 0^+, \tag{8.10}$$

and thus in particular

$$\lim_{z\to 0^+} y_\nu(z) = 0. \tag{8.11}$$

Hankel's asymptotic formula (8.3), or also (8.5), imply in particular that the line $y = 1$ is a horizontal asymptote on $(0,\infty)$ for the function $z \to y_\nu(z)$, i.e.,

$$\lim_{z\to\infty} y_\nu(z) = 1. \tag{8.12}$$

Since as we have observed above one has $I_\nu(z) > 0$ for every $z > 0$ and $\nu > -1$, we infer that the function $z \to y_\nu(z)$ is continuous on $[0,\infty)$. Therefore, (8.12) and (8.11) imply that for every $\nu > -1$ there exists $M_\nu \in (0,\infty)$ such that

$$\sup_{z>0} y_\nu(z) = M_\nu. \tag{8.13}$$

The functions $I_\nu$ and $y_\nu$ are connected by the following result.

PROPOSITION 8.2. *For every $\nu > -1$ and $z > 0$ one has*

$$I_\nu(z) = \frac{1}{\Gamma(\nu+1)} \left(\frac{z}{2}\right)^\nu \exp\left(\int_0^z y_\nu(t)dt\right).$$

PROOF. If we take the logarithmic derivative of the function $I_\nu$, we find

$$\frac{d}{dz} \log I_\nu(z) = \frac{I'_\nu(z)}{I_\nu(z)} = \frac{I_{\nu+1}(z)}{I_\nu(z)} + \frac{\nu}{z} = y_\nu(z) + \frac{d}{dz} \log z^\nu, \tag{8.14}$$

where in the second to the last equality we have used the second equation in (8.8) above. We rewrite (8.14) in the following form

$$\frac{d}{dt} \log \frac{I_\nu(t)}{t^\nu} = y_\nu(t), \qquad t > 0, \tag{8.15}$$

and integrate it between $\varepsilon$ and $z$, obtaining

$$\log\frac{I_\nu(z)}{z^\nu}-\log\frac{I_\nu(\varepsilon)}{\varepsilon^\nu}=\int_\varepsilon^z y_\nu(t)dt.$$

Taking the limit as $\varepsilon\to 0^+$, using (8.2), and exponentiating, we reach the desired conclusion. □

LEMMA 8.3. *For every $\nu>-1$ the Bessel quotient $y(z)=y_\nu(z)$ satisfies the following Cauchy problem for the Riccati equation*

$$y'(z)=1-y(z)^2-\frac{2\nu+1}{z}y(z),\qquad y(0)=0. \tag{8.16}$$

PROOF. The initial condition $y(0)=0$ is nothing but (8.11). The equation (8.16) is derived in the following way. Recall the recurrence relations (8.8) above. Since the rule for differentiating a quotient gives

$$y'(z)=\frac{I'_{\nu+1}(z)}{I_\nu(z)}-\frac{I'_\nu(z)}{I_\nu(z)}\frac{I_{\nu+1}(z)}{I_\nu(z)},$$

substituting (8.8) into the latter expression we easily obtain the differential equation in (8.16). □

One has the asymptotic formula on $(0,\infty)$.

PROPOSITION 8.4. *Let $\nu>-1$. One has*

$$\lim_{z\to\infty} z\,[1-y_\nu(z)]=\frac{2\nu+1}{2}. \tag{8.17}$$

PROOF. To prove (8.17) we use the asymptotic expansions (8.3) and (8.5), which give

$$\begin{aligned}1-y_\nu(z)&=\frac{I_\nu(z)-I_{\nu+1}(z)}{I_\nu(z)}\\&=\frac{\frac{e^z}{(2\pi z)^{1/2}}}{\frac{e^z}{(2\pi z)^{1/2}}\left(1+O(z^{-1})\right)}\left[1-\frac{(\nu,1)}{(2z)}-1+\frac{(\nu+1,1)}{(2z)}+O\left(\frac{1}{z^2}\right)\right]\\&=\frac{1}{1+O(z^{-1})}\left(\frac{-(\nu,1)+(\nu+1,1)}{(2z)}+O\left(\frac{1}{z^2}\right)\right)\\&=\left[\frac{-(\nu,1)+(\nu+1,1)}{2z}+O\left(\frac{1}{z^2}\right)\right]\left(1+O\left(\frac{1}{z}\right)\right).\end{aligned}$$

Using (8.4) we find

$$\begin{aligned}-(\nu,k)+(\nu+1,1)&=\frac{\Gamma(1/2-\nu+1)\Gamma(1/2+\nu+1)}{\Gamma(2)\Gamma(1/2-\nu)\Gamma(1/2+\nu)}\\&\quad-\frac{\Gamma(1/2-(\nu+1)+1)\Gamma(1/2+(\nu+1)+1)}{\Gamma(2)\Gamma(1/2-(\nu+1))\Gamma(1/2+(\nu+1))}\\&=(1/2-\nu)(1/2+\nu)-((1/2-\nu)-1)((1/2+\nu)+1)\\&=-(1/2-\nu)+(1/2+\nu)+1=2\nu+1.\end{aligned}$$

The latter two formulas prove that as $z\to\infty$

$$z(1-y_\nu(z))=\left(\frac{2\nu+1}{2}+O(z^{-1})\right)(1+O(z^{-1}))=\frac{2\nu+1}{2}+O(z^{-1}). \tag{8.18}$$

□

REMARK 8.5. We emphasize that (8.17) implies, in particular, that for large $z > 0$ we must have

$$\begin{cases} 1 - y_\nu(z) < 0, & \text{when } -1 < \nu < -1/2, \\ 1 - y_\nu(z) > 0, & \text{when } -1/2 < \nu. \end{cases} \tag{8.19}$$

The equation (8.19) shows that the function $y_\nu(z)$ approaches the asymptotic line $y = 1$ from above when $-1 < \nu < -1/2$, and from below when $\nu > -1/2$.

In fact, when $\nu \geq -1/2$ we have the following global information.

PROPOSITION 8.6. *For every $\nu \geq -\frac{1}{2}$ and for every $z > 0$ one has*

$$y_\nu(z) < 1. \tag{8.20}$$

PROOF. This result was proved by Soni in [**56**] when $\nu > -1/2$. Since it is not easy to obtain Soni's original paper, we provide its short proof in what follows. By (8.6) above, we have

$$I_\nu(z) = \frac{2(z/2)^\nu}{\sqrt{\pi}\Gamma(\nu + 1/2)} \int_0^1 \cosh(zt)(1 - t^2)^{\nu - 1/2} dt, \qquad \Re\nu > -\frac{1}{2}.$$

From this formula we obtain

$$I_{\nu+1}(z) = \frac{2^{-\nu} z^{\nu+1}}{\sqrt{\pi}(\nu + 1/2)\Gamma(\nu + 1/2)} \int_0^1 \cosh(zt)(1 - t^2)^{\nu + 1/2} dt, \qquad \Re\nu > -\frac{1}{2}.$$

Integrating by parts in the integral in the right-hand side we find

$$I_{\nu+1}(z) = \frac{2(z/2)^\nu}{\sqrt{\pi}\Gamma(\nu + 1/2)} \int_0^1 t \sinh(zt)(1 - t^2)^{\nu - 1/2} dt, \qquad \Re\nu > -\frac{1}{2}.$$

These formulas give for $\nu > -\frac{1}{2}$

$$\begin{aligned} I_\nu(z) - I_{\nu+1}(z) &= \frac{2(z/2)^\nu}{\sqrt{\pi}\Gamma(\nu + 1/2)} \int_0^1 [\cosh(zt) - t\sinh(zt)](1 - t^2)^{\nu - 1/2} dt \\ &= \frac{2(z/2)^\nu}{\sqrt{\pi}\Gamma(\nu + 1/2)} \int_0^1 [\coth(zt) - t] \sinh(zt)(1 - t^2)^{\nu - 1/2} dt. \end{aligned}$$

Since $t \to \sinh(zt)(1-t^2)^{\nu-1/2}$ is $> 0$ on the interval $[0, 1]$, except at $t = 0$, we infer that (8.20) will be true if for every $z > 0$ and $0 \leq t \leq 1$ one has $\coth(zt) - t \geq 0$. Now, $\partial_t(\coth(zt) - t) = -\frac{z}{\sinh^2(zt)} - 1 < 0$ for every $0 < t \leq 1$. Therefore, for every $z > 0$ the function $h(t) = \coth(zt) - t$ is strictly decreasing on $(0, 1)$. Since $h(1) = \coth(z) - 1 > 0$, we conclude that $h(t) > 0$ for every $0 \leq t \leq 1$. This proves Soni's inequality (8.20) when $\nu > -1/2$.

Later, Nasell observed in [**45**] that the statement is also true when $\nu = -1/2$. To see this note that from (2.11) above we find

$$I_{-1/2}(z) - I_{1/2}(z) = \left(\frac{2}{\pi z}\right)^{1/2} e^{-z}. \tag{8.21}$$

Dividing by $I_{-1/2}(z)$ in (8.21), we find

$$1 - \frac{I_{1/2}(z)}{I_{-1/2}(z)} = \left(\frac{2}{\pi z}\right)^{1/2} \frac{e^{-z}}{I_{-1/2}(z)} = \frac{2e^{-z}}{e^z + e^{-z}} = \frac{2}{e^{2z} + 1},$$

or equivalently

$$\frac{I_{1/2}(z)}{I_{-1/2}(z)} = 1 - \frac{2}{e^{2z}+1} < 1. \tag{8.22}$$

Notice that in agreement with (8.17) in Proposition 8.4, we have

$$\lim_{z\to\infty} z\left[1 - \frac{I_{1/2}(z)}{I_{-1/2}(z)}\right] = 0.$$

Although (8.22) shows that for every $z > 0$ we have

$$\left(\frac{I_{1/2}(z)}{I_{-1/2}(z)}\right)^2 - 1 < 0, \tag{8.23}$$

the estimate does prove that the bound with 1 is not optimal, since in fact the more precise formula

$$\left(\frac{I_{1/2}(z)}{I_{-1/2}(z)}\right)^2 - 1 = -\frac{4}{e^{2z}+1}\left(1 - \frac{1}{e^{2z}+1}\right) \tag{8.24}$$

is available. □

REMARK 8.7. We mention that the limitation $\nu \geq -1/2$ in the global estimate in Proposition 8.6 is responsible for the limitation $a \geq 0$ in Theorems 1.1 and 1.2 above.

We take the occasion here to correct an oversight in the formula (A.3) in [**61**], where the value of the limit (8.17) is stated to be $2\nu+1$. We were in fact indirectly led to rediscovering Proposition 8.4, after realizing that (A.3) in [**61**] cannot possibly be correct since it would lead to the contradictory conclusion that, when $-1 < \nu < -\frac{1}{2}$, the function

$$f(z) = f_\nu(z) = z^2(y_\nu^2(z) - 1),$$

be bounded on the whole line. While, if true, such result would be great in settling the question raised in Remark 4.5 above, unfortunately, the function $f(z)$ cannot possibly be bounded on $(0,\infty)$ since, as a consequence of (8.17), we know that

$$\lim_{z\to\infty} f(z) = \infty. \tag{8.25}$$

To see (8.25), we note that (8.12) gives

$$\lim_{z\to\infty}(1 + y(z)) = 2.$$

If we write

$$f(z) = -z^2(1 - y(z))(1 + y(z)),$$

then (8.25) follows, since we have by (8.17)

$$\lim_{z\to\infty}\{-z(1 - y(z))(1 + y(z))\} > 0.$$

We now have

$$f'(z) = -2z(1 - y^2(z)) + 2y(z)y'(z)z^2.$$

Using the differential equation (8.16), we find

$$f'(z) = -2z\left(1 - y^2(z)\right) + 2y(z)z^2\left(1 - y(z)^2 - \frac{2\nu+1}{z}y(z)\right).$$

We rewrite this equation as follows

$$\frac{f'(z)}{z} = -2\left(1 - y^2(z)\right) + 2y(z)z\left(1 - y(z)^2\right) - 2(2\nu+1)y(z)^2.$$

By (8.12) we see that $-2(1-y^2(z)) \to 0$ and that $-2(2\nu+1)y(z)^2 \to -2(2\nu+1)$ as $z \to \infty$. On the other hand, if instead of (8.17) we had (A.3) in [**61**], then such fact and (8.12) would give $2y(z)z\left(1-y(z)^2\right) \to 4(2\nu+1)$ as $z \to \infty$. We would conclude that

$$\lim_{z\to\infty} \frac{f'(z)}{z} = 4(2\nu+1) - 2(2\nu+1) = 2(2\nu+1) < 0.$$

This limit relation would imply, in particular, that $f(z)$ is decreasing in a neighborhood of infinity, and since $f(z) \geq 0$ at infinity by (8.17), we would reach the conclusion that $f(z)$ be bounded at infinity. But this is a contradiction with (8.25).

Having said this, we note that the correct limiting relation (8.17) above implies that, in fact, what one has is

$$\lim_{z\to\infty} \frac{f'(z)}{z} = 0,$$

and this leads to no contradiction.

We close this section with the second critical property of the Bessel quotient $y_\nu$ of interest in this paper, namely, its monotonicity. It is certainly well-known to most experts of special functions but the only proof we could locate is embedded in the discussion in the Appendix of [**61**]. Since we need this result in the proof of Theorems 1.5 and 1.7, for the reader's convenience we provide its proof.

PROPOSITION 8.8. *When $\nu \geq -1/2$ the Bessel quotient $y_\nu$ strictly increases on $(0,\infty)$ from $y_\nu(0) = 0$ to its asymptotic value $y_\nu(\infty) = 1$. If instead $-1 < \nu < -1/2$, then $y_\nu$ first increases to its absolute maximum $> 1$, and then it becomes strictly decreasing to its asymptotic value $y_\nu(\infty) = 1$.*

PROOF. When $\nu = -1/2$ the desired conclusion follows from the explicit expression in (8.22) above. We thus assume $\nu \neq -1/2$, or equivalently $2\nu - 1 \neq 0$. Differentiating (8.16) we obtain

$$y''(z) = \frac{2\nu+1}{z^2} y(z) - \left(\frac{2\nu+1}{z} + 2y(z)\right) y'(z). \tag{8.26}$$

At a stationary point $z_0 > 0$ we have $y'(z_0) = 0$, and thus (8.26) gives

$$y''(z_0) = \frac{2\nu+1}{z_0^2} y(z_0).$$

Since $y(z) > 0$ for every $z > 0$, it ensues that the sign of $y''(z_0)$ is the same as that of $2\nu+1$. Therefore, $y$ can only have strict local minima if $\nu > -1/2$, strict local maxima if $-1 < \nu < -1/2$.

On the other hand, the equation (8.16) implies that $y'$ cannot change sign when $\nu > -1/2$. Otherwise, there would be a point $z_0 > 0$ at which $y'(z_0) = 0$. From what has been said, $y$ must have a strict local minimum at $z_0$, and therefore $y'(z) \leq 0$ in a left neighborhood of $z_0$, whereas $y'(z) \geq 0$ in a small right neighborhood of $z_0$. Since by (8.16) we obtain

$$y'(0) = \frac{1}{2\nu+2} > 0,$$

we infer that there must be a point $0 < z_1 < z_0$ where $y'(z_1) = 0$ and where $y$ attains a local maximum. Since this is impossible from what has been said above, we conclude that $y'$ cannot change sign when $\nu > -1/2$, and therefore $y' > 0$ for such values of $\nu$. When instead $-1 < \nu < -1/2$, then (8.17) above shows that in a neighborhood of infinity we must have $y > 1$. Since $y(\infty) = 1$ there must be a point $z_0 > 0$ where $y'(z_0) < 0$. Since as we have observed $y'(0) > 0$, there must be a point $z_1 \in (0, z_0)$ where $y'(z_1) = 0$. From what we have said, $y$ can only have a strict local maximum at $z_1$. Since the equation (8.16) gives

$$0 = y'(z_1) = 1 - y(z_1)^2 - \frac{2\nu + 1}{z_1} y(z_1),$$

we conclude that $y(z_1) > 1$. Since $y'$ can only change sign once, we finally infer that $y(z_1)$ is not only a strict local maximum, but also a global □

## Acknowledgments

I thank Camelia Pop for her interest in the present work, for her kindness (and patience!) in educating me in the probabilistic aspects of the Bessel semigroup, and for bringing to my attention the references [**21**], [**22**], [**23**] and [**24**]. In Section 5 below I discuss the connection between Theorem 1.2 above and the results in [**23**], [**24**] and [**18**]. I also mention that Proposition 2.6 below came up in a conversation with Giulio Tralli and I thank him for an interesting exchange. I am also grateful to Javier Segura and Bettina Grün who have kindly corresponded with me. In an early stage of this article, Javier first provided numerical evidence that Proposition 8.6 fails in the range $-1 < \nu < -1/2$. Finally, I thank Charles Epstein and Rafe Mazzeo for their kind feedback.

Last, but not least, I thank the organizers of the AMS Special Session at Northeastern University for their gracious invitation to contribute to this volume.

## References

[1] Milton Abramowitz and Irene A. Stegun, *Handbook of mathematical functions with formulas, graphs, and mathematical tables*, National Bureau of Standards Applied Mathematics Series, vol. 55, For sale by the Superintendent of Documents, U.S. Government Printing Office, Washington, D.C., 1964. MR0167642

[2] Frederick J. Almgren Jr., *Dirichlet's problem for multiple valued functions and the regularity of mass minimizing integral currents*, Minimal submanifolds and geodesics (Proc. Japan-United States Sem., Tokyo, 1977), North-Holland, Amsterdam-New York, 1979, pp. 1–6. MR574247

[3] D. E. Amos, *Computation of modified Bessel functions and their ratios*, Math. Comp. **28** (1974), 239–251, DOI 10.2307/2005830. MR333287

[4] Ioannis Athanasopoulos, Luis Caffarelli, and Emmanouil Milakis, *On the regularity of the non-dynamic parabolic fractional obstacle problem*, J. Differential Equations **265** (2018), no. 6, 2614–2647, DOI 10.1016/j.jde.2018.04.043. MR3804726

[5] A. Audrito & S. Terracini, *On the nodal set of solutions to a class of nonlocal parabolic reaction-diffusion equations*, preprint.

[6] D. Bakry and Michel Émery, *Diffusions hypercontractives* (French), Séminaire de probabilités, XIX, 1983/84, Lecture Notes in Math., vol. 1123, Springer, Berlin, 1985, pp. 177–206, DOI 10.1007/BFb0075847. MR889476

[7] Dominique Bakry, Ivan Gentil, and Michel Ledoux, *Analysis and geometry of Markov diffusion operators*, Grundlehren der Mathematischen Wissenschaften [Fundamental Principles of Mathematical Sciences], vol. 348, Springer, Cham, 2014. MR3155209

[8] A. Banerjee, D. Danielli, N. Garofalo & A. Petrosyan, *The structure of the singular set in the thin obstacle problem for degenerate parabolic equations*, manuscript in preparation.

[9] Agnid Banerjee and Nicola Garofalo, *Monotonicity of generalized frequencies and the strong unique continuation property for fractional parabolic equations*, Adv. Math. **336** (2018), 149–241, DOI 10.1016/j.aim.2018.07.021. MR3846151

[10] A. Banerjee, N. Garofalo, I. Munive & D. M. Nhieu, *The Harnack inequality for a class of nonlocal parabolic equations*, ArXiv:1911.05619.

[11] Fabrice Baudoin, Michel Bonnefont, and Nicola Garofalo, *A sub-Riemannian curvature-dimension inequality, volume doubling property and the Poincaré inequality*, Math. Ann. **358** (2014), no. 3-4, 833–860, DOI 10.1007/s00208-013-0961-y. MR3175142

[12] Fabrice Baudoin, Michel Bonnefont, Nicola Garofalo, and Isidro H. Munive, *Volume and distance comparison theorems for sub-Riemannian manifolds*, J. Funct. Anal. **267** (2014), no. 7, 2005–2027, DOI 10.1016/j.jfa.2014.07.030. MR3250358

[13] F. Baudoin & N. Garofalo, *Generalized Bochner formulas and Ricci lower bounds for sub-Riemannian manifolds of rank two*, Arxiv preprint, (2009) arXiv:0904 1623

[14] Fabrice Baudoin and Nicola Garofalo, *A note on the boundedness of Riesz transform for some subelliptic operators*, Int. Math. Res. Not. IMRN **2** (2013), 398–421, DOI 10.1093/imrn/rnr271. MR3010694

[15] Fabrice Baudoin and Nicola Garofalo, *Curvature-dimension inequalities and Ricci lower bounds for sub-Riemannian manifolds with transverse symmetries*, J. Eur. Math. Soc. (JEMS) **19** (2017), no. 1, 151–219, DOI 10.4171/JEMS/663. MR3584561

[16] Kamil Bogus and Jacek Małecki, *Sharp estimates of transition probability density for Bessel process in half-line*, Potential Anal. **43** (2015), no. 1, 1–22, DOI 10.1007/s11118-015-9461-x. MR3361787

[17] Kamil Bogus and Jacek Małecki, *Heat kernel estimates for the Bessel differential operator in half-line*, Math. Nachr. **289** (2016), no. 17-18, 2097–2107, DOI 10.1002/mana.201500163. MR3583258

[18] Linan Chen and Daniel W. Stroock, *The fundamental solution to the Wright-Fisher equation*, SIAM J. Math. Anal. **42** (2010), no. 2, 539–567, DOI 10.1137/090764207. MR2607921

[19] Filippo Chiarenza and Raul Serapioni, *A remark on a Harnack inequality for degenerate parabolic equations*, Rend. Sem. Mat. Univ. Padova **73** (1985), 179–190. MR799906

[20] Donatella Danielli, Nicola Garofalo, Arshak Petrosyan, and Tung To, *Optimal regularity and the free boundary in the parabolic Signorini problem*, Mem. Amer. Math. Soc. **249** (2017), no. 1181, v + 103, DOI 10.1090/memo/1181. MR3709717

[21] Charles L. Epstein and Rafe Mazzeo, *Wright-Fisher diffusion in one dimension*, SIAM J. Math. Anal. **42** (2010), no. 2, 568–608, DOI 10.1137/090766152. MR2607922

[22] Charles L. Epstein and Rafe Mazzeo, *Degenerate diffusion operators arising in population biology*, Annals of Mathematics Studies, vol. 185, Princeton University Press, Princeton, NJ, 2013. MR3202406

[23] Charles L. Epstein and Rafe Mazzeo, *Harnack inequalities and heat kernel estimates for degenerate diffusion operators arising in population biology*, Appl. Math. Res. Express. AMRX **2** (2016), 217–280, DOI 10.1093/amrx/abw002. MR3551776

[24] Charles L. Epstein and Camelia A. Pop, *The Feynman-Kac formula and Harnack inequality for degenerate diffusions*, Ann. Probab. **45** (2017), no. 5, 3336–3384, DOI 10.1214/16-AOP1138. MR3706745

[25] A. Erdélyi, W. Magnus, F. Oberhettinger, and F. G. Tricomi, *Tables of integral transforms. Vol. II*, McGraw-Hill Book Company, Inc., New York-Toronto-London, 1954. Based, in part, on notes left by Harry Bateman. MR0065685

[26] Eugene B. Fabes, Carlos E. Kenig, and Raul P. Serapioni, *The local regularity of solutions of degenerate elliptic equations*, Comm. Partial Differential Equations **7** (1982), no. 1, 77–116, DOI 10.1080/03605308208820218. MR643158

[27] William Feller, *The parabolic differential equations and the associated semi-groups of transformations*, Ann. of Math. (2) **55** (1952), 468–519, DOI 10.2307/1969644. MR47886

[28] Ronald Fisher, *Dispersion on a sphere*, Proc. Roy. Soc. London. Ser. A. **217** (1953), 295–305, DOI 10.1098/rspa.1953.0064. MR0056866

[29] Nicola Garofalo, *Fractional thoughts*, New developments in the analysis of nonlocal operators, Contemp. Math., vol. 723, Amer. Math. Soc., Providence, RI, 2019, pp. 1–135, DOI 10.1090/conm/723/14569. MR3916700

[30] Nicola Garofalo, *Some properties of sub-Laplaceans*, Proceedings of the International Conference "Two nonlinear days in Urbino 2017", Electron. J. Differ. Equ. Conf., vol. 25, Texas State Univ.–San Marcos, Dept. Math., San Marcos, TX, 2018, pp. 103–131. MR3883632

[31] N. Garofalo & G. Tralli, *A class of nonlocal hypoelliptic operators and their extensions.* To appear in Indiana Univ. Math. J.

[32] I. S. Gradshteyn & I. M. Ryzhik, *Tables of integrals, series, and products*, Translated from the fourth Russian edition. Fifth edition. Translation edited and with a preface by Alan Jeffrey. Academic Press, Inc., Boston, MA, 1994. xlviii+1204 pp.

[33] J. Hadamard, *Extension à l'équation de la chaleur d'un théorème de A. Harnack* (French), Rend. Circ. Mat. Palermo (2) **3** (1954), 337–346 (1955), DOI 10.1007/BF02849264. MR68713

[34] Kurt Hornik and Bettina Grün, *Amos-type bounds for modified Bessel function ratios*, J. Math. Anal. Appl. **408** (2013), no. 1, 91–101, DOI 10.1016/j.jmaa.2013.05.070. MR3079949

[35] Kurt Hornik and Bettina Grün, *On maximum likelihood estimation of the concentration parameter of von Mises–Fisher distributions*, Comput. Statist. **29** (2014), no. 5, 945–957, DOI 10.1007/s00180-013-0471-0. MR3266042

[36] Samuel Karlin and Howard M. Taylor, *A second course in stochastic processes*, Academic Press, Inc. [Harcourt Brace Jovanovich, Publishers], New York-London, 1981. MR611513

[37] Charles Tier and Joseph B. Keller, *Asymptotic analysis of diffusion equations in population genetics*, SIAM J. Appl. Math. **34** (1978), no. 3, 549–576, DOI 10.1137/0134044. MR484539

[38] P. Langevin, *Magnétisme et théorie des électrons*, Ann. Chim. Phys., **5**, 71-127.

[39] N. N. Lebedev, *Special functions and their applications*, Dover Publications, Inc., New York, 1972. Revised edition, translated from the Russian and edited by Richard A. Silverman; Unabridged and corrected republication. MR0350075

[40] Peter Li and Shing-Tung Yau, *On the parabolic kernel of the Schrödinger operator*, Acta Math. **156** (1986), no. 3-4, 153–201, DOI 10.1007/BF02399203. MR834612

[41] Kanti V. Mardia and Peter E. Jupp, *Directional statistics*, Wiley Series in Probability and Statistics, John Wiley & Sons, Ltd., Chichester, 2000. Revised reprint of *Statistics of directional data* by Mardia [ MR0336854 (49 #1627)]. MR1828667

[42] S. A. Molčanov, *Martin boundary for invariant Markov processes on a solvable group* (Russian, with English summary), Teor. Verojatnost. i Primenen **12** (1967), 358–362. MR0219144

[43] S. A. Molčanov and E. Ostrovskiĭ, *Symmetric stable processes as traces of degenerate diffusion processes.* (Russian, with English summary), Teor. Verojatnost. i Primenen. **14** (1969), 127–130. MR0247668

[44] B. Muckenhoupt and E. M. Stein, *Classical expansions and their relation to conjugate harmonic functions*, Trans. Amer. Math. Soc. **118** (1965), 17–92, DOI 10.2307/1993944. MR199636

[45] Ingemar Nåsell, *Inequalities for modified Bessel functions*, Math. Comp. **28** (1974), 253–256, DOI 10.2307/2005831. MR333288

[46] K. Nyström and O. Sande, *Extension properties and boundary estimates for a fractional heat operator*, Nonlinear Anal. **140** (2016), 29–37, DOI 10.1016/j.na.2016.02.027. MR3492726

[47] Bruno Pini, *Sulla soluzione generalizzata di Wiener per il primo problema di valori al contorno nel caso parabolico* (Italian), Rend. Sem. Mat. Univ. Padova **23** (1954), 422–434. MR0065794

[48] Chi-Cheung Poon, *Unique continuation for parabolic equations*, Comm. Partial Differential Equations **21** (1996), no. 3-4, 521–539, DOI 10.1080/03605309608821195. MR1387458

[49] Jim Pitman and Marc Yor, *A decomposition of Bessel bridges*, Z. Wahrsch. Verw. Gebiete **59** (1982), no. 4, 425–457, DOI 10.1007/BF00532802. MR656509

[50] A. P. Prudnikov, Yu. A. Brychkov, and O. I. Marichev, *Integrals and series. Vol. 2*, 2nd ed., Gordon & Breach Science Publishers, New York, 1988. Special functions; Translated from the Russian by N. M. Queen. MR950173

[51] Daniel Revuz and Marc Yor, *Continuous martingales and Brownian motion*, 3rd ed., Grundlehren der Mathematischen Wissenschaften [Fundamental Principles of Mathematical Sciences], vol. 293, Springer-Verlag, Berlin, 1999. MR1725357

[52] Diego Ruiz-Antolín and Javier Segura, *A new type of sharp bounds for ratios of modified Bessel functions*, J. Math. Anal. Appl. **443** (2016), no. 2, 1232–1246, DOI 10.1016/j.jmaa.2016.06.011. MR3514344

[53] Geert Schou, *Estimation of the concentration parameter in von Mises-Fisher distributions*, Biometrika **65** (1978), no. 2, 369–377, DOI 10.1093/biomet/65.2.369. MR513935

[54] Javier Segura, *Bounds for ratios of modified Bessel functions and associated Turán-type inequalities*, J. Math. Anal. Appl. **374** (2011), no. 2, 516–528, DOI 10.1016/j.jmaa.2010.09.030. MR2729238

[55] Henry C. Simpson and Scott J. Spector, *Some monotonicity results for ratios of modified Bessel functions*, Quart. Appl. Math. **42** (1984), no. 1, 95–98, DOI 10.1090/qam/736509. MR736509

[56] R. P. Soni, *On an inequality for modified Bessel functions*, J. Math. and Phys. **44** (1965), 406–407. MR185164

[57] Pablo Raúl Stinga and José Luis Torrea, *Extension problem and Harnack's inequality for some fractional operators*, Comm. Partial Differential Equations **35** (2010), no. 11, 2092–2122, DOI 10.1080/03605301003735680. MR2754080

[58] Pablo Raúl Stinga and José L. Torrea, *Regularity theory and extension problem for fractional nonlocal parabolic equations and the master equation*, SIAM J. Math. Anal. **49** (2017), no. 5, 3893–3924, DOI 10.1137/16M1104317. MR3709888

[59] Michael Struwe, *On the evolution of harmonic maps in higher dimensions*, J. Differential Geom. **28** (1988), no. 3, 485–502. MR965226

[60] R. von Mises, *Über die "Ganzzahligkeit" del Atomgewichte und verwandte Fragen*, Phys. Z. **19** (1918), 490-500.

[61] Lin Yuan and John D. Kalbfleisch, *On the Bessel distribution and related problems*, Ann. Inst. Statist. Math. **52** (2000), no. 3, 438–447, DOI 10.1023/A:1004152916478. MR1794244

Dipartimento di Ingegneria Civile, Edile e Ambientale (DICEA), Università di Padova, 35131 Padova, Italy

*Email address*, Nicola Garofalo: nicola.garofalo@unipd.it

Contemporary Mathematics
Volume **748**, 2020
https://doi.org/10.1090/conm/748/15057

# On the existence of dichromatic single element lenses

Cristian E. Gutiérrez and Ahmad Sabra

Abstract. Due to dispersion, light with different wavelengths, or colors, is refracted at different angles. Our purpose is to determine when is it possible to design a lens made of a single homogeneous material so that it refracts light superposition of two colors into a desired fixed final direction. Two problems are considered: one is when light emanates in a parallel beam and the other is when light emanates from a point source. For the first problem, and when the direction of the parallel beam is different from the final desired direction, we show that such a lens does not exist; otherwise we prove the solution is trivial, i.e., the lens is confined between two parallel planes. For the second problem we prove that is impossible to design such a lens when the desired final direction is not among the set of incident directions. Otherwise, solving an appropriate system of functional differential equations we show that a local solution exists.

## Contents



2010 *Mathematics Subject Classification.* Primary 78A05, 39B72, 34K10.

*Key words and phrases.* geometric optics, functional differential equations, fixed-point theorems.

The first author was partially supported by NSF grant DMS–1600578.

The second author was partially supported by Research Grant 2015/19/P/ST1/02618 from the National Science Centre, Poland, entitled "Variational Problems in Optical Engineering and Free Material Design".

## 1. Introduction

We showed in [**GS16**] that given a function $u$ in $\Omega \subset \mathbb{R}^2$ and a unit direction $w \in S^2$ there exist surfaces $S_C$ depending on a real parameter $C$, such that the lens sandwiched by $u$ and $S_C$, made of an homogeneous material, refracts monochromatic light emanating vertically from $\Omega$ into the direction $w$. In the earlier paper [**Gut13**], a similar result is proved when light emanates from a point source. The purpose of this paper is to study if it is possible to design simple lenses doing similar refracting jobs for non monochromatic light. By a simple (or single element) lens we mean a domain in $\mathbb{R}^3$ bounded by two smooth surfaces that is filled with an homogeneous material.

To do this we need to deal with dispersion: since the index of refraction of a material depends on the wavelength of the radiation, a non monochromatic light ray after passing through a lens splits into several rays having different refraction angles and wavelengths. Therefore, when white light is refracted by a single lens each color comes to a focus at a different distance from the objective. This is called chromatic aberration and plays a vital role in lens design, see [**KJ10**, Chapter 5]. Materials have various degrees of dispersion, and low dispersion ones are used in the manufacturing of photographic lenses, see [**Can**]. A way to correct chromatic aberration is to build lenses composed of various simple lenses made of different materials. Also chromatic aberration has recently being handled numerically using demosaicing algorithms, see [**ima**]. The way in which the refractive index depends of the wavelength is given by a formula for the dispersion of light due to A. Cauchy: the refractive index $n$ in terms of the wavelength $\lambda$ is given by

$$n(\lambda) = A_1 + \frac{A_2}{\lambda^2} + \frac{A_4}{\lambda^4} + \cdots ,$$

where $A_i$ are constants depending on the material [**Cau36**]. The validity of this formula is in the visible wavelength range; see [**BW59**, pp. 99-102] for its accurateness in various materials. A more accurate formula was derived by Sellmeier, see [**JW01**, Section 23.5].

A first result related to our question is that there is no single lens bounded by two spherical surfaces that refracts non monochromatic radiation from a point into a fixed direction; this was originally stated by K. Schwarzschild [**Sch05**]. The question of designing a single lens, non spherical, that focuses one point into a fixed direction for light containing only two colors, i.e., for two refractive indices $n \neq \bar{n}$, is considered in [**Sch83**] in the plane; but no mathematically rigorous proof is given. In fact, by tracing back and forth rays of both colors, the author describes how a finite number of points should be located on the faces of the desired lens and he claims, without proof, that the desired surfaces can be completed by interpolating adjacent points with third degree polynomials. Such an interpolation will give an undesired refracting behavior outside the fixed number of points considered. For the existence of rotationally symmetric lenses capable of focusing one point into two points for two different refractive indices see [**vBO92**], [**vB94**]. The results of all these authors require size conditions on $n, \bar{n}$. The monochromatic case is due to Friedman and McLeod [**FM87**] and Rogers [**Rog88**]. The solutions obtained are analytic functions. These results are all two dimensional and therefore concern only to rotationally symmetric lenses.

In view of all this, we now state the problems that are considered and solved in this paper.

Problem A: is there a single lens sandwiched by a surface $L$ given by the graph of a function $u$ in $\Omega$, the lower surface of the lens, and a surface $S$, the top part of the lens, such that each ray of dichromatic light (superposition of two colors) emanating in the vertical direction $e$ from $(x, 0)$ for $x \in \Omega$ is refracted by the lens into the direction $w$? We denote such a lens by the pair $(L, S)$. Notice that when a dichromatic ray enters the lens, due to chromatic dispersion, it splits into two rays having different directions and colors, say red and blue, that they both travel inside the lens until they exit it at points on the surface $S$ and then both continue traveling in the direction $w$; see Figure 1(a). In other words, from the result mentioned at

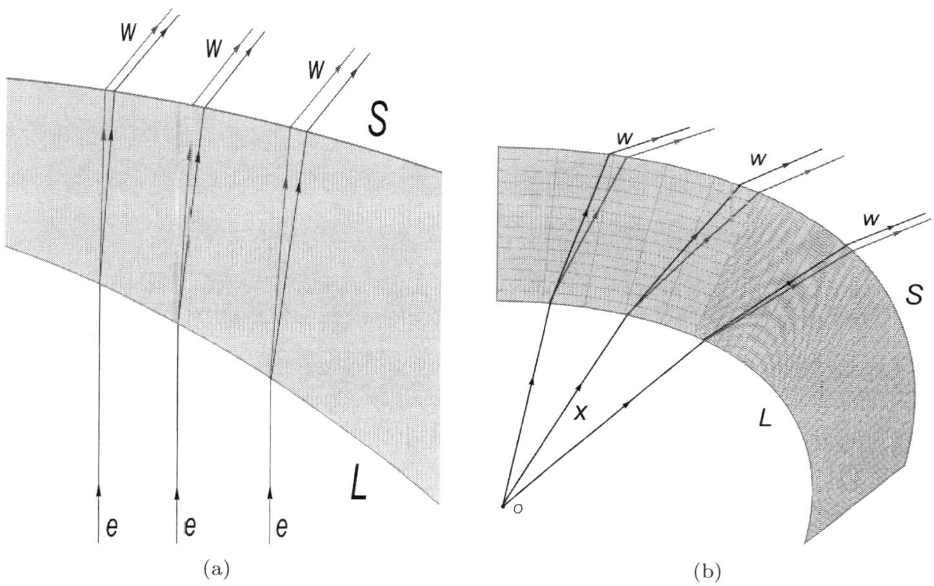


FIGURE 1. Problems A and B

the beginning of this introduction, there are lenses $(L, S_{C_b})$ refracting light with color $b$ into $w$, and the question is then to see if we can choose appropriately $u$ and the parameter $C_b$ so that the very same lens $(L, S_{C_b})$ refracts also incident light superposition now of two colors $b$ and $r$ both into the direction $w$; see (3.2).

Problem B: a similar question is when the rays emanate from a point source $O$ and we ask if a single lens $(L, S)$ exists such that all rays are refracted into a fixed given direction $w$. Now $L$ is given parametrically by $\rho(x)x$ for $x \in \Omega \subset S^2$; see Figure 1(b).

We will show in Section 3 using Brouwer fixed point theorem that Problem A has no solution if $w \neq e$. In case $w = e$, the unique solution to Problem A is the trivial solution $L$ and $S$ are contained in two horizontal planes. This is the contents of Theorem 3.1. On the other hand, since Problem A is solvable for monochromatic light and for each given lower surface $L$, we obtain two single lenses, one for each color. We then show in Section 3.1 that the difference between the upper surfaces of these two lenses can be estimated by the difference between the refractive indices for each color.

Concerning Problem B, we prove in Theorem 5.2, also using Brouwer fixed point theorem, that if $w \notin \Omega$ then Problem $B$ has no solution. The case when $w \in \Omega$ requires a more elaborate and long approach which we can implement only in dimension two. In fact, we show in Sections 5.2 and 5.3, that the solvability of Problem B is equivalent to solve a system of first order functional differential equations of one variable. For this we need an existence theorem for these type of equations that was introduced by Rogers in [**Rog88**]. We provide in Section 4 a simpler proof of this existence and uniqueness result of local solutions using the Banach fixed point theorem, Theorems 4.1 and 4.7. Section 4 is self contained and has independent interest. The existence of local solutions to Problem B in the plane is then proved in Section 5.5 by application of Theorem 4.1. For this it is necessary to assume conditions on the ratio between the thickness of the lens and its distance to the point source, Theorem 5.13. We also derive a necessary condition for the solvability of Problem B, see Corollary 5.10. To sum up our result: for $w = e \in \Omega$ and fixing two points $P$ and $Q$ on the positive $y$-axis, with $|Q| > |P|$ and letting $k = |P|/|Q - P|$, we show that if $k$ is small then there exists a unique lens $(L, S)$ local solution to Problem $B$ such that $L$ passes through the point $P$ and $S$ through $Q$; otherwise, for $k$ large no solution exists. For intermediate values of $k$ see Remark 5.14. By reversibility of optical paths, this result shows existence of single element lenses focusing dichromatic rays from infinity into a point.

The analogue of Problem B for more than two colors has no solution, i.e., if rays emitted from the origin are superposition of three or more colors, there is no simple lens refracting these rays into a unique direction $w$, see Remark 5.12.

We close this introduction mentioning that a number of results have been recently obtained for refraction of monochromatic light, these are included in the papers [**GH09**], [**GH14**], [**Gut14**], [**Kar16**], [**LGM17**], and [**GS18**].

## 2. Preliminaries

In this section we mention some consequences from the Snell law that will be used later. In an homogeneous and isotropic medium, the index of refraction depends on the wavelength of light. Suppose $\Gamma$ is a surface in $\mathbb{R}^3$ separating two media I and II that are homogeneous and isotropic. If a ray of monochromatic light having unit direction $x$ and traveling through the medium I strikes $\Gamma$ at the point $P$, then this ray is refracted in the unit direction $m$ through medium II according with the Snell law in vector form, [**Lun64**],

$$n_1(x \times \nu) = n_2(m \times \nu), \tag{2.1}$$

where $\nu$ is the unit normal to the surface to $\Gamma$ at $P$ going towards the medium II, and $n_1, n_2$ are the refractive indices for the corresponding monochromatic light.

This has several consequences:

(a) the vectors $x, m, \nu$ are all on the same plane, called plane of incidence;
(b) the well known Snell law in scalar form

$$n_1 \sin \theta_1 = n_2 \sin \theta_2,$$

where $\theta_1$ is the angle between $x$ and $\nu$ (the angle of incidence), $\theta_2$ the angle between $m$ and $\nu$ (the angle of refraction).

From (2.1), with $\kappa = n_2/n_1$,

$$x - (n_2/n_1)\, m = \lambda \nu, \tag{2.2}$$

with

$$\lambda = x \cdot \nu - \sqrt{\kappa^2 - 1 + (x \cdot \nu)^2} = \Phi_\kappa(x \cdot \nu). \tag{2.3}$$

Notice that $\lambda > 0$ when $\kappa < 1$, and $\lambda < 0$ if $\kappa > 1$. When $\kappa < 1$ total reflection occurs, unless $x \cdot \nu \geq \sqrt{1 - \kappa^2}$, see [**BW59**] and [**Gut14**, Sec. 2].

The following lemmas will be used in the remaining sections of the paper.

LEMMA 2.1. *Assume we have monochromatic light. Let $\Gamma_1$ and $\Gamma_2$ be two surfaces enclosing a lens with refractive index $n_2$, and the outside of the lens is a medium with refractive index $n_1$ with $n_1 \neq n_2$.*

*Suppose an incident ray with unit direction $x$ strikes $\Gamma_1$ at $P$, the ray propagates inside the lens and is refracted at $Q \in \Gamma_2$ into the unit direction $w$. Then $w = x$ if and only if the unit normals $\nu_1(P) = \nu_2(Q)$.*

PROOF. From the Snell law applied at $P$ and $Q$

$$x - (n_2/n_1)\, m = \lambda_1\, \nu_1(P), \qquad m - (n_1/n_2)\, w = \lambda_2\, \nu_2(Q),$$

then

$$x - w = \lambda_1\, \nu_1(P) + (n_2/n_1)\, \lambda_2\, \nu_2(Q). \tag{2.4}$$

If $x = w$, since $\lambda_1$ and $-(n_2/n_1)\, \lambda_2$ have the same sign and the normals are unit vectors, we conclude

$$\nu_1(P) = \nu_2(Q).$$

Conversely, if $\nu_1(P) = \nu_2(Q) := \nu$, then from (2.4) $x - w = (\lambda_1 + (n_1/n_2)\, \lambda_2)\, \nu$. Notice that $m \cdot \nu = (n_1/n_2)\,(x \cdot \nu - \lambda_1) = (n_1/n_2)\sqrt{(n_2/n_1)^2 - 1 + (x \cdot \nu)^2}$. Hence from (2.3)

$$\lambda_1 + (n_2/n_1)\, \lambda_2 = x\cdot\nu - (n_2/n_1)\, m\cdot\nu \;1 + (n_2/n_1)\left(m\cdot\nu - \sqrt{(n_1/n_2)^2 - 1 + (m\cdot\nu)^2}\right) = 0.$$

□

Let us now consider the case of dichromatic light, i.e., a mix of two colors b and r. That is, if a ray with direction $x$ in vacuum strikes a surface $\Gamma$ at $P$ separating two media, then the ray splits into two rays one with color b and direction $m_b$, and another with color r and direction $m_r$. Here $m_r$ satisfies (2.1) with $n_1 = 1$ and $n_2 = n_r$ (the refractive index for the color r) and $m_b$ satisfies (2.1) with $n_1 = 1$ and $n_2 = n_b$ (the refractive index for the color b). Notice $m_b, m_r$ are both in the plane of incidence containing $P$, the vector $x$, and $\nu(P)$ the normal to $\Gamma$ at $P$. Assuming $n_b > n_r$, i.e., rays with color r travel faster than rays with color b, then for a given incidence angle $\theta$ by the Snell law the angles of refraction satisfy $\theta_b \leq \theta_r$. In fact

$$\sin\theta = n_b \sin\theta_b = n_r \sin\theta_r,$$

obtaining the following Lemma.

LEMMA 2.2. *Suppose a dichromatic ray with unit direction $x$ strikes a surface $\Gamma$ at a point $P$ having normal $\nu$. Then $m_b = m_r$ if and only if $x = \nu$.*

## 3. The collimated case: Problem A

In this section we consider the following set up. We are given $\Omega \subseteq \mathbb{R}^2$ a compact and convex set with nonempty interior, and $w$ a unit vector in $\mathbb{R}^3$. Dichromatic rays with colors b and r are emitted from $(t,0)$, with $t \in \Omega$, into the vertical direction $e = (0,0,1)$. By application of the results from [**GS16**] with $n_1 = n_3 = 1$ and $n_2 = n_r$, we have the following. Given $u \in C^2$, there exist surfaces parametrized by $f_r(t) = (t,u(t)) + d_r(t)m_r(t)$, with $m_r(t) = \dfrac{1}{n_r}(e - \lambda_r \nu_u(t))$ where $\lambda_r = \Phi_{n_r}(e \cdot \nu_u(t))$ from (2.3), $\nu_u(t) = \dfrac{(-\nabla u(t), 1)}{\sqrt{1+|\nabla u(t)|^2}}$ the unit normal at $(t,u(t))$, and

$$d_r(t) = \frac{C_r - (e-w)\cdot(t,u(t))}{n_r - w\cdot m_r(t)} \tag{3.1}$$

from [**GS16**, Formula (3.14)], such that lens bounded between $u$ and $f_r$ refracts the rays with color r into $w$. Here the constant $C_r$ is chosen so that $d_r(t) > 0$ and $f_r$ has a normal vector at every point. This choice is possible from [**GS18**, Theorem 3.2 and Corollary 3.3]. In addition, given $u$, the corresponding surface parametrized by $f_r$ is unique when passing through a given point, which determines the constant $C_r$.

Likewise there exist surfaces parametrized by $f_b(t) = (t,u(t)) + d_b(t)m_b(t)$, with similar quantities as before with r replaced by b, such that lens bounded between $u$ and $f_b$ refracts the rays with color b into $w$.

We assume that $n_b > n_r > 1$, where $n_b, n_r$ are the refractive indices of the material of the lens corresponding to monochromatic light having colors b or r, and the medium surrounding the lens is vacuum.

To avoid total reflection compatibility conditions between $u$ and $w$ are needed, see [**GS16**, condition (3.4)] which in our case reads

$$\lambda_r\,\nu_u(t)\cdot w \le e\cdot w - 1, \text{ and } \lambda_b\,\nu_u(t)\cdot w \le e\cdot w - 1.$$

If $w = e$, these two conditions are automatically satisfied because $\lambda_r, \lambda_b$ are both negative and $\nu_u(t)\cdot e = \dfrac{1}{\sqrt{1+|\nabla u(t)|^2}} > 0$.

The problem we consider in this section is to determine if there exist $u$ and corresponding surfaces $f_r$ and $f_b$ for each color such that $f_r$ can be obtained by a reparametrization of $f_b$. That is, if there exist a positive function $u \in C^2(\Omega)$, real numbers $C_r$ and $C_b$, and a continuous map $\varphi : \Omega \to \Omega$ such that the surfaces $f_r$ and $f_b$, corresponding to $u, C_r, C_b$, have normals at each point and

$$f_r(t) = f_b(\varphi(t)) \qquad \forall t \in \Omega, \tag{3.2}$$

we refer to this as *Problem A*. Notice that if a solution exists then $f_r(\Omega) \subseteq f_b(\Omega)$. From an optical point of view, this means that the lens sandwiched between $u$ and $f_b$ refracts both colors into $w$. Notice that there could be points in $f_b(\Omega)$ that are not reached by red rays.

The answer to Problem A is given in the following theorem.

THEOREM 3.1. *If $w \neq e$, then Problem A has no solution, and if $w = e$ the only solutions to Problem A are lenses enclosed by two horizontal planes.*

To prove this theorem we need the following lemma.

LEMMA 3.2. *Given a surface described by $u \in C^2(\Omega)$ and the unit direction $w$, let $f_r$ and $f_b$ be the surfaces parametrized as above. If $f_r(t) = f_b(t)$ for some $t \in \Omega$, then $\nu_u(t) = e$, the unit normal vector to $u$ at $(t, u(t))$.*

PROOF. Since

$$f_b(t) = (t, u(t)) + d_b(t)\, m_b(t) = f_r(t) = (t, u(t)) + d_r(t)\, m_r(t)$$

we get $d_b(t)\, m_b(t) = d_r(t)\, m_r(t)$, and since $m_r, m_b$ are unit, $d_b(t) = d_r(t)$. Therefore $m_b(t) = m_r(t)$ which from Lemma 2.2 implies that $\nu_u(t) = e$. □

PROOF OF THEOREM 3.1. To show the first part of the theorem, suppose by contradiction that Problem A has a solution with $w \neq e$. Since $\Omega$ is compact and convex, by Brouwer fixed point theorem [**Dug78**, Sect. 2, Chap. XVI] there is $t_0 \in \Omega$ such that $\varphi(t_0) = t_0$, and so from (3.2) $f_r(t_0) = f_b(t_0)$. Hence from Lemma 3.2 $\nu_u(t_0) = e$. By Snell's law at $(t_0, u(t_0))$, $m_b(t_0) = m_r(t_0) = e$. From (3.2) the normals at $f_r(t_0) = f_b(t_0)$ are parallel. Since the normals used in the Snell law go towards the target medium, $n_r \neq n_b$, and both colors with direction $e$ are refracted at $f_b(t_0) = f_r(t_0)$ into the direction $w$, it follows that $w = e$, a contradiction.

To show the second part of the Theorem, assume there exist $u$ and $\varphi : \Omega \to \Omega$ such that problem A has a solution. Let $t \in \Omega$, and $Q = f_r(t) = f_b(\varphi(t))$. Since the ray emitted from $(t, 0)$ with direction $e$ and color $r$ is refracted by $(u, f_r)$ into $e$ at $Q$, then by Lemma 2.1 $\nu_u(t) = \nu(Q)$, where $\nu(Q)$ denotes the normal to the upper face of the lens at $Q$. Similarly, applying Lemma 2.1 to the color b we have $\nu_u(\varphi(t)) = \nu(Q)$. We conclude that for every $t \in \Omega$

$$\nu_u(t) = \nu_u(\varphi(t)). \tag{3.3}$$

We will show that $u$ is constant, i.e., $\nabla u(t) = 0$, for all $t \in \Omega$ connected. Suppose by contradiction that there exists $t_0 \in \Omega$, with $\nabla u(t_0) \neq 0$. If $t_1 = \varphi(t_0)$, then $t_1 \neq t_0$. Otherwise, from Lemma 3.2 $\nu_u(t_0) = \dfrac{(-\nabla u(t_0), 1)}{\sqrt{1 + |\nabla u(t_0)|^2}} = e$, so $\nabla u(t_0) = 0$. Also from (3.3), $\nu_u(t_0) = \nu_u(t_1)$. Let $L_r(t_0)$ be the red ray from $(t_0, u(t_0))$ to $f_r(t_0)$, and let $L_b(t_1)$ be the blue ray from $(t_1, u(t_1))$ to $f_b(t_1)$. We have that $L_r(t_0)$ and $L_b(t_1)$ intersect at $Q_0 := f_r(t_0) = f_b(t_1)$. If $\Pi_r$ denotes the plane of incidence passing through $(t_0, u(t_0))$ containing the directions $e$ and $\nu_u(t_0)$, and $\Pi_b$ denotes the plane of incidence through $(t_1, u(t_1))$ containing the directions $e$ and $\nu_u(t_1)$, then $\Pi_r$ and $\Pi_b$ are parallel since $\nu_u(t_0) = \nu_u(t_1)$. Also by Snell's law $L_r(t_0) \subset \Pi_r$ and $L_b(t_1) \subset \Pi_b$, so $Q_0 \in \Pi_r \cap \Pi_b$. We then obtain $\Pi_r = \Pi_b := \Pi$.

Let $\ell$ denote the segment $\Omega \cap \Pi$. We deduce from the above that $t_0, t_1 \in \ell$. Next, let $t_2 = \varphi(t_1)$. As before, since $\nabla u(t_1) = \nabla u(t_0) \neq 0$, by Lemma 3.2 $t_2 \neq t_1$; and by (3.3) $\nu_u(t_1) = \nu_u(t_2)$. Let $\Pi_2$ be the plane through $(t_2, u(t_2))$ and containing the vectors $e$ and $\nu_u(t_2)$. We have $L_b(t_2) \subset \Pi_2$, $f_r(t_1) = f_b(t_2)$, and $f_r(t_1) \in \Pi$. Therefore $\Pi_2 = \Pi$, in particular, $t_2 \in \ell$.

Let $\ell_1$ denote the half line starting from $t_1$ and containing $t_0$. We claim that $t_2 \notin \ell_1$. In fact, we first have that $L_r(t_0)$ and $L_b(t_1)$ intersect at $Q_0$. Since $\nu_u(t_0) = \nu_u(t_1) := \nu$, $L_r(t_0)$ is parallel to $L_r(t_1)$, and $L_b(t_0)$ is parallel to $L_b(t_1)$. And since $n_b > n_r$, it follows from the Snell law that the angle of refraction $\theta_b$ for the blue ray $L_b(t_0)$ and the angle of refraction $\theta_r$ of the red ray $L_r(t_1)$ satisfy $\theta_b < \theta_r$. Hence $L_b(t_0)$ and $L_r(t_1)$ diverge. Moreover, notice that all rays are on the plane $\Pi$, and $L_b(t_2)$ is parallel to $L_b(t_1)$. Then, if $t_2 \in \ell_1$, we have $L_b(t_2)$ and

$L_r(t_1)$ diverge and cannot intersect, a contradiction since $f_b(t_2) = f_r(t_1)$ and the claim is proved; see Figure 2 illustrating that $t_2$ cannot be on $\ell_1$.

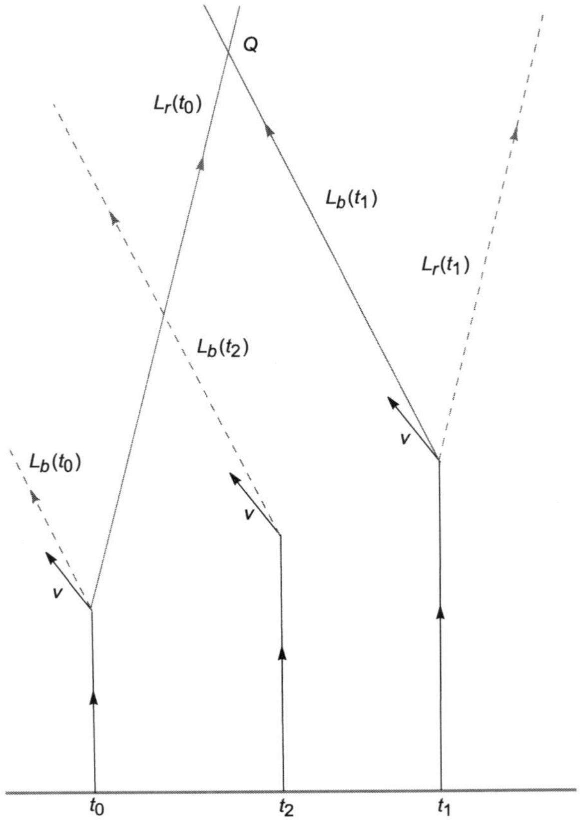


FIGURE 2. $t_2 \notin \ell_1$

Continuing in this way we construct the sequence $t_k = \varphi(t_{k-1})$. By (3.3) $\nu_u(t_k) = \nu_u(t_{k-1}) = \cdots = \nu_u(t_0) \neq 0$, and again by Lemma 3.2 $t_k \neq t_{k-1}$. By Snell's law $\{L_b(t_k)\}$ are all parallel, $\{L_r(t_k)\}$ are all parallel and arguing as before they are all contained in $\Pi$, and then $t_k \in \ell$. In addition, the angles between $L_r(t_k)$ and $L_b(t_k)$ are the same for all $k$. Also, for $k \geq 1$, if $\ell_k$ is the half line with origin $t_k$ and passing through $t_{k-1}$, then as above $t_{k+1} \notin \ell_k$. Hence the sequence $\{t_k\}$ is decreasing or increasing on the line $\ell$. Therefore $t_k$ converges to some $\hat{t} \in \ell$ so by continuity $\varphi(\hat{t}) = \hat{t}$. Hence by Lemma 3.2 $\nabla u(\hat{t}) = 0$, but since $\nabla u(t_k) = \nabla u(t_0) \neq 0$ for all $k$, and $u$ is $C^2$ we obtain a contradiction. Thus $u$ is constant in $\Omega$. Since the lower face is then contained in a horizontal plane, $m_r(t) = m_b(t) = e = (0,0,1)$. Hence from the form of the parameterizations of $f_b$ and $f_r$, and since from (3.1) $d_r$ and $d_b$ are constants, the upper face of the lens is also contained in a horizontal plane. $\square$

**3.1. Estimates of the upper surfaces for two colors.** The purpose of this section is to measure how far the surfaces $f_r$ and $f_b$ can be when $w = e$. We shall prove the following.

PROPOSITION 3.3. *Suppose $f_r(t_0) = f_b(t_0)$ at some point $t_0$, and $w = e$. Then*

$$|f_r(t) - f_b(t)| \leq \bar{C}\,|n_b - n_r|$$

*for all $t$ with a constant $\bar{C}$ depending only $t_0$ and $n_r, n_b$.*

PROOF. We begin showing an upper estimate of the difference between $m_r(t)$ and $m_b(t)$. To simplify the notation write $\nu = \nu_u$.

We have

$$m_b(t) = \frac{1}{n_b}\left(e - \lambda_b\,\nu(t)\right); \qquad \lambda_b = e\cdot\nu - \sqrt{n_b^2 - 1 + (e\cdot\nu)^2};$$

$$m_r(t) = \frac{1}{n_r}\left(e - \lambda_r\,\nu(t)\right); \qquad \lambda_r = e\cdot\nu - \sqrt{n_r^2 - 1 + (e\cdot\nu)^2}.$$

So

$$m_b(t) - m_r(t) = \left(\frac{1}{n_b} - \frac{1}{n_r}\right) e + \left(\frac{\lambda_r}{n_r} - \frac{\lambda_b}{n_b}\right)\nu(t) := A + B.$$

Notice $|A| = \dfrac{|n_b - n_r|}{n_b\,n_r}$. Next write

$$\begin{aligned}\left(\frac{\lambda_r}{n_r} - \frac{\lambda_b}{n_b}\right) &= \frac{1}{n_b\,n_r}\left(n_b\,\lambda_r - n_r\,\lambda_b\right)\\ &= \frac{1}{n_b\,n_r}\Big\{(n_b - n_r)\,(e\cdot\nu(t)) + n_r\sqrt{n_b^2 - 1 + (e\cdot\nu)^2}\\ &\qquad\qquad - n_b\sqrt{n_r^2 - 1 + (e\cdot\nu)^2}\Big\}.\end{aligned}$$

Now

$$\begin{aligned}&n_r\sqrt{n_b^2 - 1 + (e\cdot\nu)^2} - n_b\sqrt{n_r^2 - 1 + (e\cdot\nu)^2}\\ &\qquad = \frac{(n_b^2 - n_r^2)\left(1 - (e\cdot\nu)^2\right)}{n_r\sqrt{n_b^2 - 1 + (e\cdot\nu)^2} + n_b\sqrt{n_r^2 - 1 + (e\cdot\nu)^2}}.\end{aligned}$$

Hence

$$|B| \leq \frac{|n_b - n_r|}{n_b\,n_r} + \frac{1}{n_b\,n_r}\left(\frac{|n_b^2 - n_r^2|}{n_r\sqrt{n_b^2 - 1} + n_b\sqrt{n_r^2 - 1}}\right)$$

and therefore

$$|m_b(t) - m_r(t)| \leq \frac{|n_b - n_r|}{n_b\,n_r}\left(2 + \frac{n_b + n_r}{n_r\sqrt{n_b^2 - 1} + n_b\sqrt{n_r^2 - 1}}\right). \tag{3.4}$$

We next estimate $d_r(t) - d_b(t)$, where

$$d_r(t) = \frac{C_r}{n_r - m_r(t)\cdot e}, \qquad d_b(t) = \frac{C_b}{n_b - m_b(t)\cdot e},$$

by (3.1). From Lemma 3.2, since $f_r(t_0) = f_b(t_0)$, $\nu(t_0) = e = (0,0,1)$. Then by the Snell law $m_r(t_0) = m_b(t_0) = e$, and from the parametrization of $f_r$ and $f_b$, $d_r(t_0) = d_b(t_0) := d_0$. Hence

$$\frac{C_b}{n_b - 1} = \frac{C_r}{n_r - 1} = d_0.$$

We then obtain

$$\begin{aligned}
&d_r(t) - d_b(t)\\
&= \frac{d_0}{(n_r - m_r(t)\cdot e)(n_b - m_b(t)\cdot e)}\left((n_b - m_b(t)\cdot e)(n_r - 1) - (n_r - m_r(t)\cdot e)(n_b - 1)\right)\\
&= \frac{d_0}{(n_r - m_r(t)\cdot e)(n_b - m_b(t)\cdot e)}\Delta.
\end{aligned}$$

Now

$$\begin{aligned}
\Delta &= n_r - n_b + (m_b(t) - m_r(t))\cdot e + n_b\, m_r(t)\cdot e - n_r\, m_b(t)\cdot e\\
&= n_r - n_b + (m_b(t) - m_r(t))\cdot e + (n_b - n_r)\, m_r(t)\cdot e + n_r\, (m_r(t) - m_b(t))\cdot e,
\end{aligned}$$

so from (3.4)

$$|\Delta| \leq C(n_r, n_b)\, |n_r - n_b|.$$

Since $(n_r - m_r(t)\cdot e)(n_b - m_b(t)\cdot e) \geq (n_r - 1)(n_b - 1)$, we obtain

$$|d_r(t) - d_b(t)| \leq C'\, |n_r - n_b|, \tag{3.5}$$

with $C'$ depending on $d_0, n_r, n_b$.

Finally write

$$f_r(t) - f_b(t) = d_r(t) m_r(t) - d_b(t) m_b(t) = d_r(t)\,(m_r(t) - m_b(t)) - (d_b(t) - d_r(t))\, m_b(t).$$

Since $d_r(t) \leq C_r/(n_r - 1) = d_0$, then the desired estimate follows from (3.4) and (3.5). $\square$

We conclude this section analyzing the intersection of the upper surfaces of the single lenses $(u, f_r)$ and $(u, f_b)$.

Proposition 3.4. *Let $w = e$. If $\nu_u(t) \neq e$ for all $t \in \Omega$, and $\nabla u$ is injective in $\Omega$, then $S_r \cap S_b = \emptyset$, where $S_r = f_r(\Omega)$ and $S_b = f_b(\Omega)$. This means that, the upper surfaces of the single lenses $(u, f_r)$ and $(u, f_b)$ are disjoint.*

Proof. Suppose $P \in S_r \cap S_b$, i.e. there exists $t_0, t_1 \in \Omega$ such that $f_r(t_0) = f_b(t_1)$, then as in the proof of (3.3) we get $\nu_u(t_0) = \nu_u(t_1)$, and therefore $\nabla u(t_0) = \nabla u(t_1)$. Since $\nabla u$ is injective, then $t_0 = t_1$ and so $f_r(t_0) = f_b(t_0)$. Therefore from the proof of Lemma 3.2 we conclude that $\nu_u(t_0) = e$. $\square$

Remark 3.5. When $\nabla u$ is not injective the upper surfaces of the single lenses $(u, f_r)$ and $(u, f_b)$ may or may not be disjoint. We illustrate this with lenses bounded by parallel planes. In fact, by Lemma 2.1 such a lens refracts all rays blue and red into the vertical direction $e$. Notice that if the planes are sufficiently far apart, depending on the refractive indices, then $f_r(\Omega) \cap f_b(\Omega) = \emptyset$. This is illustrated in Figure 3: if the lens in between the planes $A$ and $B$, then $f_r(\Omega) \cap f_b(\Omega) \neq \emptyset$; and if the lens is between the planes $A$ and $C$, then $f_r(\Omega) \cap f_b(\Omega) = \emptyset$.

## 4. First order functional differential equations

In this section we give a new and simpler proof of an existence theorem for functional differential equations originally due to J. Rogers [**Rog88**, Sec. 2]. It will be used in Section 5 to show the existence of a dichromatic lens when rays emanate from one point source. Indeed, the existence of such dichromatic lens in two dimensions is equivalent to solve a system of functional differential equations of one variable.

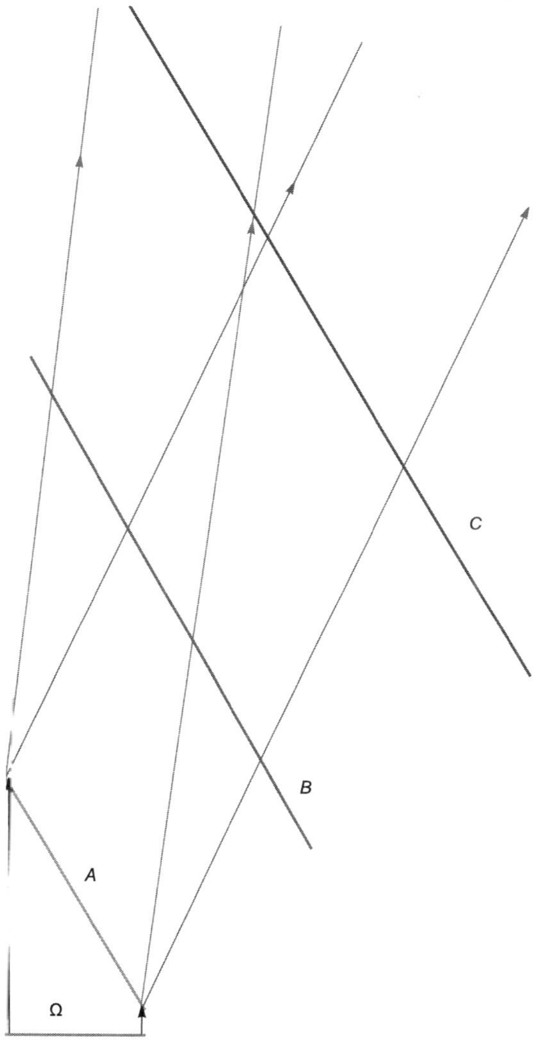


FIGURE 3. The lens is between $A$ and $B$ or between $A$ and $C$.

Developing an extension of Picard's iteration method for functional equations, Van-Brunt and Ockendon gave another proof of Roger's theorem, [**vBO92**]. We present here a topological proof, that we believe has independent interest, using Banach's fixed point theorem.

Let $H$ be a continuous map defined in an open domain in $\mathbb{R}^{4n+1}$ with values in $\mathbb{R}^n$ given by

$$H(X) = (h_1(X), h_2(X), \cdots, h_n(X)), \tag{4.1}$$

with $X := \left(t; \zeta^0, \zeta^1; \xi^0, \xi^1\right), t \in \mathbb{R}; \zeta^0, \zeta^1, \xi^0, \xi^1 \in \mathbb{R}^n$.

We are interested in solving the following system of functional differential equations

$$Z'(t) = H\left(t; Z(t), Z(z_1(t)); Z'(t), Z'(z_1(t))\right) \tag{4.2}$$
$$Z(0) = 0,$$

with $Z(t) = (z_1(t), \cdots, z_n(t))$.

THEOREM 4.1. *Let $\|\cdot\|$ be a norm in $\mathbb{R}^n$. Assume that the system*

$$P = H\left(0; \mathbf{0}, \mathbf{0}; P, P\right), \tag{4.3}$$

*has a solution $P = (p_1, p_2, \cdots, p_n)$ such that*

$$|p_1| \leq 1. \tag{4.4}$$

*Let $\mathcal{P} = (0; \mathbf{0}, \mathbf{0}; P, P) \in R^{4n+1}$, and let*

$$N_\varepsilon(\mathcal{P}) = \left\{\left(t; \zeta^0, \zeta^1; \xi^0, \xi^1\right) : |t| + \|\zeta^0\| + \|\zeta^1\| + \|\xi^0 - P\| + \|\xi^1 - P\| \leq \varepsilon\right\}$$

*be a neighborhood of $\mathcal{P}$ such that*

(i) *$H$ is uniformly Lipschitz in the variable $t$, i.e., there exists $\Lambda > 0$ such that*

$$\left\|H\left(\bar{t}; \zeta^0, \zeta^1; \xi^0, \xi^1\right) - H\left(t; \zeta^0, \zeta^1; \xi^0, \xi^1\right)\right\| \leq \Lambda\left|\bar{t} - t\right|. \tag{4.5}$$

*for all $\left(\bar{t}; \zeta^0, \zeta^1; \xi^0, \xi^1\right), \left(t; \zeta^0, \zeta^1; \xi^0, \xi^1\right) \in N_\varepsilon(\mathcal{P})$;*

(ii) *$H$ is uniformly Lipschitz in the variables $\zeta^0$ and $\zeta^1$, i.e., there exist positive constants $L_0$ and $L_1$ such that*

$$\left\|H\left(t; \bar{\zeta}^0, \bar{\zeta}^1; \xi^0, \xi^1\right) - H\left(t; \zeta^0, \zeta^1; \xi^0, \xi^1\right)\right\| \leq L_0\left\|\bar{\zeta}^0 - \zeta^0\right\| + L_1\left\|\bar{\zeta}^1 - \zeta^1\right\|, \tag{4.6}$$

*for all $\left(t; \bar{\zeta}^0, \bar{\zeta}^1; \xi^0, \xi^1\right), \left(t; \zeta^0, \zeta^1; \xi^0, \xi^1\right) \in N_\varepsilon(\mathcal{P})$;*

(iii) *$H$ is a uniform contraction in the variables $\xi^0$ and $\xi^1$, i.e., there exists constants $C_0$ and $C_1$ such that*

$$\left\|H\left(t; \zeta^0, \zeta^1; \bar{\xi}^0, \bar{\xi}^1\right) - H\left(t; \zeta^0, \zeta^1; \xi^0, \xi^1\right)\right\| \leq C_0\left\|\bar{\xi}^0 - \xi^0\right\| + C_1\left\|\bar{\xi}^1 - \xi^1\right\|, \tag{4.7}$$

*for all $\left(t; \zeta^0, \zeta^1; \bar{\xi}^0, \bar{\xi}^1\right), \left(t; \zeta^0, \zeta^1; \xi^0, \xi^1\right) \in N_\varepsilon(\mathcal{P})$, with*

$$C_0 + C_1 < 1; \tag{4.8}$$

(iv) *For all $X \in N_\varepsilon(\mathcal{P})$*

$$|h_1(X)| \leq 1. \tag{4.9}$$

*Under these assumptions, there exists $\delta > 0$ and $Z \in C^1[-\delta, \delta]$ with*

$$\left(t; Z(t), Z(z_1(t)); Z'(t), Z'(z_1(t))\right) \in N_\varepsilon(\mathcal{P})$$

*and $Z$ solving the system*

$$\begin{cases} Z'(t) = H\left(t; Z(t), Z(z_1(t)); Z'(t), Z'(z_1(t))\right) \\ Z(0) = 0, \end{cases} \tag{4.10}$$

*for $|t| \leq \delta$ and satisfying in addition that $Z'(0) = P$.*

PROOF. Since $H$ is continuous, let

$$\alpha = \max\{\|H(X)\| : X \in N_\varepsilon(\mathcal{P})\}. \tag{4.11}$$

From (4.3)

$$\|P\| = \|H(\mathcal{P})\| \leq \alpha. \tag{4.12}$$

Let $\mu$ be a constant such that

$$\mu \geq \frac{\Lambda + (L_0 + L_1)\,\alpha}{1 - C_0 - C_1}. \tag{4.13}$$

For any map $Z : \mathbb{R} \to \mathbb{R}^n$, we define the vector

$$V_Z(t) = \left(t; Z(t), Z\left(z_1(t)\right); (Z)'\,(t), (Z)'\left(z_1(t)\right)\right).$$

DEFINITION 4.2. Let $C^1[-\delta,\delta]$ denote the class of all functions $Z : [-\delta,\delta] \to \mathbb{R}^n$ that are $C^1$ equipped with the norm $\|Z\|_{C^1[-\delta,\delta]} = \max_{[-\delta,\delta]}\|Z(t)\| + \max_{[-\delta,\delta]}\|Z'(t)\|$. We define the set $\mathcal{C} = \mathcal{C}(\delta)$ as follows: $Z \in \mathcal{C}$ if and only if

(1) $Z \in C^1[-\delta,\delta]$,
(2) $Z(0) = 0$, $Z'(0) = P$,
(3) $|z_1(t)| \leq |t|$,
(4) $\|Z(t) - Z(\bar{t})\| \leq \alpha\,|t - \bar{t}|$,
(5) $|z_1(t) - z_1(\bar{t})| \leq |t - \bar{t}|$,
(6) $\|Z'(t) - Z'(\bar{t})\| \leq \mu\,|t - \bar{t}|$,
(7) $V_Z(t) \in N_\varepsilon(\mathcal{P})$, for all $|t| \leq \delta$.

Define a map $T$ on $\mathcal{C}$ as follows:

$$T\,Z(t) = \int_0^t H(s; Z(s), Z(z_1(s)); Z'(s), Z'(z_1(s)))\,ds.$$

Our goal is to show that $T : \mathcal{C} \to \mathcal{C}$, for $\delta$ sufficiently small, and therefore from Banach's fixed point theorem, $T$ has a unique fixed point $Z \in \mathcal{C}$ and so $Z$ solves (4.2).

We will prove the theorem in a series of steps.

STEP 1. There exists $\delta_0 > 0$ such that $\mathcal{C}(\delta)$ is non empty for $\delta \leq \delta_0$; in fact, the function $Z^0(t) = tP \in \mathcal{C}$.

PROOF. Obviously, $Z^0(0) = 0$, $\left(Z^0\right)' = P$, and from (4.4) $\left|z_1^0(t)\right| = |p_1|\,|t| \leq |t|$. Also from (4.3) and (4.4)

$$\|Z^0(t) - Z^0(\bar{t})\| \leq \|P\|\,|t - \bar{t}| = \|H(\mathcal{P})\|\,|t - \bar{t}| \leq \alpha\,|t - \bar{t}|,$$

$$|z_1^0(t) - z_1^0(\bar{t})| = |p_1|\,|t - \bar{t}| \leq |t - \bar{t}|,$$

and $\|\left(Z^0\right)'(t) - \left(Z^0\right)'(\bar{t})\| = 0 \leq \mu\,|t - \bar{t}|$.

It remains to show that $V_{Z^0}(t) \in N_\varepsilon(\mathcal{P})$. By definition $V_{Z^0}(t) = (t; t\,P, t\,p_1\,P; P, P)$, and so $V_{Z^0}(t) \in N_\varepsilon\,(\mathcal{P})$ if and only if $|t| + |t|\,\|P\| + |t|\,|p_1|\,\|P\| \leq \varepsilon$, which is equivalent to

$$|t| \leq \frac{\varepsilon}{1 + \|P\| + |p_1|\,\|P\|} := \delta_0. \tag{4.14}$$

□

STEP 2. $\mathcal{C}(\delta)$ is complete for every $\delta > 0$.

PROOF. Let $Z^k$ be a Cauchy sequence in $\mathcal{C}$. Since $C^1[-\delta,\delta]$ is complete, $Z^k$ converges uniformly to a function $Z \in C^1[-\delta,\delta]$, and $\left(Z^k\right)'$ converges uniformly to $Z'$. Since $Z^k$ satisfy properties (1)-(7) in Definition 4.2, then Step 2 follows by uniform convergence. $\square$

STEP 3. If $Z \in \mathcal{C}$ and $W = TZ = (w_1, \cdots, w_n)$, then

$$|w_1(t)| \leq |t|.$$

PROOF. From (4.9) and since $V_Z(t) \in N_\varepsilon(\mathcal{P})$ for $|t| \leq \delta$, it follows that

$$|w_1(t)| = \left| \int_0^t h_1(V_Z(s))\, ds \right| \leq |t|.$$

$\square$

STEP 4. If $Z \in \mathcal{C}$ and $W = TZ$, then

$$\|W(t) - W(\bar{t})\| \leq \alpha |t - \bar{t}|,$$

and

$$|w_1(t) - w_1(\bar{t})| \leq |t - \bar{t}|$$

for every $t, \bar{t} \in [-\delta, \delta]$.

PROOF. Since $V_Z(t) \in N_\varepsilon(\mathcal{P})$, by (4.11) for all $|t| \leq \delta$

$$\|W(t) - W(\bar{t})\| = \left\| \int_{\bar{t}}^t H(V_Z(s))\, ds \right\| \leq \alpha\, |t - \bar{t}|,$$

and from (4.9), we get similarly the desired estimate for $w_1(t) - w_1(\bar{t})$. $\square$

STEP 5. If $Z \in \mathcal{C}(\delta)$ and $W = TZ$, then

$$\|W'(t) - W'(\bar{t})\| \leq \mu |t - \bar{t}|,$$

for every $t, \bar{t} \in [-\delta, \delta]$.

PROOF. From the Lipschitz estimates for $H$

$$\begin{aligned}
\|W'(t) - W'(\bar{t})\| &= \|H(V_Z(t)) - H(V_Z(\bar{t}))\| \\
&\leq \Lambda\, |t - \bar{t}| + L_0\, \|Z(t) - Z(\bar{t})\| + L_1\, \|Z(z_1(t)) - Z(z_1(\bar{t}))\| \\
&\qquad + C_0\, \|Z'(t) - Z'(\bar{t})\| + C_1\, \|Z'(z_1(t)) - Z'(z_1(\bar{t}))\|.
\end{aligned}$$

Since $|z_1(t)| \leq |t| \leq \delta$ and $|z_1(\bar{t})| \leq |\bar{t}| \leq \delta$, we get from the Lipschitz properties of $Z$, $z_1$, and $Z'$

$$\begin{aligned}
\|W'(t) - W'(\bar{t})\| &\leq \Lambda\, |t - \bar{t}| + L_0\, \alpha |t - \bar{t}| + L_1\, \alpha |z_1(t) - z_1(\bar{t})| + C_0\, \mu |t - \bar{t}| \\
&\qquad + C_1\, \mu |z_1(t) - z_1(\bar{t})| \\
&\leq \left(\Lambda + (L_0 + L_1)\alpha + (C_0 + C_1)\mu\right)\, |t - \bar{t}|.
\end{aligned}$$

From (4.13), $\Lambda + (L_0 + L_1)\alpha \leq (1 - C_0 - C_1)\mu$, then Step 5 follows. $\square$

STEP 6. For $\delta$ sufficiently small, $W = TZ \in \mathcal{C}$ for each $Z \in \mathcal{C}$.

PROOF. From the previous steps, it remains to show that $V_W(t) \in N_\varepsilon(\mathcal{P})$. Define

$$\begin{aligned} S_Z(t) &= |t| + \|Z(t)\| + \|Z(z_1(t))\| + \|Z'(t) - P\| + \|Z'(z_1(t)) - P\| \\ S_W(t) &= |t| + \|W(t)\| + \|W(w_1(t))\| + \|W'(t) - P\| + \|W'(w_1(t)) - P\| . \end{aligned}$$

Since $V_Z(t) \in N_\varepsilon(\mathcal{P})$, we have $S_Z(t) \le \varepsilon$. We shall prove that $S_W(t) \le \varepsilon$ by choosing $\delta$ sufficiently small. In fact, from (4.11) for every $|t| \le \delta$

$$\|W(t)\| = \left\| \int_0^t H(V_Z(s))\, ds \right\| \le \alpha\, |t| \le \alpha\, \delta, \tag{4.15}$$

and from (4.9)

$$|w_1(t)| = \left| \int_0^t h_1(V_Z(s))\, ds \right| \le |t| \le \delta.$$

Hence $\|W(w_1(t))\| \le \alpha\,\delta$. We next estimate $\|W'(t) - P\|$. Using the Lipschitz properties of $H$ we write

$$\begin{aligned} \|W'(t) - P\| &= \|H(V_Z(t)) - H(\mathcal{P})\| \\ &\le \Lambda|t| + L_0\, \|Z(t)\| + L_1\, \|Z(z_1(t))\| + C_0\, \|Z'(t) - P\| \\ &\qquad\qquad\qquad + C_1\, \|Z'(z_1(t)) - P\| . \end{aligned}$$

Notice that $\|Z(t)\| = \|Z(t) - Z(0)\| \le \alpha|t| \le \alpha\delta$, and since $|z_1(t)| \le |t| \le \delta$ then $\|Z(z_1(t))\| \le \alpha\delta$. Also

$$\|Z'(t) - P\| = \|Z'(t) - Z'(0)\| \le \mu|t| \le \mu\delta.$$

and

$$\|Z'(z_1(t)) - P\| \le \mu\delta.$$

Therefore, for $|t| \le \delta$

$$\|W'(t) - P\| \le [\Lambda + \alpha(L_0 + L_1)\delta + \mu(C_0 + C_1)]\,\delta,$$

and since $|w_1(t)| \le \delta$, we also get

$$\|W'(w_1(t)) - P\| \le [\Lambda + \alpha(L_0 + L_1)\delta + \mu(C_0 + C_1)]\,\delta.$$

We conclude that

$$S_W(t) \le \delta\, \left(1 + 2\Lambda + 2\alpha(1 + L_0 + L_1) + 2\mu(C_0 + C_1)\right),$$

so choosing $\delta \le \dfrac{\varepsilon}{1 + 2\Lambda + 2\alpha(1 + L_0 + L_1) + 2\mu(C_0 + C_1)}$ Step 6 follows. $\square$

It remains to show that $T$ is a contraction.

STEP 7. If $Z^1, Z^2 \in \mathcal{C}(\delta)$ with $\delta$ small enough from the previous steps, then

$$\left\|TZ^1 - TZ^2\right\|_{C^1[-\delta,\delta]} \le q\, \left\|Z^1 - Z^2\right\|_{C^1[-\delta,\delta]},$$

for some $q < 1$.

Proof. Let $W^1(t) = TZ^1(t)$, $W^2(t) = TZ^2(t)$. By the fundamental theorem of calculus we have for every $|t| \leq \delta$

$$\begin{aligned} \|W^1(t) - W^2(t)\| &\leq \left| \int_0^t \left\| \left(W^1\right)'(s) - \left(W^2\right)'(s) \right\| ds \right| \\ &\leq \delta \sup_{|t|\leq\delta} \left\| \left(W^1\right)'(t) - \left(W^2\right)'(t) \right\| \\ &\leq \delta \|W^1 - W^2\|_{C^1[-\delta,\delta]}, \end{aligned}$$

and similarly $\|Z^1(t) - Z^2(t)\| \leq \delta \|Z^1 - Z^2\|_{C^1[-\delta,\delta]}$. From the Lipschitz properties of $H$, for every $|t| \leq \delta$,

$$\begin{aligned} \left\| \left(W^1\right)'(t) - \left(W^2\right)'(t) \right\| &\leq \|H\left(V_{Z^1}(t)\right) - H\left(V_{Z^2}(t)\right)\| \\ &\leq L_0 \|Z^1(t) - Z^2(t)\| + L_1 \|Z^1\left(z_1^1(t)\right) - Z^2\left(z_1^2(t)\right)\| \\ &\quad + C_0 \left\| \left(Z^1\right)'(t) - \left(Z^2\right)'(t) \right\| \\ &\quad + C_1 \left\| \left(Z^1\right)'\left(z_1^1(t)\right) - \left(Z^2\right)'\left(z_1^2(t)\right) \right\|. \end{aligned}$$

We have

$$\begin{aligned} \|Z^1(t) - Z^2(t)\| &\leq \delta \|Z^1 - Z^2\|_{C^1[-\delta,\delta]} \\ \|Z^1(z_1^1(t)) - Z^2(z_1^2(t))\| &\leq \|Z^1(z_1^1(t)) - Z^2(z_1^1(t))\| \\ &\quad + \|Z^2(z_1^1(t)) - Z^2(z_1^2(t))\| \\ &\leq \delta \|Z^1 - Z^2\|_{C^1[-\delta,\delta]} + \alpha |z_1^1(t) - z_1^2(t)| \\ &\leq \delta \|Z^1 - Z^2\|_{C^1[-\delta,\delta]} + \alpha C_{\|\cdot\|} \|Z^1(t) - Z^2(t)\| \\ &\leq \delta \left(\alpha C_{\|\cdot\|} + 1\right) \|Z^1 - Z^2\|_{C^1[\delta,\delta]} \\ \left\| \left(Z^1\right)'(t) - \left(Z^2\right)'(t) \right\| &\leq \|Z^1 - Z^2\|_{C^1[-\delta,\delta]} \\ \left\| \left(Z^1\right)'\left(z_1^1(t)\right) - \left(Z^2\right)'\left(z_1^2(t)\right) \right\| &\leq \left\| \left(Z^1\right)'\left(z_1^1(t)\right) - \left(Z^2\right)'\left(z_1^1(t)\right) \right\| \\ &\quad + \left\| \left(Z^2\right)'\left(z_1^1(t)\right) - \left(Z^2\right)'\left(z_1^2(t)\right) \right\| \\ &\leq \left\|Z^1 - Z^2\right\|_{C^1[-\delta,\delta]} + \mu |z_1^1(t) - z_1^2(t)| \\ &\leq \left\|Z^1 - Z^2\right\|_{C^1[-\delta,\delta]} + \mu C_{\|\cdot\|} \left\|Z^1(t) - Z^2(t)\right\| \\ &\leq \left(\mu C_{\|\cdot\|} \delta + 1\right) \left\|Z^1 - Z^2\right\|_{C^1[-\delta,\delta]}; \end{aligned}$$

here $C_{\|\cdot\|}$ is a constant larger than 1, depending only on the choice of the norm in $\mathbb{R}^n$, since all norms in $\mathbb{R}^n$ are equivalent such constant exists. Combining the above inequalities, we obtain

$$\begin{aligned} &\left\| \left(W^1\right)'(t) - \left(W^2\right)'(t) \right\| \\ &\leq \left(L_0 \delta + L_1 \delta \left(\alpha C_{\|\cdot\|} + 1\right) + C_0 + C_1 \left(\mu C_{\|\cdot\|} \delta + 1\right)\right) \left\|Z^1 - Z^1\right\|_{C^1} \\ &\qquad := \left(M \delta + C_0 + C_1\right) \left\|Z^1 - Z^2\right\|_{C^1}, \end{aligned}$$

and from the fundamental theorem of calculus

$$\|W^1(t) - W^2(t)\| \leq \delta \sup_{|t|\leq\delta} \left\| \left(W^1\right)'(t) - \left(W^2\right)'(t) \right\| \leq \delta \left(M\delta + C_0 + C_1\right) \left\|Z^1 - Z^2\right\|_{C^1}.$$

We conclude that

$$\left\|W^1 - W^2\right\|_{C^1[-\delta,\delta]} \leq (1+\delta)\,(M\,\delta + C_0 + C_1)\,\left\|Z^1 - Z^2\right\|_{C^1[-\delta,\delta]}.$$

Since $C_0 + C_1 < 1$, then choosing $\delta$ sufficiently small Step 7 follows. $\square$

We conclude that there exists $\delta^* > 0$ small such that for $0 < \delta \leq \delta^*$, the map $T : \mathcal{C}(\delta) \to \mathcal{C}(\delta)$ is a contraction and hence by the Banach fixed point theorem there is a unique $Z \in \mathcal{C}(\delta)$ such that

$$Z(t) = TZ(t) = \int_0^t H(V_Z(s))\,ds.$$

Differentiating with respect to $t$, we get that $Z$ solves (4.10) for $|t| \leq \delta$. $\square$

We make the following observations about the assumptions in Theorem 4.1.

REMARK 4.3. We show that even for $H$ smooth, satisfying (4.3) with (4.4), and (4.9), the system (4.2) might not have any real solutions in a neighborhood of $t = 0$. In fact, consider for example the following ode:

$$z'(t) = z'(t)^2 + \frac{1-t}{4}, \qquad z(0) = 0. \tag{4.16}$$

In this case $n = 1$ and $H : \mathbb{R}^5 \to \mathbb{R}$ with

$$H\left(t; \zeta^0, \zeta^1; \xi^0, \xi^1\right) = \left(\xi^0\right)^2 + \frac{1-t}{4}$$

analytic. The system $P = H(0;0,0;P,P)$ has a unique solution $P = 1/2$, and so (4.3) and (4.4) hold. Let $\mathcal{P} = (0;0,0;1/2,1/2)$. Since $H(\mathcal{P}) < 1$, (4.9) holds in a small neighborhood of $\mathcal{P}$. On the other hand, (4.16) cannot have real solutions for $t < 0$ and so in any neighborhood of $t = 0$. This shows that there cannot exist a norm $\|\cdot\|$ in $\mathbb{R}$ so that the contraction condition (4.8) is satisfied. In particular, this shows that the conclusion of [**vBO92**, Lemma 2.2] is in error.

REMARK 4.4. Let $H$ be a map from a domain in $\mathbb{R}^{4n+1}$ with values in $\mathbb{R}^n$, and let $P$ be a solution to the system $P = H(0;\mathbf{0},\mathbf{0};P,P)$ satisfying (4.4). Assume $H$ is $C^1$ in a neighborhood of $\mathcal{P} = (0;\mathbf{0},\mathbf{0};P,P)$. Given a norm $\|\cdot\|$ in $\mathbb{R}^n$, let $|||\cdot|||$ be the induced norm on the space of $n \times n$ matrices, i.e., for a $n \times n$ matrix $A$

$$|||A||| = \max\{\|Av\| : v \in \mathbb{R}^n, \|v\| = 1\},$$

see [**HJ85**, Section 5.6]. Since $H$ is $C^1$ then there exists a neighborhood $N_\varepsilon(\mathcal{P})$ as defined in Theorem 4.1 such that (4.5), (4.6), (4.7) are satisfied.

The following proposition shows estimates for $C_0 + C_1$ in (4.7).

PROPOSITION 4.5. *We define the $n \times n$ matrices*

$$\nabla_{\xi^0} H = \left(\frac{\partial h_i}{\partial \xi_j^0}\right)_{1\leq i,j\leq n}, \qquad \nabla_{\xi^1} H = \left(\frac{\partial h_i}{\partial \xi_j^1}\right)_{1\leq i,j\leq n}.$$

*If (4.7) holds for some $C_0, C_1$, then*

$$C_0 \geq \left|\left|\left|\nabla_{\xi^0} H(\mathcal{P})\right|\right|\right|, \quad \textit{and } C_1 \geq \left|\left|\left|\nabla_{\xi^0} H(\mathcal{P})\right|\right|\right|. \tag{4.17}$$

*In addition, (4.7) holds with $C_0 = \max_{N_\varepsilon(\mathcal{P})} \left|\left|\left|\nabla_{\xi^0} H\right|\right|\right|$ and $C_1 = \max_{N_\varepsilon(\mathcal{P})} \left|\left|\left|\nabla_{\xi^1} H\right|\right|\right|$.*

PROOF. We first prove (4.17). Let $v \in \mathbb{R}^n$ with $\|v\| = 1$; for $s > 0$ small, the vector $(0; \mathbf{0}, \mathbf{0}; P + s\,v, P) \in N_\varepsilon(\mathcal{P})$. From (4.7), we get

$$\|H(0; \mathbf{0}, \mathbf{0}; P + s\,v, P) - H(0; \mathbf{0}, \mathbf{0}; P, P)\| \leq C_0\,|s|.$$

Dividing by $|s|$ and letting $s \to 0$ we obtain, by application of the mean value theorem on each component, that for every $v \in \mathbb{R}^n$ with $\|v\| = 1$

$$C_0 \geq \left\|\nabla_{\xi^0} H(\mathcal{P})\, v^t\right\|.$$

Taking the supremum over all $v \in \mathbb{R}^n$ with $\|v\| = 1$, we obtain $C_0 \geq \left|\left|\left|\nabla_{\xi^0} H(\mathcal{P})\right|\right|\right|$. Similarly we get $C_1 \geq \left|\left|\left|\nabla_{\xi^1} H(\mathcal{P})\right|\right|\right|$.

To show the second part of the proposition, by the fundamental theorem of calculus, we have for $\left(t; \zeta^0, \zeta^1; \xi^0, \xi^1\right), \left(t; \zeta^0, \zeta^1; \bar{\xi}^0, \bar{\xi}^1\right) \in N_\varepsilon(\mathcal{P})$ that

$$\begin{aligned}
&H\left(t; \zeta^0, \zeta^1; \xi^0, \xi^1\right) - H\left(t; \zeta^0, \zeta^1; \bar{\xi}^0, \bar{\xi}^1\right) \\
&= \int_0^1 DH\left((1-s)\left(t; \zeta^0, \zeta^1; \xi^0, \xi^1\right) + s\left(t; \zeta^0, \zeta^1; \bar{\xi}^0, \bar{\xi}^1\right)\right) \left(0; \mathbf{0}, \mathbf{0}; \xi^0 - \bar{\xi}^0, \xi^1 - \bar{\xi}^1\right)^t\, ds \\
&= \int_0^1 \nabla_{\xi^0} H\left((1-s)\left(t; \zeta^0, \zeta^1; \xi^0, \xi^1\right) + s\left(t; \zeta^0, \zeta^1; \bar{\xi}^0, \bar{\xi}^1\right)\right) \left(\xi^0 - \bar{\xi}^0\right)^t\, ds \\
&\qquad + \int_0^1 \nabla_{\xi^1} H\left((1-s)\left(t; \zeta^0, \zeta^1; \xi^0, \xi^1\right) + s\left(t; \zeta^0, \zeta^1; \bar{\xi}^0, \bar{\xi}^1\right)\right) \left(\xi^1 - \bar{\xi}^1\right)^t\, ds,
\end{aligned}$$

where $DH$ is the $n \times (4n+1)$ matrix of the first derivatives of $H$ with respect to all variables. Then

$$\begin{aligned}
&\left\|H\left(t; \zeta^0, \zeta^1; \xi^0, \xi^1\right) - H\left(t; \zeta^0, \zeta^1; \bar{\xi}^0, \bar{\xi}^1\right)\right\| \\
&\leq \int_0^1 \left|\left|\left|\nabla_{\xi^0} H\left((1-s)\left(t; \zeta^0, \zeta^1; \xi^0, \xi^1\right) + s\left(t; \zeta^0, \zeta^1; \bar{\xi}^0, \bar{\xi}^1\right)\right)\right|\right|\right|\, \left\|\xi^0 - \bar{\xi}^0\right\|\, ds \\
&\qquad + \int_0^1 \left|\left|\left|\nabla_{\xi^1} H\left((1-s)\left(t; \zeta^0, \zeta^1; \xi^0, \xi^1\right) + s\left(t; \zeta^0, \zeta^1; \bar{\xi}^0, \bar{\xi}^1\right)\right)\right|\right|\right|\, \left\|\xi^1 - \bar{\xi}^1\right\|\, ds \\
&\leq \max_{N_\varepsilon(\mathcal{P})} \left|\left|\left|\nabla_{\xi^0} H\right|\right|\right|\, \left\|\bar{\xi}^0 - \xi^0\right\| + \max_{N_\varepsilon(\mathcal{P})} \left|\left|\left|\nabla_{\xi^1} H\right|\right|\right|\, \left\|\bar{\xi}^1 - \xi^1\right\|.
\end{aligned}$$

The proof is then complete. □

Given an $n \times n$ matrix $A$, let $R_A$ be its spectral radius, i.e., $R_A$ is the largest absolute value of the eigenvalues of $A$. From [**HJ85**, Theorem 5.6.9] we have $R_A \leq |||A|||$ for any matrix norm $|||\cdot|||$. Then from (4.17) we get the following corollary that shows that the possibility of choosing a norm in $\mathbb{R}^n$ for which the contraction property (4.8) holds depends on the spectral radii of the matrices $\nabla_{\xi^0} H(\mathcal{P}), \nabla_{\xi^1} H(\mathcal{P})$.

COROLLARY 4.6. *Let $H$ be as above. Denote by $R_{\xi^0}$ and $R_{\xi^1}$ the spectral radii of the matrices $\nabla_{\xi^0} H(\mathcal{P})$, and $\nabla_{\xi^1} H(\mathcal{P})$ respectively. Then for any norm in $\mathbb{R}^n$ and any $C_0, C_1$ satisfying* (4.7) *we have*

$$C_0 + C_1 \geq R_{\xi^0} + R_{\xi^1}.$$

To apply Theorem 4.1, we need $H$ to satisfy (4.7) together with the contraction condition (4.8) for some norm $\|\cdot\|$ in $\mathbb{R}^n$. It might not be possible to find such a norm. In fact, if an $n \times n$ matrix $A$ has spectral radius $R_A \geq 1$, then for each norm $\|\cdot\|$ in $\mathbb{R}^n$ the induced matrix norm satisfies $|||A||| \geq 1$; see [Theorem 5.6.9 and Lemma 5.6.10][**HJ85**]. So if the sum of the spectral radii of the Jacobian matrices

$\nabla_{\xi^0} H(\mathcal{P})$, $\nabla_{\xi^1} H(\mathcal{P})$ is bigger than one, then from Corollary 4.6 it is not possible to find a norm $\|\cdot\|$ in $\mathbb{R}^n$ for which (4.8) holds.

**4.1. Uniqueness of solutions.** In this section we show the following uniqueness theorem.

THEOREM 4.7. *Under the assumptions of Theorem* 4.1, *the local solution to* (4.10) *with* $Z'(0) = P$ *is unique.*

The theorem is a consequence of the following lemma.

LEMMA 4.8. *Let* $\mathcal{C}(\delta)$ *be the set in Definition* 4.2. *If there exists* $\delta > 0$ *such that* $W$ *solves* (4.10) *for* $|t| \leq \delta$, *with* $W'(0) = P$, *and* $V_W(t) \in N_\varepsilon(\mathcal{P})$ *for* $|t| \leq \delta$, *then* $W \in \mathcal{C}(\delta)$.

PROOF. Since $w_1'(t) = h_1(V_W(t))$, then (4.9) implies

$$|w_1(t)| = \left|\int_0^t h_1(V_W(s))\,ds\right| \leq |t|. \tag{4.18}$$

Also for $|t|, |\bar{t}| \leq \delta$ since $W$ is a solution to (4.10) then from (4.11)

$$\|W(t) - W(\bar{t})\| = \left\|\int_{\bar{t}}^t H(V_W(s))\,ds\right\| \leq \alpha|t - \bar{t}| \tag{4.19}$$

and by (4.9)

$$|w_1(t) - w_1(\bar{t})| = \left|\int_{\bar{t}}^t h_1(V_W(s))\,ds\right| \leq |t - \bar{t}|. \tag{4.20}$$

It remains to show the Lipschitz estimate on $W'$. Let $|t|, |\bar{t}| \leq \delta$, then by the Lipschitz properties of $H$

$$\begin{aligned}\|W'(t) - W'(\bar{t})\| &= \|H(V_W(t)) - H(V_W(\bar{t}))\| \\ &\leq \Lambda|t - \bar{t}| + L_0\,\|W(t) - W(\bar{t})\| + L_1\,\|W(w_1(t)) - W(w_1(\bar{t}))\| \\ &\quad + C_0\,\|W'(t) - W'(\bar{t})\| + C_1\,\|W'(w_1(t)) - W'(w_1(\bar{t}))\|.\end{aligned}$$

Using (4.19), (4.18), and (4.20), we get that for every $|t|, |\bar{t}| \leq \delta$

$$\begin{aligned}\|W'(t) - W'(\bar{t})\| \leq (\Lambda + (L_0 + L_1)\alpha)|t - \bar{t}| + C_0\,\|W'(t) - W'(\bar{t})\| \\ + C_1\,\|W'(w_1(t)) - W'(w_1(\bar{t}))\|.\end{aligned} \tag{4.21}$$

Fix $t$ and $\bar{t}$ and let $r = |t - \bar{t}|$. Let $\tau$, and $\bar{\tau}$ be such that $|\tau|, |\bar{\tau}| \leq \delta$ and $|\tau - \bar{\tau}| \leq r$, then by (4.20)

$$|w_1(\tau) - w_1(\bar{\tau})| \leq |\tau - \bar{\tau}| \leq r.$$

Hence applying (4.21) for $\tau$ and $\bar{\tau}$ we get for $|\tau|, |\bar{\tau}| \leq \delta$, $|\tau - \bar{\tau}| \leq r$ that

$$\begin{aligned}\|W'(\tau) - W'(\bar{\tau})\| &\leq (\Lambda + (L_0 + L_1)\,\alpha)|\tau - \bar{\tau}| + C_0\,\|W'(\tau) - W'(\bar{\tau})\| \\ &\qquad + C_1\,\|W'(w_1(\tau)) - W'(w_1(\bar{\tau}))\| \\ &\leq (\Lambda + (L_0 + L_1)\,\alpha)\,r + (C_0 + C_1) \sup_{|\tau|,|\bar{\tau}|\leq\delta, |\tau-\bar{\tau}|\leq r} \|W'(\tau) - W'(\bar{\tau})\|.\end{aligned}$$

Hence taking the supremum on the left hand side of the inequality, and using (4.13) we get

$$\sup_{|\tau|,|\bar{\tau}|\leq\delta, |\tau-\bar{\tau}|\leq r} \|W'(\tau) - W'(\bar{\tau})\| \leq \frac{\Lambda + (L_0 + L_1)\,\alpha}{1 - C_0 - C_1}\,r \leq \mu\, r$$

so for every $|t|, |\bar{t}| \leq \delta$

$$\|W'(t) - W'(\bar{t})\| \leq \sup_{|\tau|,|\bar{\tau}|\leq\delta, |\tau-\bar{\tau}|\leq|t-\bar{t}|} \|W'(\tau) - W'(\bar{\tau})\| \leq \mu\,|t-\bar{t}|,$$

and the lemma follows. □

PROOF OF THEOREM 4.7. Let $\delta_1, \delta_2 \leq \delta^*$ and let $W^i$ solving (4.10) for $|t| \leq \delta_i$, with $\left(W^i\right)'(0) = P$, and $V_{W^i}(t) \in N_\varepsilon(\mathcal{P})$ for $|t| \leq \delta_i$ for $i = 1, 2$. From Lemma 4.8 $W^i \in \mathcal{C}(\delta_i)$, and since they solve (4.10) we have $TW^i(t) = W^i(t)$ for $|t| \leq \delta_i$, $i = 1, 2$. Let $\delta = \min\{\delta_1, \delta_2\}$. We have $\mathcal{C}(\delta_i) \subset \mathcal{C}(\delta)$, $i = 1, 2$. Since $\delta \leq \delta^*$, from the proof of the existence theorem $T$ has a unique fixed point in $\mathcal{C}(\delta)$. But, $TW^i = W^i$ for $|t| \leq \delta$ and so $W^1 = W^2$ for $|t| \leq \delta$. □

REMARK 4.9. If the vector $P$ solution to the system (4.3) does not satisfy (4.4), then the system (4.10) may have infinitely many solutions. This goes back to the paper by Kato and McLeod [**KM71**, Thm. 2] about single functional differential equations. We refer to [**FMOT71**, Equation (1.7)] for representation formulas for infinitely many solutions.

REMARK 4.10. Theorem 4.1 has an extension to more variables and the proof is basically the same. In fact, the set up in this case and the result are as follows.

We set $Z(t) = (z_1(t), \cdots, z_n(t))$ where $z_i(t)$ are real valued functions of one variable, and $Z'(t) = (z_1'(t), \cdots, z_n'(t))$. Let $H = (h_1, \cdots, h_n)$ where

$$h_i = h_i\left(t; \zeta^0, \zeta^1, \cdots, \zeta^m; \xi^0, \xi^1, \cdots, \xi^m\right)$$

are real valued functions with $\zeta^j, \xi^j \in \mathbb{R}^n$ for $0 \leq j \leq m$, so each $h_i$ has $1+2\,(m+1)\,n$ variables. Let

$$X = \left(t; \zeta^0, \zeta^1, \cdots, \zeta^m; \xi^0, \xi^1, \cdots, \xi^m\right).$$

We assume $1 \leq m \leq n$. Set $Z(z_k(t)) = (z_1(z_k(t)), \cdots, z_n(z_k(t)))$ for $1 \leq k \leq m$. We want to find $Z(t)$ as above satisfying the following functional differential equation

$$\begin{aligned} &Z'(t) = H\left(t; Z(t), Z(z_1(t)), \cdots, Z(z_m(t)); Z'(t), Z'(z_1(t)), \cdots, Z'(z_m(t))\right) \\ &Z(0) = (0, \cdots, 0) \in \mathbb{R}^n, \end{aligned}$$

for $t$ in a neighborhood of 0. We assume that there exists $P = (p_1, \cdots, p_n) \in \mathbb{R}^n$ with $|p_i| \leq 1$ for $1 \leq i \leq m$ solving

$$P = H\left(0; \mathbf{0}, \cdots, \mathbf{0}; P, \cdots, P\right)$$

where dots mean $m+1$ times.

Let $\mathcal{P} = (0; \mathbf{0}, \cdots, \mathbf{0}; P, \cdots, P) \in R^{2(m+1)n+1}$, and let $\|\cdot\|$ be a norm in $\mathbb{R}^n$ with

$$\begin{aligned} N_\varepsilon(\mathcal{P}) = \big\{\left(t; \zeta^0, \cdots, \zeta^m; \xi^0, \cdots, \xi^m\right); |t| + \|\zeta^0\| + \cdots + \|\zeta^m\| + \|\xi^0 - P\| \\ + \cdots + \|\xi^m - P\| \leq \varepsilon\big\} \end{aligned}$$

a neighborhood of $\mathcal{P}$ such that

(i) $H$ is uniformly Lipschitz in the variable $t$, i.e., there exists $\Lambda > 0$ such that

$$\begin{aligned} \left\|H\left(\bar{t}; \zeta^0, \cdots, \zeta^m; \xi^0, \cdots, \xi^m\right) - H\left(t; \zeta^0, \cdots, \zeta^m; \xi^0, \cdots, \xi^m\right)\right\| \\ \leq \Lambda|\bar{t} - t|. \end{aligned}$$

for all $\left(\bar{t}; \zeta^0, \cdots, \zeta^m; \xi^0, \cdots, \xi^m\right), \left(t; \zeta^0, \cdots, \zeta^m; \xi^0, \cdots, \xi^m\right) \in N_\varepsilon(\mathcal{P})$;

(ii) $H$ is uniformly Lipschitz in the variables $\zeta^0, \cdots, \zeta^m$ i.e., there exist positive constants $L_0, \cdots, L_m$ such that

$$\left\|H\left(t; \bar{\zeta}^0, \cdots, \bar{\zeta}^m; \xi^0, \cdots, \xi^m\right) - H\left(t; \zeta^0, \cdots, \zeta^m; \xi^0, \cdots, \xi^m\right)\right\| \leq L_0 \left\|\bar{\zeta}^0 - \zeta^0\right\| + \cdots + L_m \left\|\bar{\zeta}^m - \zeta^m\right\|,$$

for all $\left(t; \bar{\zeta}^0, \cdots, \bar{\zeta}^m; \xi^0, \cdots, \xi^m\right), \left(t; \zeta^0, \cdots, \zeta^m; \xi^0, \cdots, \xi^m\right) \in N_\varepsilon(\mathcal{P})$;

(iii) $H$ is a uniform contraction in the variables $\xi^0, \cdots, \xi^m$, i.e., there exists constants $C_0, \cdots, C_m$ such that

$$\left\|H\left(t; \zeta^0, \cdots, \zeta^m; \bar{\xi}^0, \cdots, \bar{\xi}^m\right) - H\left(t; \zeta^0, \cdots, \zeta^m; \xi^0, \cdots, \xi^m\right)\right\| \leq C_0 \left\|\bar{\xi}^0 - \xi^0\right\| + \cdots + C_m \left\|\bar{\xi}^m - \xi^m\right\|,$$

for all $\left(t; \zeta^0, \cdots, \zeta^m; \bar{\xi}^0, \cdots, \bar{\xi}^m\right), \left(t; \zeta^0, \cdots, \zeta^m; \xi^0, \cdots, \xi^m\right) \in N_\varepsilon(\mathcal{P})$. with

$$C_0 + \cdots + C_m < 1;$$

(iv) For all $X \in N_\varepsilon(\mathcal{P})$

$$|h_1(X)| \leq 1.$$

Under these assumptions, there exists $\delta > 0$, such that the system

$$\begin{cases} Z'(t) = H\left(t; Z(t), Z(z_1(t)), \cdots, Z(z_m(t)); Z'(t), Z'(z_1(t)), \cdots, Z'(z_m(t))\right) \\ Z(0) = 0, \end{cases}$$

has a unique solution defined for $|t| \leq \delta$ satisfying $Z'(0) = P$.

## 5. One point source case: Problem B

The setup in this section is the following. We are given a unit vector $w \in \mathbb{R}^3$, and a compact domain $\Omega$ contained in the upper unit sphere $S^2$, such that $\Omega = x(D)$, where $D$ is a convex and compact domain in $\mathbb{R}^2$ with nonempty interior. Here $x(t)$ are for example spherical coordinates, $t \in D$. Dichromatic rays with colors b and r are now emitted from the origin with unit direction $x(t)$, $t \in D$. From the results from [**Gut13**, Section 3] with $n_1 = n_3 = 1$, $n_2 = n_r$, and $e_1 = w$ we have the following. Consider a $C^2$ surface with a given polar parametrization $\rho(t)x(t)$ for $t \in D$, and the surface parametrized by $f_r(t) = \rho(t)x(t) + d_r(t)m_r(t)$, with $m_r(t) = \dfrac{1}{n_r}\left(x(t) - \lambda_r \nu_\rho(t)\right)$, $\lambda_r(t) = \Phi_{n_r}\left(x(t) \cdot \nu_\rho(t)\right)$ from (2.3), $\nu_\rho(t)$ the outer unit normal at $\rho(t)x(t)$, and with

$$d_r(t) = \frac{C_r - \rho(t)\left(1 - w \cdot x(t)\right)}{n_r - w \cdot m_r(t)}, \tag{5.1}$$

for some constant $C_r$. Then the lens bounded between $\rho$ and $f_r$ refracts the rays with color r into the direction $w$ provided that $C_r$ is chosen so that $d_r(t) > 0$ and $f_r$ has a normal at each point. Likewise and for the color b the surface $f_b(t) = \rho(t)x(t) + d_b(t)m_b(t)$, with similar quantities as before with r replaced by b, does a similar refracting job for rays with color b.

As before, we assume $n_b > n_r > 1$, and the medium surrounding the lens is vacuum. To avoid total reflection for each color, compatibility conditions between $\rho$ and $w$ are needed, see [**Gut13**, condition (3.8)] which in our case reads

$$\lambda_r\, \nu_\rho(t) \cdot w \leq x(t) \cdot w - 1, \text{ and } \lambda_b\, \nu_\rho(t) \cdot w \leq x(t) \cdot w - 1.$$

The problem we consider in this section is to determine if there exist $\rho$ and corresponding surfaces $f_r$ and $f_b$ for each color such that $f_r$ can be obtained by a re-parametrization of $f_b$. That is, if there exist a positive function $\rho \in C^2(D)$, real numbers $C_r$ and $C_b$, and a $C^1$ map $\varphi : D \to D$ such that the surfaces $f_r$ and $f_b$, corresponding to $\rho$, $C_r, C_b$, have normals at each point and

$$f_r(t) = f_b(\varphi(t)) \qquad \forall t \in D. \tag{5.2}$$

We refer to this as *Problem B.* As in the collimated case, if a solution exists $f_r(D) \subseteq f_b(D)$. Again, from an optical point of view, this means that the lens sandwiched between $\rho$ and $f_b$ refracts both colors into $w$; however, there could be points in $f_b(D)$ that are not reached by red rays.

If $w \notin \Omega$, we will show in Theorem 5.2 that Problem B is not solvable. On the other hand, when $w \in \Omega$ we shall prove, in dimension two, that problem B is locally solvable, Theorem 5.13. Notice that by rotating the coordinates we may assume without loss of generality that $w = e$. Theorem 5.13 will follow from Theorem 4.1 on functional differential equations, assuming an initial size condition on the ratio between the thickness of the lens and its distance to the origin. By local solution we mean that there exists an interval $[-\delta, \delta] \subseteq D$, a positive function $\rho \in C^2[-\delta, \delta]$, real numbers $C_r$ and $C_b$, and $\varphi : [-\delta, \delta] \to [-\delta, \delta]$ $C^1$ such that the corresponding surfaces $f_r$ and $f_b$ have normals at every point and

$$f_r(t) = f_b(\varphi(t)) \qquad \forall t \in [-\delta, \delta].$$

We will also show a necessary condition for solvability of Problem B, Corollary 5.10.

We first state the following lemma whose proof is the same as that of Lemma 3.2.

LEMMA 5.1. *Given a surface $\rho(t)x(t)$, $t \in D$, and $w$ a unit vector in $\mathbb{R}^3$, let $f_r$ and $f_b$ be the surfaces parametrized as above. If $f_r(t) = f_b(t)$ for some $t \in D$, then $\nu_\rho(t) = x(t)$. In addition $d_b(t) = d_r(t)$.*

We next show nonexistence of solutions to Problem $B$ for $w \notin \Omega$.

THEOREM 5.2. *Let $w$ be a unit vector in $\mathbb{R}^3$. If Problem B is solvable, then $x(t) = w$ for some $t \in D$. Therefore, since $x(D) = \Omega$, Problem B has no solutions when $w \notin \Omega$.*

PROOF. Suppose there exist $\rho$ and $\varphi : D \to D$ satisfying (5.2). Since $D$ is a compact and convex domain, by Brouwer fixed point theorem $\varphi$ has a fixed point $t_0$, and from (5.2) $f_r(t_0) = f_b(t_0)$. Therefore, by Lemma 5.1 $\nu_\rho(t_0) = x(t_0)$, and by the Snell's law at $\rho(t_0)x(t_0)$ we have $m_b(t_0) = m_r(t_0) = x(t_0)$. Using Snell's law again at $f_r(t_0) = f_b(t_0)$, since $n_r \neq n_b$ and both colors with direction $x(t_0)$ are refracted at $f_r(t_0) = f_b(t_0)$ into $w$, we obtain $x(t_0) = w$. □

From now on our objective is to show that problem B is locally solvable in dimension two when $w \in \Omega$, Theorem 5.13.

### 5.1. Two dimensional case, $w \in \Omega$.

Let $w$ be a unit vector in $\mathbb{R}^2$, by rotating the coordinates we will assume that $w = e = (0, 1)$. Let $\Omega$ be a compact domain of the upper circle, such that $\Omega = x(D)$ where $D$ is a closed interval in $(-\pi/2, \pi/2)$, and $x(t) = (\sin t, \cos t)$.

We will use the following expression for the normal to a parametric curve.

LEMMA 5.3. *If a curve is given by the polar parametrization $\rho(t)x(t) = \rho(t)(\sin t, \cos t)$, with $\rho \in C^1$, then the unit outer normal is*

$$\nu(t) = \frac{1}{\sqrt{\rho^2(t) + \rho'(t)^2}} \left(\rho(t)\sin t - \rho'(t)\cos t, \rho'(t)\sin t + \rho(t)\cos t\right).$$

PROOF. The tangent vector to the curve at the point $\rho(t)\,x(t)$, with $x(t) = (\sin t, \cos t)$, equals

$$(\rho(t)\,x(t))' = \rho'(t)x(t) + \rho(t)x'(t) = \left(\rho'(t)\sin t + \rho(t)\cos t, \rho'(t)\cos t - \rho(t)\sin t\right).$$

Thus

$$|(\rho(t)\,x(t))'|^2 = \rho(t)^2 + \rho'(t)^2.$$

Hence the unit normal

$$\nu(t) = \pm\frac{1}{\sqrt{\rho^2(t) + \rho'(t)^2}} \left(\rho(t)\sin t - \rho'(t)\cos t, \rho'(t)\sin t + \rho(t)\cos t\right).$$

Since $\nu(t)$ is outer, i.e. $x(t)\cdot\nu(t) \geq 0$, so we take the positive sign above and the lemma follows. □

As a consequence, we obtain the following important lemma.

LEMMA 5.4. *Assume Problem B is solvable in the plane when $w = e$. Then $0 \in D$,*

$$\varphi(0) = 0, \qquad d_b(0) = d_r(0), \qquad \text{and } \rho'(0) = 0.$$

PROOF. Using the proof of Theorem 5.2, there exists $t_0 \in D$ such that $\varphi(t_0) = t_0$ and $x(t_0) = e = (0,1)$, then $(\sin t_0, \cos t_0) = (0,1)$, and $t_0 = 0$. By Lemma 5.1, we get $d_b(0) = d_r(0)$, and $\nu_\rho(0) = e$. Therefore, Lemma 5.3 yields

$$(0,1) = \frac{1}{\sqrt{\rho^2(0) + \rho'(0)^2}}(-\rho'(0), \rho(0)).$$

□

**5.2. Derivation of a system of functional equations from the solvability of problem B in the plane.** Assume Problem B has a solution refracting rays of both colors b and r into the direction e, and recall Lemma 5.4. [1]

We set $\rho(0) = \rho_0$ and $d_b(0) = d_r(0) = d_0$, and prove the following theorem.

THEOREM 5.5. *Suppose there exist $\rho$ and $\varphi$ solving Problem B in an interval $D$. Let $Z(t) = (z_1(t), z_2(t), z_3(t), z_4(t), z_5(t)) \in \mathbb{R}^5$ with*

(5.3)
$$z_1(t) = \varphi(t), \quad z_2(t) = v_1(t) + \rho_0, \quad z_3(t) = v_2(t), \quad z_4(t) = v_1'(t), \quad z_5(t) = v_2'(t) - \rho_0,$$

*where $v_1(t) = -\rho(t)\cos t$, and $v_2(t) = \rho(t)\sin t$.*

[1] We are assuming that 0 is an interior point of $D$. Otherwise, in our statements the interval $[-\delta, \delta]$ has to be replaced by either $[-\delta, 0]$ or $[0, \delta]$.

*Let $\mathcal{Z} = (0;\mathbf{0},\mathbf{0};Z'(0),Z'(0)) \in \mathbb{R}^{21}$.*[2] *There exists a neighborhood of $\mathcal{Z}$ and a map $H$ defined and smooth in that neighborhood with*

$$H := H(t;\zeta^0,\zeta^1;\xi^0,\xi^1) = (h_1,\cdots,h_5)$$

*where $\zeta^0,\zeta^1,\xi^0,\xi^1 \in \mathbb{R}^5$, $\zeta^i = \left(\zeta_1^i,\zeta_2^i,\cdots,\zeta_5^i\right), \xi^i = \left(\xi_1^i,\xi_2^i,\cdots,\xi_5^i\right)$, and with the functions $h_1,\cdots,h_5$ given by (5.24), (5.26), (5.28), (5.30), and (5.32), respectively, such that $Z$ is a solution to the system of functional differential equations*

$$\begin{aligned} Z'(t) &= H\left(t; Z(t), Z(z_1(t)); Z'(t), Z'(z_1(t))\right) \\ Z(0) &= \mathbf{0} \end{aligned} \tag{5.4}$$

*for $t$ in a neighborhood of 0. The map $H$ depends on the values $\rho_0$ and $d_0$.*

PROOF. From Lemma 5.4, $Z(0) = \mathbf{0}$. We will derive the expressions for $h_i(t;\zeta^0,\zeta^1;\xi^0,\xi^1)$, $1 \le i \le 5$, so that

$$z_i'(t) = h_i\left(t; Z(t), Z(z_1(t)), Z'(t), Z'(z_1(t))\right).$$

The first step is to express the quantities involved as functions of

$$\left(t; Z(t), Z(z_1(t)); Z'(t), Z'(z_1(t))\right).$$

From the Snell law, a ray emitted from the origin with color r and direction $x(t) = (\sin t, \cos t)$ refracts by a curve $\rho(t)x(t)$ into a medium with refractive index $n_r$ into the direction $m_r(t)$ such that $x(t) - n_r m_r(t) = \Phi_{n_r}(x\cdot\nu(t))\,\nu(t)$, where $\nu(t)$ is the outward unit normal to $\rho$ at $\rho(t)x(t)$. From (2.3)

$$\Phi_{n_r}(s) = s - \sqrt{n_r^2 - 1 + s^2} = \frac{1-n_r^2}{s + \sqrt{n_r^2 - 1 + s^2}},$$

and from Lemma 5.3 $\nu(t) = \dfrac{v'(t)}{|v'(t)|}$, and $x(t)\cdot\nu(t) = \dfrac{\rho(t)}{\sqrt{\rho^2(t)+\rho'(t)^2}} = \dfrac{|v(t)|}{|v'(t)|}$, with $v(t) = (v_1(t), v_2(t))$. So

$$\Phi_{n_r}(x(t)\cdot\nu(t)) = \frac{1-n_r^2}{\dfrac{|v(t)|}{|v'(t)|} + \sqrt{n_r^2 - 1 + \dfrac{|v(t)|^2}{|v'(t)|^2}}} = \frac{(1-n_r^2)|v'(t)|}{|v(t)| + \sqrt{|v(t)|^2 + (n_r^2-1)|v'(t)|^2}}.$$

Hence

$$\begin{aligned} m_r(t) &= \frac{1}{n_r}\left[x(t) - \Phi_{n_r}(x(t)\cdot\nu(t))\nu(t)\right] = \frac{1}{n_r}\left[x(t) - A_r\left(v(t),v'(t)\right)v'(t)\right] \\ &:= (m_{1r}(t), m_{2r}(t)) \end{aligned} \tag{5.5}$$

with

$$A_r(v(t),v'(t)) = \frac{1-n_r^2}{|v(t)| + \sqrt{|v(t)|^2 + (n_r^2-1)|v'(t)|^2}}. \tag{5.6}$$

[2] $Z'(0) = (\varphi'(0), 0, \rho_0, -\rho''(0) + \rho_0, 0)$.

Rewriting the last expressions in terms of the variables $z_i(t)$ introduced in (5.3), and omitting the dependance in $t$ to simplify the notation, we obtain

(5.7)

$$A_r(v(t), v'(t)) = \frac{1-n_r^2}{|(z_2-\rho_0, z_3)| + \sqrt{|(z_2-\rho_0, z_3)|^2 + (n_r^2-1)\,|(z_4, z_5+\rho_0)|^2}} := \mathcal{A}_r(Z(t)),$$

(5.8)

$$m_{1r}(t) = \frac{1}{n_r}\left[\sin t - \mathcal{A}_r(Z)\, z_4\right] := \mu_r\,(t, Z(t)),$$

(5.9)

$$m_{2r}(t) = \frac{1}{n_r}\left[\cos t - \mathcal{A}_r(Z)\,(z_5+\rho_0)\right] := \tau_r\,(t, Z(t)).$$

Notice that $\tau_r\,(0, Z(0)) = \tau_r\,(0, \mathbf{0}) = 1$.

If for each $t$, the ray with direction $m_r(t)$ is refracted by the upper face of the lens into the direction $e = (0,1)$, then the upper face is parametrized by the vector

$$f_r(t) = \rho(t)x(t) + d_r(t)m_r(t) := (f_{1r}(t), f_{2r}(t)).$$

From (5.1), and (5.9)

$$(5.10)\qquad d_r(t) = \frac{C_r - \rho(t)(1-\cos t)}{n_r - m_{2r}(t)} = \frac{C_r - |v(t)| - v_1(t)}{n_r - \tau_r(t, Z(t))}$$
$$= \frac{C_r - |(z_2-\rho_0, z_3)| - z_2 + \rho_0}{n_r - \tau_r\,(t, Z(t))} := D_r\,(t, Z(t)),$$

with $C_r$ a constant. Since $\rho$ and $\varphi$ solve Problem B, from Lemma 5.4 we have $\rho'(0) = 0$, and $v(0) = (0,1)$. So $m_r(0) = (0,1)$ and from (5.1) we get $C_r = (n_r - 1)\,d_0$. Hence

$$(5.11)\qquad f_{1r}(t) = z_3 + D_r(t, Z)\,\mu_r(t, Z) := F_{1r}(t, Z(t))$$

$$(5.12)\qquad f_{2r}(t) = -z_2 + \rho_0 + D_r(t, Z)\,\tau_r(t, Z) := F_{2r}(t, Z(t)).$$

In addition, by the Snell law at $f_r(t)$, $m_r(t) - \frac{1}{n_r}e = \lambda_{2,r}(t)\,\nu_r(t)$, where $\nu_r$ is the normal to the upper surface at the point $f_r(t)$ (we are using here that the normal to $f_r$ exists since we are assuming Problem B is solvable). Since $n_r > 1$, then $\lambda_{2,r} > 0$ and so taking absolute values in the last expression yields

$$(5.13)\qquad \lambda_{2,r}(t) = \left|m_r(t) - \frac{1}{n_r}e\right| = \sqrt{1 + \frac{1}{n_r^2} - \frac{2}{n_r}m_{2r}(t)}$$
$$= \sqrt{1 + \frac{1}{n_r^2} - \frac{2}{n_r}\tau_r(t, Z)} := \Lambda_{2,r}(t, Z(t)).$$

For $t \in \mathbb{R}$ and $\zeta = (\zeta_1, \cdots, \zeta_5) \in \mathbb{R}^5$ we let

(5.14)

$$\begin{cases} \mathcal{A}_r(\zeta) = \dfrac{1-n_r^2}{|(\zeta_2-\rho_0,\zeta_3)| + \sqrt{|(\zeta_2-\rho_0,\zeta_3)|^2 + (n_r^2-1)\,|(\zeta_4,\zeta_5+\rho_0)|^2}} \\ \mu_r(t,\zeta) = \dfrac{1}{n_r}\left[\sin t - \mathcal{A}_r(\zeta)\,\zeta_4\right], \qquad \tau_r(t,\zeta) = \dfrac{1}{n_r}\left[\cos t - \mathcal{A}_r(\zeta)\,(\zeta_5+\rho_0)\right] \\ D_r(t,\zeta) = \dfrac{C_r - |(\zeta_2-\rho_0,\zeta_3)| - \zeta_2 + \rho_0}{n_r - \tau_r(t,\zeta)}, \qquad \text{with } C_r = (n_r-1)\,d_0 \\ F_{1r}(t,\zeta) = \zeta_3 + D_r(t,\zeta)\,\mu_r(t,\zeta), \qquad F_{2r}(t,\zeta) = -\zeta_2 + \rho_0 + D_r(t,\zeta)\,\tau_r(t,\zeta) \\ \Lambda_{2,r}(t,\zeta) = \sqrt{1 + \dfrac{1}{n_r^2} - \dfrac{2}{n_r}\tau_r(t,\zeta)}. \end{cases}$$

The functions $\mathcal{A}_r, \mu_r, \tau_r$ are well defined and smooth for all $t \in \mathbb{R}$ and for all $\zeta = (\zeta_1, \zeta_2, \zeta_3, \zeta_4, \zeta_5)$ with $(\zeta_2, \zeta_3) \neq (\rho_0, 0)$. Since $\tau_r(0, \mathbf{0}) = 1$, then all functions in (5.14) are well defined and smooth in a neighborhood of $t = 0$ and $\zeta = \mathbf{0}$. Notice that the definitions of $\mathcal{A}_r(\zeta), \mu_r(t,\zeta), \tau_r(t,\zeta)$ and $\Lambda_{2,r}(t,\zeta)$ depend on the value of $\rho_0$, and the definitions of $D_r(t,\zeta), F_{1r}(t,\zeta)$ and $F_{2r}(t,\zeta)$ depend on the values of $\rho_0$ and $d_0$.

To determine later the functions $h_i$ we next calculate the derivatives of $A_r, m_{1r}, m_{2r}, d_r, F_{1r}, F_{2r}$, and $\lambda_{2,r}$ with respect to $t$. Differentiating (5.7), (5.8), (5.9), (5.10), (5.11), and (5.12) with respect to $t$ yields

$$(5.15)\quad \frac{d}{dt}A_r(v(t), v'(t))$$
$$= \frac{n_r^2-1}{\left(|(z_2-\rho_0,z_3)| + \sqrt{|(z_2-\rho_0,z_3)|^2 + (n_r^2-1)\,|(z_4,z_5+\rho_0)|^2}\right)^2}$$
$$\left[\frac{(z_2-\rho_0,z_3)\cdot(z_2',z_3')}{|(z_2-\rho_0,z_3)|} + \frac{(z_2-\rho_0,z_3)\cdot(z_2',z_3') + (n_r^2-1)(z_4,z_5+\rho_0)\cdot(z_4',z_5')}{\sqrt{|(z_2-\rho_0,z_3)|^2 + (n_r^2-1)\,|(z_4,z_5+\rho_0)|^2}}\right]$$
$$= \frac{\mathcal{A}_r(Z)^2}{n_r^2-1}\left[\frac{(z_2-\rho_0,z_3)\cdot(z_2',z_3')}{|(z_2-\rho_0,z_3)|} + \frac{(z_2-\rho_0,z_3)\cdot(z_2',z_3') + (n_r^2-1)(z_4,z_5+\rho_0)\cdot(z_4',z_5')}{\sqrt{|(z_2-\rho_0,z_3)|^2 + (n_r^2-1)\,|(z_4,z_5+\rho_0)|^2}}\right]$$
$$:= \widetilde{\mathcal{A}}_r(Z(t), Z'(t)),$$

(5.16)

$$\frac{d}{dt}m_{1r}(t) = \frac{1}{n_r}\left[\cos t - \mathcal{A}_r(Z)\,z_4' - \widetilde{\mathcal{A}}_r(Z,Z')\,z_4\right] := \widetilde{\mu}_r\left(t, Z(t), Z'(t)\right),$$

(5.17)

$$\frac{d}{dt}m_{2r}(t) = \frac{1}{n_r}\left[-\sin t - \mathcal{A}_r(Z)\,z_5' - \widetilde{\mathcal{A}}_r(Z,Z')\,(z_5+\rho_0)\right] := \widetilde{\tau}_r\left(t, Z(t), Z'(t)\right),$$

(5.18)

$$\frac{d}{dt}d_r(t)$$
$$= \frac{\left(-\dfrac{(z_2-\rho_0,z_3)\cdot(z_2',z_3')}{|(z_2-\rho_0,z_3)|} - z_2'\right)(n_r - \tau_r(t,Z)) + \widetilde{\tau}_r(t,Z,Z')\,(C_r - |(z_2-\rho_0,z_3)| - z_2 + \rho_0)}{(n_r-\tau_r(t,Z))^2}$$
$$= \frac{-\dfrac{(z_2-\rho_0,z_3)\cdot(z_2',z_3')}{|(z_2-\rho_0,z_3)|} - z_2' + \widetilde{\tau}_r(t,Z,Z')\,D_r(t,Z)}{n_r-\tau_r(t,Z)} := \widetilde{D}_r(t, Z(t), Z'(t))$$

$$\frac{d}{dt}f_{1r}(t) = z_3' + D_r(t,Z)\,\widetilde{\mu}_r(t,Z,Z') + \widetilde{D}_r(t,Z,Z')\,\mu_r(t,Z) := \widetilde{F}_{1r}(t,Z(t),Z'(t)) \tag{5.19}$$

$$\frac{d}{dt}f_{2r}(t) = -z_2' + D_r(t,Z)\,\widetilde{\tau}_r(t,Z,Z') + \widetilde{D}_r(t,Z,Z')\,\tau_r(t,Z) := \widetilde{F}_{2r}(t,Z(t),Z'(t)) \tag{5.20}$$

$$\frac{d}{dt}\lambda_{2,r}(t) = -\frac{\frac{1}{n_r}\widetilde{\tau}_r(t,Z,Z')}{\Lambda_{2,r}(t,Z)} := \widetilde{\Lambda}_{2,r}(t,Z(t),Z'(t)). \tag{5.21}$$

For $t\in\mathbb{R}$ and $\zeta,\xi\in\mathbb{R}^5$ we let

(5.22)

$$\begin{cases}
\widetilde{\mathcal{A}}_r(\zeta,\xi) = \dfrac{\mathcal{A}_r(\zeta)^2}{n_r^2-1}\left[\dfrac{(\zeta_2-\rho_0,\zeta_3)\cdot(\xi_2,\xi_3)}{|(\zeta_2-\rho_0,\zeta_3)|}+\dfrac{(\zeta_2-\rho_0,\zeta_3)\cdot(\xi_2,\xi_3)+(n_r^2-1)(\zeta_4,\zeta_5+\rho_0)\cdot(\xi_4,\xi_5)}{\sqrt{|(\zeta_2-\rho_0,\zeta_3)|^2+(n_r^2-1)|(\zeta_4,\zeta_5+\rho_0)|^2}}\right]\\
\widetilde{\mu}_r(t,\zeta,\xi) = \dfrac{1}{n_r}\left[\cos t - \mathcal{A}_r(\zeta)\xi_4 - \widetilde{\mathcal{A}}_r(\zeta,\xi)\zeta_4\right]\\
\widetilde{\tau}_r(t,\zeta,\xi) = \dfrac{1}{n_r}\left[-\sin t - \mathcal{A}_r(\zeta)\xi_5 - \widetilde{\mathcal{A}}_r(\zeta,\xi)(\zeta_5+\rho_0)\right]\\
\widetilde{D}_r(t,\zeta,\xi) = \dfrac{-\dfrac{(\zeta_2-\rho_0,\zeta_3)\cdot(\xi_2,\xi_3)}{|(\zeta_2-\rho_0,\zeta_3)|} - \xi_2 + \widetilde{\tau}_r(t,\zeta,\xi)\,D_r(t,\zeta)}{n_r-\tau_r(t,\zeta)}\\
\widetilde{F}_{1r}(t,\zeta,\xi) = \xi_3 + D_r(t,\zeta)\widetilde{\mu}_r(t,\zeta,\xi) + \widetilde{D}_r(t,\zeta,\xi)\mu_r(t,\zeta)\\
\widetilde{F}_{2r}(t,\zeta,\xi) = -\xi_2 + D_r(t,\zeta)\widetilde{\tau}_r(t,\zeta,\xi) + \widetilde{D}_r(t,\zeta,\xi)\tau_r(t,\zeta)\\
\widetilde{\Lambda}_{2,r}(t,\zeta,\xi) = -\dfrac{\frac{1}{n_r}\widetilde{\tau}_r(t,\zeta,\xi)}{\Lambda_{2,r}(t,\zeta)}.
\end{cases}$$

As for (5.14), all functions in (5.22) are well defined and smooth in a neighborhood of $t=0$ and $\zeta=\mathbf{0}$, and for any $\xi$.

We mention the following remark that will be used later.

REMARK 5.6. Using the formulas in (5.14) and (5.22), notice that for any differentiable map $U:V\to R^5$ with $V$ a neighborhood of $t=0$ and with $U(0)=\mathbf{0}$, we have the following formulas valid for $t$ in a neighborhood $V'$ of $t=0$ (possibly smaller than $V$):

$$\begin{aligned}
\frac{d}{dt}[\mathcal{A}_r(U(t))] &= \widetilde{\mathcal{A}}_r(U(t),U'(t)), & \frac{d}{dt}[\mu_r(t,U(t))] &= \widetilde{\mu}_r(t,U(t),U'(t)),\\
\frac{d}{dt}[\tau_r(t,U(t))] &= \widetilde{\tau}_r(t,U(t),U'(t)), & \frac{d}{dt}[D_r(t,U(t))] &= \widetilde{D}_r(t,U(t),U'(t)),\\
\frac{d}{dt}[F_{1r}(t,U(t))] &= \widetilde{F}_{1r}(t,U(t),U'(t)), & \frac{d}{dt}[F_{2r}(t,U(t))] &= \widetilde{F}_{2r}(t,U(t),U'(t)),\\
\frac{d}{dt}[\Lambda_{2,r}(t,U(t))] &= \widetilde{\Lambda}_{2,r}(t,U(t),U'(t)).
\end{aligned}$$

We also obtain the same formulas for the color $b$ with $n_r$ replaced by $n_b$.

We are now ready to calculate $h_i$, one by one, for $i=1,\cdots,5$ to obtain the system of functional differential equations satisfied by $Z(t)=(\varphi(t),v_1(t)+\rho_0,v_2(t),v_1'(t),v_2'(t)-\rho_0)$. Recall $h_i := h_i\left(t;\zeta^0,\zeta^1;\xi^0,\xi^1\right)$, with $\zeta^0,\zeta^1,\xi^0,\xi^1\in\mathbb{R}^5$, $\zeta^i=\left(\zeta_1^i,\cdots,\zeta_5^i\right)$, $\xi^i=\left(\xi_1^i,\cdots,\xi_5^i\right)$, $i=0,1$.

**Calculation of $h_1$.** Since $z_1 = \varphi$ satisfies $f_r(t) = f_b(\varphi(t))$, taking the first components and differentiating with respect to $t$ yields

$$f'_{1r}(t) = \varphi'(t)\, f'_{1b}(\varphi(t)). \tag{5.23}$$

We claim that $f'_{1b}(\varphi(t)) \neq 0$ in a neighborhood of $t = 0$.

PROOF OF THE CLAIM. By continuity of $f'_{1b} \circ \varphi$ and since $\varphi(0) = 0$ by Lemma 5.4, it is equivalent to show that $f'_{1b}(0) \neq 0$.

Recall that $f_{1r}(t) = \rho(t)\sin t + d_r(t) m_{1r}(t)$, $\rho(0) = \rho_0$, $d_r(0) = d_0$, and $m_r(0) = (0,1)$. Then

$$f'_{1r}(0) = \rho_0 + d_0\, m'_{1r}(0).$$

Also from (5.16), since $Z(0) = \mathbf{0}$

$$m'_{1r}(0) = \frac{1}{n_r}\left(1 - \mathcal{A}_r(\mathbf{0}) z'_4(0)\right).$$

Notice that $z'_4(t) = v''_1(t) = \rho(t)\cos t - \rho''(t)\cos t + 2\rho'(t)\sin t$, then $z'_4(0) = \rho_0 - \rho''(0)$. Also from (5.7)

$$\mathcal{A}_r(\mathbf{0}) = \frac{1 - n_r^2}{(1 + n_r)\rho_0} = \frac{1 - n_r}{\rho_0}.$$

We conclude that

$$m'_{1r}(0) = \frac{1}{n_r}\left(1 + \frac{n_r - 1}{\rho_0}(\rho_0 - \rho''(0))\right) = 1 - \frac{n_r - 1}{n_r}\frac{\rho''(0)}{\rho_0}.$$

Similarly, $f_{1b}(t) = \rho(t)\sin t + d_b(t) m_b(t)$, $d_b(0) = d_0$, $m_b(0) = (0,1)$, and we get $f'_{1b}(0) = \rho_0 + d_0\, m'_{1b}(0)$, with

$$m'_{1b}(0) = 1 - \frac{n_b - 1}{n_b}\frac{\rho''(0)}{\rho_0}.$$

Suppose by contradiction that $f'_{1b}(0) = 0$. Then by (5.23) $f'_{1r}(0) = 0$. Hence from the calculations above $m'_{1b}(0) = m'_{1r}(0) = -\rho_0/d_0$. Since $n_r \neq n_b$, $m'_{1b}(0) = m'_{1r}(0)$ implies $\rho''(0) = 0$. So

$$m'_{1b}(0) = m'_{1r}(0) = 1 = -\rho_0/d_0 < 0,$$

a contradiction. □

Since $z_1 = \varphi$ we then conclude that $f'_{1b}(z_1(t)) \neq 0$ in a neighborhood of $t = 0$, and

$$z'_1(t) = \varphi'(t) = \frac{f'_{1r}(t)}{f'_{1b}(z_1(t))}.$$

Applying formula (5.19) for both $b$ and $r$ yields

$$z'_1(t) = \varphi'(t) = \frac{\widetilde{F}_{1r}(t, Z(t), Z'(t))}{\widetilde{F}_{1b}(z_1(t), Z(z_1(t)), Z'(z_1(t)))} := h_1(t; Z(t), Z(z_1(t)); Z'(t), Z'(z_1(t)))$$

with

$$h_1(t; \zeta^0, \zeta^1; \xi^0, \xi^1) = \frac{\widetilde{F}_{1r}(t, \zeta^0, \xi^0)}{\widetilde{F}_{1b}(\zeta^0_1, \zeta^1, \xi^1)}; \tag{5.24}$$

$\widetilde{F}_{1r}$ is defined explicitly in (5.22), and $\widetilde{F}_{1b}$ has a similar expression with $n_r$ replaced by $n_b$.

We next verify that $h_1$ is smooth in a neighborhood of $\mathcal{Z} = (0; \mathbf{0}, \mathbf{0}; Z'(0), Z'(0))$; $Z'(0) = (\varphi'(0), 0, \rho_0, -\rho''(0) + \rho_0, 0)$. From (5.14), $\mathcal{A}_r$ is smooth in a neighborhood of $\mathbf{0} \in R^5$, $\mu_r, \tau_r, D_r, F_{1r}, F_{2r}, \Lambda_{2,r}$ are smooth in a neighborhood of $(0, \mathbf{0}) \in \mathbb{R}^6$. Also, from (5.22), $\widetilde{\mathcal{A}}_r$ is smooth in a neighborhood of $(\mathbf{0}, \mathbf{Z}'(\mathbf{0})) \in R^{10}$, and $\widetilde{\mu}_r, \widetilde{\tau}_r, \widetilde{D}_r, \widetilde{F}_{1r}, \widetilde{F}_{2r}, \widetilde{\Lambda}_{2,r}$ are smooth in a neighborhood of $(0, \mathbf{0}, \mathbf{Z}'(\mathbf{0})) \in R^{11}$. Similarly, we have the same smoothness for the functions corresponding to $n_b$. Therefore, from (5.24), to show that $h_1$ is smooth in a neighborhood of $\mathcal{Z}$, it is enough to prove that $\widetilde{F}_{1b}(0, \mathbf{0}, Z'(0)) \neq 0$. In fact, since $Z(0) = \mathbf{0}$ we obtain from (5.19) for $n_b$ and the claim above that

$$\widetilde{F}_{1b}(0, \mathbf{0}, Z'(0)) = \widetilde{F}_{1b}(0, Z(0), Z'(0)) = f'_{1b}(0) \neq 0.$$

**Calculation of $h_2$ and $h_3$.** We have

$$z'_2(t) = v'_1(t) = z_4(t) := h_2(t; Z(t), Z(z_1(t)); Z'(t), Z'(z_1(t))) \tag{5.25}$$

with,

$$h_2(t; \zeta^0, \zeta^1; \xi^0, \xi^1) = \zeta^0_4. \tag{5.26}$$

Similarly,

$$z'_3(t) = v'_2(t) = z_5(t) + \rho_0 := h_3(t; Z(t), Z(z_1(t)); Z'(t), Z'(z_1(t))) \tag{5.27}$$

with

$$h_3(t; \zeta^0, \zeta^1; \xi^0, \xi^1) = \zeta^0_5 + \rho_0. \tag{5.28}$$

Trivially, $h_2$ and $h_3$ are smooth everywhere.

**Calculation of $h_4$.**

The rays $m_r(t)$ and $m_b(\varphi(t))$ are both refracted into $e = (0, 1)$ at $f_r(t)$ (since Problem B is solvable $f_r$ has a normal vector), then by the Snell law

$$m_r(t) - \frac{1}{n_r} e = \lambda_{2,r}(t) \nu_S(t)$$
$$m_b(\varphi(t)) - \frac{1}{n_b} e = \lambda_{2,b}(\varphi(t)) \nu_S(t)$$

with $\nu_S$ the outer unit normal to $S = \{f_r(D)\}$. If $\alpha_S$ denotes the first component of $\nu_S$, then

$$m_{1r}(t) = \lambda_{2,r}(t) \alpha_S(t), \qquad m_{1b}(\varphi(t)) = \lambda_{2,b}(\varphi(t)) \alpha_S(t).$$

Solving in $\alpha_S(t)$ and using (5.8) and (5.13) yields

$$\mu_r(t, Z(t))\, \Lambda_{2,b}(z_1(t), Z(z_1(t))) = \mu_b(z_1(t), Z(z_1(t)))\, \Lambda_{2,r}(t, Z(t)).$$

Differentiating the last expression with respect to $t$, Remark 5.6 yields

$$\begin{aligned}
&\widetilde{\mu}_r(t, Z, Z')\, \Lambda_{2,b}(z_1, Z(z_1)) + z'_1\, \widetilde{\Lambda}_{2,b}(z_1, Z(z_1), Z'(z_1))\, \mu_r(t, Z) \\
&\quad = z'_1\, \widetilde{\mu}_b(z_1, Z(z_1), Z'(z_1))\, \Lambda_{2,r}(t, Z) + \mu_b(z_1, Z(z_1))\, \widetilde{\Lambda}_{2,r}(t, Z, Z').
\end{aligned}$$

Replacing (5.16) in the above expression

$$\begin{aligned}
&\frac{1}{n_r}\left(\cos t - \mathcal{A}_r(Z)\, z'_4 - \widetilde{\mathcal{A}}_r(Z, Z')\, z_4\right) \Lambda_{2,b}(z_1, Z(z_1)) \\
&\quad = z'_1\left[\widetilde{\mu}_b(z_1, Z(z_1), Z'(z_1)) \Lambda_{2,r}(t, Z) - \widetilde{\Lambda}_{2,b}(z_1, Z(z_1), Z'(z_1)) \mu_r(t, Z)\right] \\
&\qquad\qquad + \mu_b(z_1, Z(z_1)) \widetilde{\Lambda}_{2,r}(t, Z, Z').
\end{aligned}$$

From (5.7) and (5.13), $\mathcal{A}_r(Z)\Lambda_{2,b}(z_1, Z(z_1)) < 0$. Then solving the last equation in $z_4'$ yields

(5.29)

$$z_4' = -\left(\frac{z_1'\left[\widetilde{\mu}_b(z_1, Z(z_1), Z'(z_1))\Lambda_{2,r}(t,Z) - \widetilde{\Lambda}_{2,b}(z_1, Z(z_1), Z'(z_1))\mu_r(t,Z)\right] + \mu_b(z_1, Z(z_1))\widetilde{\Lambda}_{2,r}(t,Z,Z')}{\frac{1}{n_r}\mathcal{A}_r(Z)\Lambda_{2,b}(z_1, Z(z_1))}\right.$$
$$\left. - \frac{\frac{1}{n_r}\left(\cos t - \widetilde{\mathcal{A}}_r(Z,Z')\, z_4\right)\Lambda_{2,b}(z_1, Z(z_1))}{\frac{1}{n_r}\mathcal{A}_r(Z)\Lambda_{2,b}(z_1, Z(z_1))}\right)$$

$$:= h_4(t; Z(t), Z(z_1(t)); Z'(t), Z'(z_1(t))),$$

so

(5.30)

$$h_4(t;\zeta^0,\zeta^1;\xi^0,\xi^1)$$
$$= -\left(\frac{\xi_1^0\left[\widetilde{\mu}_b(\zeta_1^0,\zeta^1,\xi^1)\Lambda_{2,r}(t,\zeta^0) - \widetilde{\Lambda}_{2,b}(\zeta_1^0,\zeta^1,\xi^1)\mu_r(t,\zeta^0)\right] + \mu_b(\zeta_1^0,\zeta^1)\widetilde{\Lambda}_{2,r}(t,\zeta^0,\xi^0)}{\frac{1}{n_r}\mathcal{A}_r(\zeta^0)\Lambda_{2,b}(\zeta_1^0,\zeta^1)}\right.$$
$$\left. - \frac{\frac{1}{n_r}\left(\cos t - \widetilde{\mathcal{A}}_r(\zeta^0,\xi^0)\zeta_4^0\right)\Lambda_{2,b}(\zeta_1^0,\zeta^1)}{\frac{1}{n_r}\mathcal{A}_r(\zeta^0)\Lambda_{2,b}(\zeta_1^0,\zeta^1)}\right).$$

$h_4$ is smooth in a neighborhood of $\mathcal{Z}$, since $\mathcal{A}_r(\mathbf{0})\Lambda_{2,b}(0,\mathbf{0}) < 0$, and as shown before all the functions appearing in the expression for $h_4$ are smooth in a neighborhood of $\mathcal{Z}$ from the comments after (5.14) and (5.22).

**Calculation of $h_5$.** We have $v_2(t) = -(\tan t)\, v_1(t)$. Differentiating twice we get

$$v_2''(t) = -(\tan t)\, v_1''(t) - 2\,\frac{v_1'(t)}{\cos^2 t} - 2\frac{\sin t}{\cos^3 t}\, v_1(t).$$

Then

$$z_5'(t) = -(\tan t)\, z_4'(t) - \frac{2}{\cos^2 t}\, z_4(t) - \frac{2\sin t}{\cos^3 t}\,(z_2(t) - \rho_0) \tag{5.31}$$
$$:= h_5(t; Z(t), Z(z_1(t)); Z'(t), Z'(z_1(t))),$$

and so

$$h_5(t;\zeta^0,\zeta^1;\xi^0,\xi^1) = -(\tan t)\,\xi_4^0 - \frac{2}{\cos^2 t}\,\zeta_4^0 - \frac{2\sin t}{\cos^3 t}\,(\zeta_2^0 - \rho_0). \tag{5.32}$$

Since $t \in D \subset (-\pi/2, \pi/2)$, then $h_5$ is smooth in $D \times \mathbb{R}^{20}$.

The proof of Theorem 5.5 is then complete. □

**5.3. Solutions of** (5.4)**yield local solutions to the optical problem.** In this section, we show how to obtain a local solution to Problem B by solving the system of functional differential equations (5.4).

THEOREM 5.7. *Let $\rho_0, d_0 > 0$ be given, and suppose that $H$ is the corresponding map defined in Theorem 5.5. Assume $P = (p_1, \cdots, p_n)$ is a solution to the system*

$$P = H(0; \mathbf{0}, \mathbf{0}; P, P),$$

*such that $H$ is smooth in a neighborhood of $\mathcal{P} := (0; \mathbf{0}, \mathbf{0}; P, P)$, and*

$$0 < |p_1| \leq 1. \tag{5.33}$$

*Let $Z(t) = (z_1(t), \cdots, z_5(t))$ be a $C^1$ solution to the system (5.4) in some open interval $I$ containing 0, with $Z'(0) = P$ and*

$$|z_1(t)| \leq |t|. \tag{5.34}$$

*Define*

$$\rho(t) = -\frac{z_2(t) - \rho_0}{\cos t}, \tag{5.35}$$

*and $\varphi(t) = z_1(t)$.*

*Then there is $\delta > 0$ sufficiently small, so that $\varphi : [-\delta, \delta] \to [-\delta, \delta]$ and*

$$f_r(t) = f_b(\varphi(t)).$$

*Here, $f_r(t) = \rho(t)x(t) + d_r(t)m_r(t)$, and $f_b(t) = \rho(t)x(t) + d_b(t)m_b(t)$ where*

$$d_r(t) = \frac{C_r - \rho(t)(1 - \cos t)}{n_r - e \cdot m_r(t)}, \qquad d_b(t) = \frac{C_b - \rho(t)(1 - \cos t)}{n_b - e \cdot m_b(t)}$$

*with $C_r = (n_r - 1)\, d_0$, $C_b = (n_b - 1)\, d_0$; $m_r(t)$ and $m_b(t)$ are the refracted directions of the rays $x(t)$ by the curve $\rho(t)x(t)$ corresponding to each color $r$ and $b$. In addition, for $t \in [-\delta, \delta]$, $f_r$ and $f_b$ have normal vectors for every $t$ and*

$$\rho(t) > 0, \quad d_r(t), d_b(t) > 0, \quad m_r(t) \cdot e \geq 1/n_r, \quad m_b(t) \cdot e \geq 1/n_b. \tag{5.36}$$

Notice that (5.36) implies the lens defined with $\rho(t)x(t)$, $f_r$ and $f_b$ is well defined, and moreover total internal reflection is avoided.

PROOF. We obtain the theorem by proving a series of steps.

STEP 1. If $\rho$ is from (5.35), then

$$z_3(t) = \rho(t) \sin t. \tag{5.37}$$

PROOF. Since $z_2(0) = 0$, from (5.35)

$$\rho(0) = \rho_0. \tag{5.38}$$

Since $Z$ is a solution to (5.4), $z_2'(t) = z_4(t)$ by the definition of $h_2$ in (5.26). Hence and from the definition of $h_5$ in (5.32),

$$\begin{aligned} z_5'(t) &= -(\tan t)\, z_4'(t) - \frac{2}{\cos^2 t}\, z_4(t) - \frac{2 \sin t}{\cos^3 t}\, (z_2(t) - \rho_0) \\ &= -(\tan t)\, z_2''(t) - \frac{2}{\cos^2 t}\, z_2'(t) - \frac{2 \sin t}{\cos^3 t}\, (z_2(t) - \rho_0). \end{aligned}$$

By (5.35), we have $z_2(t) = -\rho(t) \cos t + \rho_0$, then

$$\begin{aligned} z_2'(t) &= -\rho'(t) \cos t + \rho(t) \sin t \\ z_2''(t) &= -\rho''(t) \cos t + 2\rho'(t) \sin t + \rho(t) \cos t. \end{aligned}$$

Replacing in the formula for $z_5'$ we get

$$\begin{aligned} z_5'(t) &= (\tan t)(\rho''(t) \cos t - 2\rho'(t) \sin t - \rho(t) \cos t) + \frac{2}{\cos^2 t}(\rho'(t) \cos t - \rho(t) \sin t) \\ &\quad + \frac{2 \sin t}{\cos^3 t} \rho(t) \cos t = \rho''(t) \sin t + 2\rho'(t) \cos t - \rho(t) \sin t \\ &= \frac{d^2}{dt^2} \left[\rho(t) \sin t\right]. \end{aligned}$$

Hence $z_5(t) = \dfrac{d}{dt}\left[\rho(t)\sin t\right] + c = \rho'(t)\sin t + \rho(t)\cos t + c$. Since $z_5(0) = 0$, then by (5.38) $c = -\rho_0$, and therefore

$$z_5(t) = \rho'(t)\sin t + \rho(t)\cos t - \rho_0. \tag{5.39}$$

By the definition of $h_3$ in (5.28), we have

$$z_3'(t) = z_5(t) + \rho_0 = \rho'(t)\sin t + \rho(t)\cos t = \frac{d}{dt}\left[\rho(t)\sin t\right],$$

and since $z_3(0) = 0$, we conclude (5.37). □

STEP 2. For each $t \in I$

$$F_{1r}(t, Z(t)) = F_{1b}(z_1(t), Z(z_1(t))),$$

with $F_{1r}, F_{1b}$ from (5.14).

PROOF. From the definition of $h_1$ in (5.24)

$$\widetilde{F}_{1r}(t, Z(t), Z'(t)) = z_1'(t)\widetilde{F}_{1b}(z_1(t), Z(z_1(t)), Z'(z_1(t))).$$

Hence by Remark 5.6, integrating the above equality yields

$$F_{1r}(t, Z(t)) = F_{1b}(z_1(t), Z(z_1(t))) + c.$$

Since $Z(0) = \mathbf{0}$, then from the formulas of $F_{1r}$ and $\mu_r$ in (5.14) we get $F_{1r}(0, \mathbf{0}) = F_{1b}(0, \mathbf{0}) = 0$. Hence $c = 0$ and Step 2 follows. □

STEP 3. There is $\delta > 0$ small so that (5.36) holds.

PROOF. Since $\rho(0) = \rho_0 > 0$, by continuity there is $\delta > 0$ with $[-\delta, \delta] \subseteq I$ so that $\rho(t)$ given by (5.35) is positive for $t \in [-\delta, \delta]$. Rays with colors r and b emitted from the origin with direction $x(t) = (\sin t, \cos t)$ are refracted at $\rho(t)x(t)$ into the directions $m_r(t)$, and $m_b(t)$. For the upper faces $f_r$ and $f_b$ to be able to refract the rays $m_r$ and $m_b$ into $e$, they need to have a normal vector for each $t$, $d_r(t)$ and $d_b(t)$ must be positive for $t \in [-\delta, \delta]$, and $m_r$ and $m_b$ must satisfy the conditions $m_r(t) \cdot e \geq \dfrac{1}{n_r}$, $m_b(t) \cdot e \geq \dfrac{1}{n_b}$ to avoid total reflection. Notice that from (5.35)

$$\rho'(t) = -\frac{z_2'(t)\cos t + (z_2(t) - \rho_0)\sin t}{\cos^2 t}.$$

From the definition of $h_2$ in (5.26), and since $z_4(0) = 0$ we conclude that $\rho'(0) = 0$. Hence from Lemma 5.3, the normal to $\rho$ at $\rho(0)(0,1)$ is $\nu(0) = (0,1)$. Since $x(0) = \nu(0) = (0,1)$ we obtain from Snell's law that

$$m_r(0) = m_b(0) = (0,1). \tag{5.40}$$

Thus, $e \cdot m_r(0) = e \cdot m_b(0) = 1$, and so

$$d_r(0) = \frac{C_r}{n_r - 1} = d_0 = \frac{C_b}{n_b - 1} = d_b(0) > 0. \tag{5.41}$$

So choosing $\delta$ small we get $d_r(t), d_b(t) > 0$, $e \cdot m_r(t) \geq \dfrac{1}{n_r}$ and $e \cdot m_b(t) \geq \dfrac{1}{n_b}$ in $[-\delta, \delta]$. Therefore, (5.36) holds for $\delta > 0$ sufficiently small. The fact that $f_r$ and $f_b$ have a normals for every $t$ will be proved in Step 5. □

STEP 4. For each $t \in [-\delta, \delta]$

$$f_r(t) = (F_{1r}(t, Z(t)), F_{2r}(t, Z(t))),$$

and similarly $f_b(t) = (F_{1b}(t, Z(t)), F_{2b}(t, Z(t)))$.

PROOF. We first show that

$$m_r(t) = (\mu_r(t, Z(t)), \tau_r(t, Z(t))), \tag{5.42}$$

and similarly for $m_b$. Since the direction $x(t) = (\sin t, \cos t)$ is refracted by $\rho$ into the unit direction $m_r(t)$, by (5.5)

$$m_r(t) = \frac{1}{n_r}\left[x(t) - A_r(v(t), v'(t))\, v'(t)\right]$$

with $A_r$ given in (5.6) and $v(t) = (-\rho(t)\cos t, \rho(t)\sin t)$. From (5.35) and (5.37),

$$v(t) = (z_2(t) - \rho_0, z_3(t)), \qquad v'(t) = (z_2'(t), z_3'(t)).$$

However, by the definition of $h_2$ in (5.26) and the definition of $h_3$ in (5.28), we have $z_2' = z_4$ and $z_3' = z_5 + \rho_0$, then $v'(t) = (z_4, z_5 + \rho_0)$. Hence (5.6) becomes

$$A_r(v(t), v'(t)) = \frac{1 - n_r^2}{|(z_2(t) - \rho_0, z_3(t))| + \sqrt{|(z_2(t) - \rho_0, z_3(t))|^2 + (n_r^2 - 1)|(z_4(t), z_5(t) + \rho_0)|^2}},$$

and so by (5.14), $A_r(v(t), v'(t)) = \mathcal{A}_r(Z(t))$.

We conclude that

$$m_r(t) = \frac{1}{n_r}\left[(\sin t, \cos t) - \mathcal{A}_r(Z(t))\,(z_4(t), z_5(t) + \rho_0)\right],$$

and hence once again from (5.14), the identity (5.42) follows.

We now show that $d_r(t) = D_r(t, Z(t))$. In fact, by definition of $v$ we have

$$\rho(t) = |v(t)| = |(z_2(t) - \rho_0, z_3(t))|.$$

Then by (5.42), (5.35) and the formula for $D_r$ in (5.14) we get

$$d_r(t) = \frac{C_r - \rho(t) + \rho(t)\cos t}{n_r - m_{2r}(t)} = \frac{C_r - |(z_2(t) - \rho_0, z_3(t))| - z_2(t) + \rho_0}{n_r - \tau_r(t, Z(t))} = D_r(t, Z(t)).$$

Hence from (5.35), (5.37), (5.42), and the formulas of $F_{1r}, F_{2r}$ in (5.14), we conclude

$$\begin{aligned} f_r(t) &= \rho(t)(\sin t, \cos t) + d_r(t) m_r(t) \\ &= (z_3(t), -z_2(t) + \rho_0) + D_r(t, Z(t))\,(\mu_r(t, Z(t)), \tau_r(t, Z(t))) \\ &= (F_{1r}(t, Z(t)), F_{2r}(t, Z(t))). \end{aligned}$$

A similar result holds for $f_b$. □

By assumption $|z_1(t)| \le |t|$ for $t \in I$, then $z_1(t) \in [-\delta, \delta]$ for every $t \in [-\delta, \delta]$.

STEP 5. For $\delta$ small, $f_{1r}'(t) \neq 0$ and $f_{1b}'(t) \neq 0$, and hence $f_r$ and $f_b$ have normal vectors for each $t \in [-\delta, \delta]$.

PROOF. By continuity of $f_{1r}'$ and $f_{1b}'$, it is enough to show that $f_{1r}'(0) \neq 0$ and $f_{1b}'(0) \neq 0$. Since $H$ is well defined at $\mathcal{P}$, then from the definition of $h_1$ in (5.24), $\widetilde{F}_{1b}(0, \mathbf{0}, P) \neq 0$. We have $Z'(0) = P$, then from Remark 5.6 and Step 4

$$f_{1b}'(0) = \widetilde{F}_{1b}(0, \mathbf{0}, Z'(0)) = \widetilde{F}_{1b}(0, \mathbf{0}, P) \neq 0.$$

We next prove that $f_{1r}'(0) \neq 0$. From Steps 2 and 4, $f_{1r}(t) = f_{1b}(z_1(t))$. Differentiating and letting $t = 0$ yields

$$f_{1r}'(0) = z_1'(0) f_{1b}'(0).$$

Since $f_{1b}'(0) \neq 0$ and from (5.33) $z_1'(0) = p_1 \neq 0$, we obtain $f_{1r}'(0) \neq 0$. □

STEP 6. The vectors $m_r(t) - \dfrac{1}{n_r} e$ and $m_b(z_1(t)) - \dfrac{1}{n_b} e$ are colinear for every $t \in [-\delta, \delta]$.

PROOF. Since $z_4'(t) = h_4(t; Z(t), Z(z_1(t)); Z'(t), Z'(z_1(t)))$, from (5.30) it follows that

$$\begin{aligned}
\frac{1}{n_r}&\left(\cos t - \mathcal{A}_r(Z(t)) z_4'(t) - \widetilde{\mathcal{A}}_r(Z(t), Z'(t)) z_4(t)\right) \Lambda_{2,b}(z_1(t), Z(z_1(t))) \\
&= z_1'(t) \left[\widetilde{\mu}_b(z_1(t), Z(z_1(t)), Z'(z_1(t))) \Lambda_{2,r}(t, Z(t)) \right. \\
&\quad \left. - \widetilde{\Lambda}_{2,b}(z_1(t), Z(z_1(t)), Z'(z_1(t))) \mu_r(t, Z(t)) \right] \\
&\quad + \mu_b(z_1(t), Z(z_1(t))) \widetilde{\Lambda}_{2,r}(t, Z(t), Z'(t)).
\end{aligned}$$

Hence from the formula of $\widetilde{\mu}_r$ in (5.22), we obtain

$$\begin{aligned}
&\widetilde{\mu}_r(t, Z(t), Z'(t))\, \Lambda_{2,b}(z_1(t), Z(z_1(t))) + z_1'(t)\, \widetilde{\Lambda}_{2,b}(z_1(t), Z(z_1(t)))\, \mu_r(t, Z(t)) \\
&\quad = z_1'(t)\, \widetilde{\mu}_b(z_1(t), Z(z_1(t))) \Lambda_{2,r}(t, Z(t)) + \mu_b(z_1(t), Z(z_1(t))) \widetilde{\Lambda}_{2,r}(t, Z(t), Z'(t)).
\end{aligned}$$

Integrating the resulting identity using Remark 5.6, and that $\mu_r(0, \mathbf{0}) = \mu_b(0, \mathbf{0}) = 0$ from (5.14), we obtain

$$\mu_r(t, Z(t))\, \Lambda_{2,b}\left(z_1(t), Z(z_1(t))\right) = \mu_b\left(z_1(t), Z(z_1(t))\right)\, \Lambda_{2,r}(t, Z(t)). \tag{5.43}$$

On the other hand, from (5.42)

$$\mu_r(t, Z(t))^2 + \tau_r(t, Z(t))^2 = 1, \tag{5.44}$$

then from (5.14), $\Lambda_{2,r}$ can be written as follows

$$\begin{aligned}
\Lambda_{2,r}(t, Z(t)) &= \sqrt{1 + \frac{1}{n_r^2} - \frac{2}{n_r} \tau_r(t, Z(t))} = \sqrt{1 - \tau_r(t, Z(t))^2 + \left(\frac{1}{n_r} - \tau_r(t, Z(t))\right)^2} \\
&= \sqrt{\mu_r(t, Z(t))^2 + \left(\frac{1}{n_r} - \tau_r(t, Z(t))\right)^2},
\end{aligned}$$

and similarly for $\Lambda_{2,b}$. Squaring (5.43) and using the above identity for $n_r$ and $n_b$ we get

$$\begin{aligned}
&\mu_r(t, Z(t))^2 \left[\mu_b(z_1(t), Z(z_1(t)))^2 + \left(\frac{1}{n_b} - \tau_b(z_1(t), Z(z_1(t))\right)^2\right] \\
&= \mu_b(z_1(t), Z(z_1(t)))^2 \left[\mu_r(t, Z(t))^2 + \left(\frac{1}{n_r} - \tau_r(t, Z(t))\right)^2\right].
\end{aligned}$$

Hence

$$\mu_r(t, Z(t))^2 \left(\frac{1}{n_b} - \tau_b(z_1(t), Z(z_1(t)))\right)^2 = \mu_b(z_1(t), Z(z_1(t)))^2 \left(\frac{1}{n_r} - \tau_r(t, Z(t))\right)^2,$$

and taking square roots

$$\left|\mu_r(t,Z(t))\left(\frac{1}{n_b}-\tau_b(z_1(t),Z(z_1(t)))\right)\right|=\left|\mu_b(z_1(t),Z(z_1(t)))\left(\frac{1}{n_r}-\tau_r(t,Z(t))\right)\right|.$$

From (5.42), since $\delta$ is chosen so that (5.36) holds in $[-\delta,\delta]$, and $z_1(t)\in[-\delta,\delta]$, we have that $1/n_b-\tau_b(z_1(t),Z(z_1(t)))=1/n_b-e\cdot m_b(z_1(t))\leq 0$, and $1/n_r-\tau_r(t,Z(t))=1/n_r-e\cdot m_r(t)\leq 0$. Moreover, since the functions $\Lambda_{2,b}$ and $\Lambda_{2,r}$ are both positive, then by (5.43) $\mu_r(t,Z(t))$ and $\mu_b(z_1(t),Z(z_1(t)))$ have the same sign obtaining

$$\mu_r(t,Z(t))\left(\tau_b(z_1(t),Z(z_1(t)))-\frac{1}{n_b}\right)=\mu_b(z_1(t),Z(z_1(t)))\left(\tau_r(t,Z(t))-\frac{1}{n_r}\right).$$

We conclude that the vectors

$$\left(\mu_r(t,Z(t)),\tau_r(t,Z(t))-\frac{1}{n_r}\right),\quad\left(\mu_b\left(z_1(t),Z(z_1(t))\right),\tau_b\left(z_1(t),Z(z_1(t))\right)-\frac{1}{n_b}\right)$$

are colinear and the claim follows from (5.42). □

By Steps 2 and 4, we obtain that $f_{1r}(t)=f_{1b}(z_1(t))$, so to conclude the proof of Theorem 5.7, it remains to show the following.

STEP 7. We have for $|t|\leq\delta$ that

$$f_{2r}(t)=f_{2b}(z_1(t)).$$

PROOF. From Step 5, for $|t|\leq\delta$, $f_r$ and $f_b$ have normals at every point. From (5.36) and the definitions of $f_r$ and $f_b$, $m_r(t)$ and $m_b(t)$ are refracted into the direction $e$ at $f_r(t)$ and $f_b(t)$, for each $t$, respectively. So the Snell law at $f_r(t)$ and $f_b(t)$ implies that $m_r(t)-\dfrac{1}{n_r}e$ is orthogonal to the tangent vector $f_r'(t)$, and $m_b(z_1(t))-\dfrac{1}{n_b}e$ is orthogonal to $f_b'(z_1(t))$. Hence by Step 6, $f_r'(t)$ and $f_b'(z_1(t))$ are parallel. From Remark 5.6 and Steps 2 and 4 we also obtain that $f_{1r}'(t)=z_1'(t)f_{1b}'(z_1(t))$. From Step 5 and assumption (5.34), $f_{1r}'(t)\neq 0$ and $f_{1b}'(z_1(t))\neq 0$ for all $t\in[-\delta,\delta]$, then

$$f_{2r}'(t)=z_1'(t)f_{2b}'(z_1(t)).$$

Integrating the last identity, we obtain $f_{2r}(t)=f_{2b}(z_1(t))+c$. On the other hand, $z_1(0)=0$, $f_{2r}(0)=\rho(0)+d_r(0)m_{2r}(0)$, and $f_{2b}(0)=\rho(0)+d_b(0)m_{2b}(0)$. Hence from (5.40) and (5.41), $f_{2r}(0)=f_{2b}(0)$ and so $c=0$, and Step 7 follows. □

This completes the proof Theorem 5.7. □

REMARK 5.8. Notice that $f_b$ has no self intersections in the interval $[-\delta,\delta]$. Because if there would exist $t_1,t_2\in[-\delta,\delta]$ such that $f_b(t_1)=f_b(t_2)$, then by Rolle's theorem applied to $f_{1b}$ we would get $f_{1b}'(t_0)=0$ for some $t_0\in[t_1,t_2]$, a contradiction with Step 5. The issue of self-intersections in the monochromatic case is discussed in detail in [**GS18**]. This implies that $f_b$ is injective, and similarly $f_r$ is injective. We also deduce that $\varphi$ is injective: in fact, if $\varphi(t_1)=\varphi(t_2)$ then $f_r(t_1)=f_b(\varphi(t_1))=f_b(\varphi(t_2))=f_r(t_2)$ and so $t_1=t_2$.

**5.4. On the solvability of the algebraic system** (4.3). In this section, we analyze for what values of $\rho_0$, $d_0$ the algebraic system $P = H(0;\mathbf{0},\mathbf{0};P,P)$ has a solution, where $H$ is given in Theorem 5.5. This analysis will be used to apply Theorem 4.1, and to decide when Problem B has a local solution. Denote

$$k_0 = \frac{\rho_0}{d_0}, \qquad \Delta_r = \frac{n_r}{n_r - 1}, \qquad \Delta_b = \frac{n_b}{n_b - 1}.$$

THEOREM 5.9. *The algebraic system $P = H(0;\mathbf{0},\mathbf{0};P,P)$ has a solution if and only if $k_0 \leq \dfrac{(\Delta_r - \Delta_b)^2}{4\Delta_r\Delta_b}$. In case of equality, the system has only one solution $P$ with $|p_1| = 1$, and in case of strict inequality the system has two solutions $P$ and $P'$ with $0 < |p_1| < 1$ and $|p_1'| > 1$.*

PROOF. Recall that $H = H\left(t;\zeta^0,\zeta^1;\xi^0,\xi^1\right)$. Suppose $P = (p_1,\cdots,p_5)$ solves $P = H\left(0;\mathbf{0},\mathbf{0};P,P\right)$. Then from the definitions of $h_2$ in (5.26), of $h_3$ in (5.28), and of $h_5$ in (5.32) we get

(5.45)
$$p_2 = h_2\left(0;\mathbf{0},\mathbf{0};P,P\right) = 0, \quad p_3 = h_3\left(0;\mathbf{0},\mathbf{0};P,P\right) = \rho_0, \quad p_5 = h_5\left(0;\mathbf{0},\mathbf{0};P,P\right) = 0.$$

Then, from (5.14) and (5.22)

(5.46)
$$\begin{cases} \mathcal{A}_r\left(\mathbf{0}\right) = -\dfrac{n_r}{\Delta_r\,\rho_0}, \qquad \mathcal{A}_b\left(\mathbf{0}\right) = -\dfrac{n_b}{\Delta_b\,\rho_0} \\ \mu_r\left(0,\mathbf{0}\right) = \mu_b\left(0,\mathbf{0}\right) = 0 \\ \tau_r\left(0,\mathbf{0}\right) = \tau_b\left(0,\mathbf{0}\right) = 1 \\ D_r\left(0,\mathbf{0}\right) = D_b\left(0,\mathbf{0}\right) = d_0 \\ F_{1r}\left(0,\mathbf{0}\right) = F_{1b}\left(0,\mathbf{0}\right) = 0, \qquad F_{2r}\left(0,\mathbf{0}\right) = F_{2b}\left(0,\mathbf{0}\right) = \rho_0 + d_0 \\ \Lambda_{2,r}\left(0,\mathbf{0}\right) = \dfrac{1}{\Delta_r}, \qquad \Lambda_{2,b}\left(0,\mathbf{0}\right) = \dfrac{1}{\Delta_b} \\ \widetilde{\mathcal{A}}_r\left(\mathbf{0},P\right) = \widetilde{\mathcal{A}}_b\left(\mathbf{0},P\right) = 0 \\ \widetilde{\mu}_r\left(0,\mathbf{0},P\right) = \dfrac{1}{\Delta_r}\left(\dfrac{\Delta_r}{n_r} + \dfrac{p_4}{\rho_0}\right), \qquad \widetilde{\mu}_b\left(0,\mathbf{0},P\right) = \dfrac{1}{\Delta_b}\left(\dfrac{\Delta_b}{n_b} + \dfrac{p_4}{\rho_0}\right) \\ \widetilde{\tau}_r\left(0,\mathbf{0},P\right) = \widetilde{\tau}_{n_r}\left(0,\mathbf{0},P\right) = 0 \\ \widetilde{D}_r\left(0,\mathbf{0},P\right) = \widetilde{D}_b\left(0,\mathbf{0},P\right) = 0 \\ \widetilde{F}_{1r}\left(0,\mathbf{0},P\right) = \rho_0 + \dfrac{d_0}{\Delta_r}\left(\dfrac{\Delta_r}{n_r} + \dfrac{p_4}{\rho_0}\right), \qquad \widetilde{F}_{1b}\left(0,\mathbf{0},P\right) = \rho_0 + \dfrac{d_0}{\Delta_b}\left(\dfrac{\Delta_b}{n_b} + \dfrac{p_4}{\rho_0}\right) \\ \widetilde{F}_{2r}\left(0,\mathbf{0},P\right) = \widetilde{F}_{2b}\left(0,\mathbf{0},P\right) = 0 \\ \widetilde{\Lambda}_{2,r}\left(0,\mathbf{0},P\right) = \widetilde{\Lambda}_{2,b}\left(0,\mathbf{0},P\right) = 0. \end{cases}$$

Hence by the definition of $h_1$ in (5.24), and since $h_1\left(0;\mathbf{0},\mathbf{0};P,P\right)$ is well defined, we get that $\widetilde{F}_{1b}\left(0,\mathbf{0},P\right) \neq 0$. That is, from (5.46), $\rho_0 + \dfrac{d_0}{\Delta_b}\left(\dfrac{\Delta_b}{n_b} + \dfrac{p_4}{\rho_0}\right) \neq 0$ and we have

$$p_1 = h_1\left(0;\mathbf{0},\mathbf{0};P,P\right) = \frac{\widetilde{F}_{1r}\left(0,\mathbf{0},P\right)}{\widetilde{F}_{1b}\left(0,\mathbf{0},P\right)} = \frac{\rho_0 + \dfrac{d_0}{\Delta_r}\left(\dfrac{\Delta_r}{n_r} + \dfrac{p_4}{\rho_0}\right)}{\rho_0 + \dfrac{d_0}{\Delta_b}\left(\dfrac{\Delta_b}{n_b} + \dfrac{p_4}{\rho_0}\right)}. \tag{5.47}$$

From the definition of $h_4$ in (5.30)

(5.48)

$$p_4 = h_4\left(0; \mathbf{0}, \mathbf{0}; P, P\right) = \frac{\dfrac{p_1}{\Delta_b\Delta_r}\left(\dfrac{\Delta_b}{n_b}+\dfrac{p_4}{\rho_0}\right) - \dfrac{1}{n_r\Delta_b}}{\dfrac{1}{\Delta_r\Delta_b\rho_0}} = \left[p_1\left(\frac{\Delta_b}{n_b}+\frac{p_4}{\rho_0}\right) - \frac{\Delta_r}{n_r}\right]\rho_0.$$

To solve (5.48) in $p_1$, first notice that $\dfrac{\Delta_b}{n_b}+\dfrac{p_4}{\rho_0} \neq 0$. Otherwise, from (5.48) we would obtain $\dfrac{p_4}{\rho_0} = -\dfrac{\Delta_r}{n_r}$, and so $\dfrac{\Delta_r}{n_r} = \dfrac{\Delta_b}{n_b}$, hence $n_r = n_b$, a contradiction since $n_b > n_r$. Then (5.48) yields

$$p_1 = \frac{\dfrac{\Delta_r}{n_r}+\dfrac{p_4}{\rho_0}}{\dfrac{\Delta_b}{n_b}+\dfrac{p_4}{\rho_0}}. \tag{5.49}$$

Hence from (5.47) and (5.49) we get that

$$\frac{\dfrac{\Delta_r}{n_r}+\dfrac{p_4}{\rho_0}}{\dfrac{\Delta_b}{n_b}+\dfrac{p_4}{\rho_0}} = \frac{\rho_0 + \dfrac{d_0}{\Delta_r}\left(\dfrac{\Delta_r}{n_r}+\dfrac{p_4}{\rho_0}\right)}{\rho_0 + \dfrac{d_0}{\Delta_b}\left(\dfrac{\Delta_b}{n_b}+\dfrac{p_4}{\rho_0}\right)}$$

Then

(5.50)

$$\left(\frac{\Delta_r}{n_r}+\frac{p_4}{\rho_0}\right)\left(\rho_0 + \frac{d_0}{\Delta_b}\left(\frac{\Delta_b}{n_b}+\frac{p_4}{\rho_0}\right)\right) = \left(\frac{\Delta_b}{n_b}+\frac{p_4}{\rho_0}\right)\left(\rho_0 + \frac{d_0}{\Delta_r}\left(\frac{\Delta_r}{n_r}+\frac{p_4}{\rho_0}\right)\right).$$

Simplifying,

$$\rho_0\left(\frac{\Delta_r}{n_r}-\frac{\Delta_b}{n_b}\right) = \left(\frac{\Delta_r}{n_r}+\frac{p_4}{\rho_0}\right)\left(\frac{\Delta_b}{n_b}+\frac{p_4}{\rho_0}\right)\left(\frac{1}{\Delta_r}-\frac{1}{\Delta_b}\right)d_0.$$

Notice that

$$\frac{\Delta_r}{n_r}-\frac{\Delta_b}{n_b} = \frac{1}{n_r-1}-\frac{1}{n_b-1} = -1+\Delta_r+1-\Delta_b = \Delta_r-\Delta_b. \tag{5.51}$$

Then dividing by $\Delta_r - \Delta_b$[3], and using the notation $k_0 = \dfrac{\rho_0}{d_0}$ yields

$$k_0\Delta_r\Delta_b + \left(\frac{\Delta_b}{n_b}+\frac{p_4}{\rho_0}\right)\left(\frac{\Delta_r}{n_r}+\frac{p_4}{\rho_0}\right) = 0. \tag{5.52}$$

Expanding we obtain that $p_4/\rho_0$ satisfies the following quadratic equation:

$$\left(\frac{p_4}{\rho_0}\right)^2 + \left(\frac{\Delta_r}{n_r}+\frac{\Delta_b}{n_b}\right)\frac{p_4}{\rho_0} + k_0\Delta_r\Delta_b + \frac{\Delta_r\Delta_b}{n_rn_b} = 0. \tag{5.53}$$

The discriminant of (5.53) is

$$\delta = \left(\frac{\Delta_r}{n_r}+\frac{\Delta_b}{n_b}\right)^2 - 4\frac{\Delta_r\Delta_b}{n_rn_b} - 4k_0\Delta_r\Delta_b = \left(\frac{\Delta_r}{n_r}-\frac{\Delta_b}{n_b}\right)^2 - 4k_0\Delta_r\Delta_b, \tag{5.54}$$

[3] $\Delta_r - \Delta_b = \dfrac{n_b-n_r}{(n_r-1)(n_b-1)} > 0$

then (5.51) yields

$$\delta = (\Delta_r - \Delta_b)^2 - 4k_0\Delta_r\Delta_b. \tag{5.55}$$

Hence (5.53) has a real solutions if and only if $k_0 \leq \dfrac{(\Delta_r - \Delta_b)^2}{4\Delta_r\Delta_b}$. Therefore we have proved the necessity part in Theorem 5.9, and if $P$ solves the algebraic system, then

$$p_4 = \frac{-\left(\frac{\Delta_r}{n_r} + \frac{\Delta_b}{n_b}\right) - \sqrt{\delta}}{2}\rho_0, \qquad p_4' = \frac{-\left(\frac{\Delta_r}{n_r} + \frac{\Delta_b}{n_b}\right) + \sqrt{\delta}}{2}\rho_0$$

and by (5.49) and (5.51), the corresponding $p_1$ and $p_1'$ are

$$p_1 = \frac{\Delta_r - \Delta_b - \sqrt{\delta}}{\Delta_b - \Delta_r - \sqrt{\delta}}, \qquad p_1' = \frac{\Delta_r - \Delta_b + \sqrt{\delta}}{\Delta_b - \Delta_r + \sqrt{\delta}}.$$

Therefore if $P = H(0;\mathbf{0},\mathbf{0};P,P)$, then the solutions are

$$P = \left(\frac{\Delta_r - \Delta_b - \sqrt{\delta}}{\Delta_b - \Delta_r - \sqrt{\delta}}, 0, \rho_0, \frac{-\left(\frac{\Delta_r}{n_r} + \frac{\Delta_b}{n_b}\right) - \sqrt{\delta}}{2}\rho_0, 0\right) \tag{5.56}$$

$$P' = \left(\frac{\Delta_r - \Delta_b + \sqrt{\delta}}{\Delta_b - \Delta_r + \sqrt{\delta}}, 0, \rho_0, \frac{-\left(\frac{\Delta_r}{n_r} + \frac{\Delta_b}{n_b}\right) + \sqrt{\delta}}{2}\rho_0, 0\right) \tag{5.57}$$

with $\delta$ given in (5.55).

Let us now prove that if $k_0 \leq \dfrac{(\Delta_r - \Delta_b)^2}{4\Delta_r\Delta_b}$, then the system $P = H(0;\mathbf{0},\mathbf{0};P,P)$ is solvable. In fact, from the assumption on $k_0$ there is $p_4$ solving (5.53). We claim that this implies $\rho_0 + \dfrac{d_0}{\Delta_b}\left(\dfrac{\Delta_b}{n_b} + \dfrac{p_4}{\rho_0}\right) \neq 0$. Assume otherwise, since (5.53) is equivalent to (5.50) and $\rho_0 > 0$ then $p_4$ solves the system

$$\rho_0 + \frac{d_0}{\Delta_b}\left(\frac{\Delta_b}{n_b} + \frac{p_4}{\rho_0}\right) = 0$$
$$\rho_0 + \frac{d_0}{\Delta_r}\left(\frac{\Delta_r}{n_r} + \frac{p_4}{\rho_0}\right) = 0$$

Subtracting both identities we get

$$d_0\left(\frac{1}{n_b} - \frac{1}{n_r}\right) + \frac{p_4 d_0}{\rho_0}\left(\frac{1}{\Delta_b} - \frac{1}{\Delta_r}\right) = 0.$$

Since $\dfrac{1}{\Delta_b} - \dfrac{1}{\Delta_r} = \dfrac{1}{n_r} - \dfrac{1}{n_b}$ and $n_r \neq n_b$, dividing by $d_0\left(\dfrac{1}{n_b} - \dfrac{1}{n_r}\right)$ we get $1 - \dfrac{p_4}{\rho_0} = 0$, and then $p_4 = \rho_0$. Replacing in the first equation of the system yields $\rho_0 = -\dfrac{d_0}{\Delta_b}\left(\dfrac{\Delta_b}{n_b} + 1\right)$, and since $\rho_0, d_0 > 0$, we get a contradiction. Now let $p_1$ be the corresponding value to this $p_4$ given by (5.47), and $p_2, p_3, p_5$ from (5.45); and $P$ be the point with these coordinates. Hence by the formula for $\widetilde{F}_{1b}(0,\mathbf{0},P)$ in (5.46), we get that $\widetilde{F}_{1b}(0,\mathbf{0},P) \neq 0$, and therefore $h_1(0;\mathbf{0},\mathbf{0};P,P)$ is well defined, and $P$

solves the algebraic system. Then the possible values of $P$ solving the algebraic system are $P$ and $P'$ given by (5.56) and (5.57).

We now prove the last part of the theorem. If $k_0 = \dfrac{(\Delta_r - \Delta_b)^2}{4\Delta_r\Delta_b}$ then $\delta = 0$ and $P = P' = \left(-1, 0, \rho_0, \dfrac{-\left(\frac{\Delta_r}{n_r} + \frac{\Delta_b}{n_b}\right)}{2}\rho_0, 0\right)$. If $k_0 < \dfrac{(\Delta_r - \Delta_b)^2}{4\Delta_r\Delta_b}$ then the solutions $P \neq P'$. Moreover, since $\Delta_r > \Delta_b$, and from (5.55) $0 < \sqrt{\delta} < \Delta_r - \Delta_b$, Therefore

$$0 < |p_1| = \frac{\left|\Delta_r - \Delta_b - \sqrt{\delta}\right|}{\left|\Delta_b - \Delta_r - \sqrt{\delta}\right|} = \frac{\Delta_r - \Delta_b - \sqrt{\delta}}{\Delta_r - \Delta_b + \sqrt{\delta}} < 1$$

$$|p_1'| = \frac{\left|\Delta_r - \Delta_b + \sqrt{\delta}\right|}{\left|\Delta_b - \Delta_r + \sqrt{\delta}\right|} = \frac{\Delta_r - \Delta_b + \sqrt{\delta}}{\Delta_r - \Delta_b - \sqrt{\delta}} > 1.$$

□

The following corollary gives a necessary condition on $k_0$ for the existence of solutions to Problem $B$.

Corollary 5.10. *If $k_0 > \dfrac{(\Delta_r - \Delta_b)^2}{4\Delta_r\Delta_b}$ then Problem B has no local solutions.*

Proof. If Problem B has a solution then by Theorem 5.5, the vector $Z(t) = (\varphi(t), v_1(t) + \rho_0, v_2(t), v_1'(t), v_2'(t) - \rho_0)$ solves the functional system (5.4) for $t$ in a neighborhood of 0. Plugging $t = 0$ in (5.4) yields

$$(0; \mathbf{0}, \mathbf{0}; Z'(0), Z'(0)) = H(0; \mathbf{0}, \mathbf{0}; Z'(0), Z'(0)).$$

Hence $Z'(0)$ is a solution to the algebraic system $P = H(0; \mathbf{0}, \mathbf{0}; P, P)$, and from Theorem 5.9 we have $k_0 \leq \dfrac{(\Delta_r - \Delta_b)^2}{4\Delta_r\Delta_b}$. □

Corollary 5.11. *If $k_0 \leq \dfrac{(\Delta_r - \Delta_b)^2}{4\Delta_r\Delta_b}$ and a solution $\rho$ and $\varphi$ to Problem B exists, then*

$$\frac{\rho''(0)}{\rho_0} \in (\Delta_b, \Delta_r) \tag{5.58}$$

*In fact,*

$$\frac{\rho''(0)}{\rho_0} = \frac{2 + \left(\frac{\Delta_r}{n_r} + \frac{\Delta_b}{n_b}\right) \pm \sqrt{\delta}}{2}, \tag{5.59}$$

*with $\delta$ given in* (5.55).

Proof. If $\rho$ and $\varphi$ solve Problem B, then from the proof of Corollary 5.10, $Z'(0)$ solves the algebraic system, with $Z(t) = (\varphi(t), v_1(t) + \rho_0, v_2(t), v_1'(t), v_2'(t) - \rho_0)$. Using the proof of Theorem 5.9, it follows that $z_4'(0)$ satisfies (5.52) then

$$\left(\frac{\Delta_r}{n_r} + \frac{z_4'(0)}{\rho_0}\right)\left(\frac{\Delta_b}{n_b} + \frac{z_4'(0)}{\rho_0}\right) = -k_0\Delta_r\Delta_b < 0,$$

and therefore $\dfrac{z_4'(0)}{\rho_0} \in \left(-\dfrac{\Delta_r}{n_r}, -\dfrac{\Delta_b}{n_b}\right)$.

On the other hand, $z_4(t) = v_1'(t) = \rho(t)\sin t - \rho'(t)\cos t$, obtaining $z_4'(0) = \rho_0 - \rho''(0)$, so

$$\frac{\rho''(0)}{\rho_0} = 1 - \frac{z_4'(0)}{\rho_0}. \tag{5.60}$$

We conclude that

$$\frac{\rho''(0)}{\rho_0} \in \left(1 + \frac{\Delta_b}{n_b}, 1 + \frac{\Delta_r}{n_r}\right) = (\Delta_b, \Delta_r).$$

Finally, from (5.56), (5.57), and (5.60), we obtain (5.59). □

REMARK 5.12. The analogue of Problem B for three or more colors has no solution. In fact, if rays, superposition of three colors, are emitted from $O$ and $n_r < n_j < n_b$ are the refractive indices inside the lens for each color, then (5.58) must be satisfied for the pairs $n_r, n_j$ and $n_j$, $n_b$. Hence

$$\frac{\rho''(0)}{\rho(0)} \in (\Delta_j, \Delta_r) \cap (\Delta_b, \Delta_j),$$

which is impossible since the last intersection is empty.

**5.5. Existence of local solutions to** (5.4). In order to prove existence of solutions to the system (5.4), we will apply Theorem 4.1. From Theorem 5.9, we must assume that $k_0 \leq \dfrac{(\Delta_r - \Delta_b)^2}{4\Delta_r \Delta_b}$. In this case, the algebraic system $P = H(0; \mathbf{0}, \mathbf{0}; P, P)$ has a solution given by (5.56) with $|p_1| \leq 1$, and therefore (4.4) holds. Let $\mathcal{P} = (0; \mathbf{0}, \mathbf{0}; P, P)$. To show that $H$ satisfies all the hypotheses of Theorem 4.1, it remains to show there is a norm in $\mathbb{R}^5$ so that $H$ satisfies conditions (i)-(iv) with respect to this norm in a neighborhood $N_\varepsilon(\mathcal{P})$. Our result is as follows.

THEOREM 5.13. *There exists a positive constant* $C(r,b) < \dfrac{(\Delta_r - \Delta_b)^2}{4\Delta_r \Delta_b}$, *depending only on* $n_r$ *and* $n_b$, *such that for* $0 < k_0 < C(r,b)$, *the system* (5.4) *has a unique local solution* $Z(t) = (z_1(t), \cdots, z_5(t))$ *with* $Z'(0) = P$, *and* $|z_1(t)| \leq |t|$, *with* $P$ *given in* (5.56). *Hence from Theorem* 5.7 *and for those values of* $k_0$, *there exist unique* $\rho$ *and* $\varphi$ *solving Problem B.*

PROOF. Since by construction $h_i$ are all smooth in a small neighborhood of $\mathcal{P}$, $H$ is Lipschitz in that neighborhood for any norm. To apply Theorem 4.1, we need to find a norm $\|\cdot\|$ in $\mathbb{R}^5$ and a neighborhood $N_\varepsilon(\mathcal{P})$ so that $|h_1| \leq 1$, and $H$ satisfies the contraction condition (4.8).

To prove the contraction property of $H$, we first calculate the following matrices $\nabla_{\xi^0} H = \left(\dfrac{\partial h_i}{\partial \xi_j^0}\right)_{1 \leq i,j \leq 5}$, and $\nabla_{\xi^1} H = \left(\dfrac{\partial h_i}{\partial \xi_j^1}\right)_{1 \leq i,j \leq 5}$ at the point $\mathcal{P}$.

**Calculation of** $\nabla_{\xi^0} H(\mathcal{P}) = \left(\dfrac{\partial h_i}{\partial \xi_j^0}(\mathcal{P})\right)_{1 \leq i,j \leq 5}$. From (5.26) and (5.28), $h_2$ and $h_3$ do not depend on $\xi^0$, then

$$\partial_{\xi_j^0} h_2(\mathcal{P}) = \partial_{\xi_j^0} h_3(\mathcal{P}) = 0, \qquad 1 \leq j \leq 5.$$

Also from the definition of $h_5$ in (5.32)

$$\partial_{\xi_j^0} h_5(\mathcal{P}) = -\delta_j^5 (\tan t)|_{\mathcal{P}} = 0, \qquad 1 \le j \le 5.$$

We next calculate $\nabla_{\xi^0} h_1(\mathcal{P})$. From (5.24),

$$\nabla_{\xi^0} h_1(\mathcal{P}) = \frac{1}{\widetilde{F}_{1b}(0,\mathbf{0},P)} \nabla_{\xi^0} \widetilde{F}_{1r}(0,\mathbf{0},P).$$

Recall from (5.46)

$$\widetilde{F}_{1b}(0,\mathbf{0},P) = \rho_0 + \frac{d_0}{\Delta_b}\left(\frac{\Delta_b}{n_b} + \frac{p_4}{\rho_0}\right), \qquad \mu_r(0,\mathbf{0}) = 0, \qquad \mathcal{D}_r(0,\mathbf{0}) = d_0.$$

Differentiating $\widetilde{F}_{1r}$ given in (5.22) with respect to $\xi_j^0$, we then get

$$\frac{\partial \widetilde{F}_{1r}}{\partial \xi_j^0}(0,\mathbf{0},P) = \delta_j^3 + d_0 \frac{\partial \widetilde{\mu}_r}{\partial \xi_j^0}(0,\mathbf{0},P), \qquad 1 \le j \le 5.$$

From (5.46), $\mathcal{A}_r(\mathbf{0}) = \dfrac{-n_r}{\Delta_r \rho_0}$, then differentiating $\widetilde{\mu}_r$ in (5.22) with respect to $\xi_j^0$ at $(0,\mathbf{0},P)$ yields

$$\frac{\partial \widetilde{\mu}_r}{\partial \xi_j^0}(0,\mathbf{0},P) = \frac{1}{\Delta_r \rho_0}\delta_j^4. \tag{5.61}$$

Therefore

$$\nabla_{\xi^0} \widetilde{F}_{1r}(0,\mathbf{0},P) = \left(0,0,1,\frac{1}{k_0 \Delta_r},0\right). \tag{5.62}$$

We conclude that

$$\nabla_{\xi^0} h_1(\mathcal{P}) = \frac{1}{\rho_0 + \dfrac{d_0}{\Delta_b}\left(\dfrac{\Delta_b}{n_b} + \dfrac{p_4}{\rho_0}\right)} \left(0,0,1,\frac{1}{k_0 \Delta_r},0\right). \tag{5.63}$$

We next calculate $\nabla_{\xi^0} h_4(\mathcal{P})$. Recall from (5.46) that

$$\mu_b(0,\mathbf{0}) = \mu_r(0,\mathbf{0}) = 0, \quad \widetilde{\mu}_b(0,\mathbf{0},P) = \frac{1}{\Delta_b}\left(\frac{\Delta_b}{n_b} + \frac{p_4}{\rho_0}\right), \quad \Lambda_{2,r}(0,\mathbf{0}) = \frac{1}{\Delta_r},$$
$$\Lambda_{2,b}(0,\mathbf{0}) = \frac{1}{\Delta_b}, \quad \mathcal{A}_r(\mathbf{0}) = -\frac{n_r}{\Delta_r \rho_0}.$$

Then from (5.30) it follows that

$$\frac{\partial h_4}{\partial \xi_j^0} = \frac{\dfrac{1}{\Delta_b}\left(\dfrac{\Delta_b}{n_b} + \dfrac{p_4}{\rho_0}\right)\dfrac{1}{\Delta_r}}{\dfrac{1}{\Delta_r \rho_0}\dfrac{1}{\Delta_b}} \delta_j^1 = \rho_0 \left(\frac{\Delta_b}{n_b} + \frac{p_4}{\rho_0}\right)\delta_j^1, \quad 1 \le j \le 5.$$

Hence

$$\nabla_{\xi^0} h_4(\mathcal{P}) = \rho_0 \left(\frac{\Delta_b}{n_b} + \frac{p_4}{\rho_0}, 0,0,0,0\right). \tag{5.64}$$

We then conclude that
(5.65)

$$\nabla_{\xi^0} H(\mathcal{P}) = \begin{bmatrix} 0 & 0 & \dfrac{1}{\rho_0 + \dfrac{d_0}{\Delta_b}\left(\dfrac{\Delta_b}{n_b} + \dfrac{p_4}{\rho_0}\right)} & \dfrac{\dfrac{1}{k_0\Delta_r}}{\rho_0 + \dfrac{d_0}{\Delta_b}\left(\dfrac{\Delta_b}{n_b} + \dfrac{p_4}{\rho_0}\right)} & 0 \\ 0 & 0 & 0 & 0 & 0 \\ 0 & 0 & 0 & 0 & 0 \\ \rho_0\left(\dfrac{\Delta_b}{n_b} + \dfrac{p_4}{\rho_0}\right) & 0 & 0 & 0 & 0 \\ 0 & 0 & 0 & 0 & 0 \end{bmatrix}.$$

To calculate the spectral radius of the matrix $\nabla_{\xi^0} H(\mathcal{P})$, set

$$a = \frac{1}{\rho_0 + \dfrac{d_0}{\Delta_b}\left(\dfrac{\Delta_b}{n_b} + \dfrac{p_4}{\rho_0}\right)}, \qquad c = \rho_0\left(\frac{\Delta_b}{n_b} + \frac{p_4}{\rho_0}\right). \tag{5.66}$$

The eigenvalues of $\nabla_{\xi^0} H(\mathcal{P})$ are 0 (with multiplicity 3), and $\pm\sqrt{\dfrac{ac}{k_0\Delta_r}}$. Notice that from (5.56), and (5.51)

$$\begin{aligned} \frac{\Delta_r}{n_r} + \frac{p_4}{\rho_0} &= \frac{\Delta_r - \Delta_b - \sqrt{(\Delta_r - \Delta_b)^2 - 4k_0\Delta_r\Delta_b}}{2} \\ &= \frac{2k_0\Delta_r\Delta_b}{\Delta_r - \Delta_b + \sqrt{(\Delta_r - \Delta_b)^2 - 4k_0\Delta_r\Delta_b}} > 0, \end{aligned} \tag{5.67}$$

then by (5.47) and (5.49)

$$\frac{ac}{k_0\Delta_r} = \frac{\dfrac{\rho_0}{k_0\Delta_r}\left(\dfrac{\Delta_b}{n_b} + \dfrac{p_4}{\rho_0}\right)}{\rho_0 + \dfrac{d_0}{\Delta_b}\left(\dfrac{\Delta_b}{n_b} + \dfrac{p_4}{\rho_0}\right)} = \frac{\dfrac{\rho_0}{k_0\Delta_r}\left(\dfrac{\Delta_r}{n_r} + \dfrac{p_4}{\rho_0}\right)}{\rho_0 + \dfrac{d_0}{\Delta_r}\left(\dfrac{\Delta_r}{n_r} + \dfrac{p_4}{\rho_0}\right)} > 0. \tag{5.68}$$

Therefore all the eigenvalues of $\nabla_{\xi^0} H(\mathcal{P})$ are real and the spectral radius of $\nabla_{\xi^0} H(\mathcal{P})$ is $R_{\xi^0} = \sqrt{\dfrac{ac}{k_0\Delta_r}}$. We estimate $R_{\xi^0}$.

From (5.55), (5.67), and (5.68)

$$\frac{ac}{k_0\Delta_r} = \frac{\dfrac{2\Delta_b}{\Delta_r - \Delta_b + \sqrt{\delta}}}{1 + \dfrac{2\Delta_b}{\Delta_r - \Delta_b + \sqrt{\delta}}} = \frac{2\,\Delta_b}{\Delta_r + \Delta_b + \sqrt{\delta}} \leq \frac{2\,\Delta_b}{\Delta_r + \Delta_b} := \delta_0 < 1. \tag{5.69}$$

Since $\sqrt{\delta} < \Delta_r - \Delta_b$, we conclude that

$$\sqrt{\frac{\Delta_b}{\Delta_r}} < R_{\xi^0} = \sqrt{\frac{2\,\Delta_b}{\Delta_r + \Delta_b + \sqrt{\delta}}} \leq \sqrt{\frac{2\Delta_b}{\Delta_b + \Delta_r}} = \sqrt{\delta_0} < 1. \tag{5.70}$$

**Calculation of** $\nabla_{\xi^1} H(\mathcal{P}) = \left( \dfrac{\partial h_i}{\partial \xi^1_j}(\mathcal{P}) \right)_{1\leq i,j\leq 5}$. Notice from (5.26), (5.28) and (5.32) that $h_2$, $h_3$ and $h_5$ do not depend on $\xi^1$ then

$$\nabla_{\xi^1} h_2(\mathcal{P}) = \nabla_{\xi^1} h_3(\mathcal{P}) = \nabla_{\xi^1} h_5(\mathcal{P}) = \mathbf{0}.$$

We calculate $\nabla_{\xi^1} h_1(\mathcal{P})$. From (5.24), and the fact that $p_1 = h_1(\mathcal{P}) = \dfrac{\widetilde{F}_{1r}(0,\mathbf{0},P)}{\widetilde{F}_{1b}(0,\mathbf{0},P)}$ we get

$$\nabla_{\xi^1} h_1(\mathcal{P}) = -\frac{\widetilde{F}_{1r}(0,\mathbf{0},P)}{\widetilde{F}_{1b}(0,\mathbf{0},P)^2}\nabla_{\xi^1}\widetilde{F}_{1b}(0,\mathbf{0},P) = -\frac{p_1}{\widetilde{F}_{1b}(0,\mathbf{0},P)}\nabla_{\xi^1}\widetilde{F}_{1b}(0,\mathbf{0},P).$$

Recall from (5.46)

$$\widetilde{F}_{1b}(0,\mathbf{0},P) = \rho_0 + \frac{d_0}{\Delta_b}\left(\frac{\Delta_b}{n_b} + \frac{p_4}{\rho_0}\right).$$

Similarly as in (5.62)

$$\nabla_{\xi^1}\widetilde{F}_{1b}(\mathcal{P}) = \left(0,0,1,\frac{1}{k_0\Delta_b},0\right).$$

Hence

$$\nabla_{\xi^1} h_1(\mathcal{P}) = \frac{-p_1}{\rho_0 + \dfrac{d_0}{\Delta_b}\left(\dfrac{\Delta_b}{n_b} + \dfrac{p_4}{\rho_0}\right)}\left(0,0,1,\frac{1}{k_0\Delta_b},0\right).$$

We next calculate $\nabla_{\xi^1} h_4(\mathcal{P})$. Recall from (5.46)

$$\mathcal{A}_b(\mathbf{0}) = -\frac{n_b}{\Delta_b\rho_0},\quad \Lambda_{2,b}(0,\mathbf{0}) = \frac{1}{\Delta_b},\quad \Lambda_{2,r}(0,\mathbf{0}) = \frac{1}{\Delta_r},\quad \mu_r(0,\mathbf{0}) = \mu_b(0,\mathbf{0}) = 0,$$

and as in (5.61) $\nabla_{\xi^1}\widetilde{\mu}_b(0,\mathbf{0},P) = \dfrac{1}{\Delta_b\rho_0}(0,0,0,1,0)$, then from (5.30)

$$\nabla_{\xi^1} h_4(\mathcal{P}) = \frac{p_1\dfrac{1}{\Delta_r}}{\dfrac{1}{\Delta_r\rho_0}\dfrac{1}{\Delta_b}}\frac{1}{\Delta_b\rho_0}(0,0,0,1,0) = p_1(0,0,0,1,0)$$

We conclude that

$$\nabla_{\xi^1} H(\mathcal{P}) = \begin{bmatrix} 0 & 0 & \dfrac{-p_1}{\rho_0 + \dfrac{d_0}{\Delta_b}\left(\dfrac{\Delta_b}{n_b} + \dfrac{p_4}{\rho_0}\right)} & \dfrac{\dfrac{-p_1}{k_0\Delta_b}}{\rho_0 + \dfrac{d_0}{\Delta_b}\left(\dfrac{\Delta_b}{n_b} + \dfrac{p_4}{\rho_0}\right)} & 0 \\ 0 & 0 & 0 & 0 & 0 \\ 0 & 0 & 0 & 0 & 0 \\ 0 & 0 & 0 & p_1 & 0 \\ 0 & 0 & 0 & 0 & 0 \end{bmatrix}. \tag{5.71}$$

Notice that $\nabla_{\xi^1} H(\mathcal{P})$ is an upper triangular matrix with eigenvalues 0 (with multiplicity 4) and $p_1$, the spectral radius of $\nabla_{\xi^1} H(\mathcal{P})$ is

$$R_{\xi^1} = |p_1|. \tag{5.72}$$

**Choice of the norm.** We are now ready to construct a norm for which $H$ is a contraction in the last two variables. In fact, we will construct a norm denoted by $\|\cdot\|_{k_0}$ in $\mathbb{R}^5$ depending on $k_0$ such that for small $k_0$

$$|||\nabla_{\xi^0} H(\mathcal{P})|||_{k_0} + |||\nabla_{\xi^1} H(\mathcal{P})|||_{k_0} < 1 \tag{5.73}$$

where $|||\cdot|||_{k_0}$ is the matrix norm in $R^{5\times 5}$ induced by $\|\cdot\|_{k_0}$. Recall that $k_0 = \rho_0/d_0$.

We will show that for each $0 < k_0 \leq \dfrac{(\Delta_r - \Delta_b)^2}{4\Delta_r\Delta_b}$, there exist $\lambda_1, \cdots, \lambda_5$ positive depending on $k_0$ such that the norm in $\mathbb{R}^5$ having the form

$$\|x\|_{k_0} = \max\left(\lambda_1|x_1|, \lambda_2|x_2|, \lambda_3|x_3|, \lambda_4|x_4|, \lambda_5|x_5|\right),$$

satisfies

$$|||\nabla_{\xi^0} H(\mathcal{P})|||_{k_0} < 1.$$

We first choose $\lambda_1 = \lambda_2 = \lambda_5 = 1$. Assume $x \in \mathbb{R}^5$ with $\|x\|_{k_0} = 1$, which implies $|x_i| \leq \dfrac{1}{\lambda_i}$. Then from (5.65), and (5.69)

$$\begin{aligned}\left\|\nabla_{\xi^0} H(\mathcal{P})x\right\|_{k_0} &= \max\left(\left|a\,x_3 + \frac{a}{k_0\Delta_r}\,x_4\right|, \lambda_4\,|c\,x_1|\right)\\ &\leq \max\left(\frac{1}{\lambda_3}\,|a| + \frac{1}{\lambda_4}\frac{|a|}{k_0\Delta_r}, \lambda_4\,|c|\right) \leq \max\left(\frac{|a|}{\lambda_3} + \frac{\delta_0}{\lambda_4\,|c|}, \lambda_4\,|c|\right)\end{aligned}$$

with $a$ and $c$ defined in (5.66). Hence

$$|||\nabla_{\xi^0} H(\mathcal{P})|||_{k_0} = \max_{\|x\|_{k_0}=1} \left\|\nabla_{\xi^0} H(\mathcal{P})x\right\|_{k_0} \leq \max\left\{\frac{|a|}{\lambda_3} + \frac{\delta_0}{\lambda_4\,|c|}, \lambda_4\,|c|\right\}.$$

We will choose $\lambda_3$ and $\lambda_4$ so that the last maximum is less than one. Let $\delta_0 < \delta_1 < \delta_2 < 1$, with $\delta_0$ defined in (5.69), $\lambda_4 = \delta_2/|c|$ and $\lambda_3 = N\,\lambda_4$, with $N$ to be determined depending only on $n_r$ and $n_b$. Then

$$\max\left\{\frac{|a|}{\lambda_3} + \frac{\delta_0}{\lambda_4\,|c|}, \lambda_4\,|c|\right\} = \max\left\{\frac{1}{\lambda_4\,|c|}\left(\frac{a\,c}{N} + \delta_0\right), \lambda_4\,|c|\right\},$$

notice that $ac > 0$ from (5.68). From (5.69), $a\,c \leq k_0\,\Delta_r \leq \Delta_r\,\dfrac{(\Delta_r - \Delta_b)^2}{4\Delta_r\Delta_b} := B(r,b)$, we obtain

$$\max\left\{\frac{|a|}{\lambda_3} + \frac{\delta_0}{\lambda_4\,|c|}, \lambda_4\,|c|\right\} \leq \max\left\{\frac{1}{\delta_2}\left(\frac{B(r,b)}{N} + \delta_0\right), \delta_2\right\}.$$

Now pick $N$ large, depending only on $n_r$ and $n_b$, so that $\dfrac{B(r,b)}{N} + \delta_0 < \delta_1$. Therefore,

$$|||\nabla_{\xi^0} H(\mathcal{P})|||_{k_0} \leq \max\left\{\frac{\delta_1}{\delta_2}, \delta_2\right\} := s_0 < 1,$$

for all $0 < k_0 \leq \dfrac{(\Delta_r - \Delta_b)^2}{4\Delta_r\Delta_b}$.

It remains to show that with the above norm $\|\cdot\|_{k_0}$ we also have (5.73). To do this we need to choose $k_0$ sufficiently small. In fact, from (5.71) and for $\|x\|_{k_0} = 1$

we have (since $\lambda_3 = N\,\lambda_4$, $\lambda_4 = \delta_2/|c|$, and $ac < k_0\Delta_r \le B(r,b)$)

$$\begin{aligned}\left|\left|\left|\nabla_{\xi^1} H(\mathcal{P})x\right|\right|\right|_{k_0} &= |p_1|\,\max\left(\left|a\,x_3 + \frac{1}{k_0\Delta_b}a\,x_4\right|, \lambda_4\,|x_4|\right)\\ &\le |p_1|\,\max\left(\frac{|a|}{\lambda_3} + \frac{1}{\lambda_4}\frac{|a|}{k_0\Delta_b}, 1\right)\\ &= |p_1|\,\max\left(\frac{1}{\delta_2}\left(\frac{a\,c}{N} + \frac{a\,c}{k_0\,\Delta_b}\right), 1\right)\\ &\le |p_1|\,\max\left(\frac{1}{\delta_2}\left(\frac{B(r,b)}{N} + \frac{\Delta_r}{\Delta_b}\right), 1\right) := |p_1|\,s_1,\end{aligned}$$

for all $0 < k_0 \le \dfrac{(\Delta_r - \Delta_b)^2}{4\Delta_r\Delta_b}$ with $s_1$ depending only on $n_r$ and $n_b$ ($s_1 > 1$). Hence

$$\left|\left|\left|\nabla_{\xi^1} H(\mathcal{P})\right|\right|\right|_{k_0} \le |p_1|\,s_1.$$

Therefore

$$\left|\left|\left|\nabla_{\xi^0} H(\mathcal{P})\right|\right|\right|_{k_0} + \left|\left|\left|\nabla_{\xi^1} H(\mathcal{P})\right|\right|\right|_{k_0} \le s_0 + |p_1|\,s_1.$$

From (5.55) and (5.56), $p_1 \to 0$ as $k_0 \to 0$, and therefore we obtain (5.73) for $k_0$ close to 0.

**Verification of** (4.8). Let us now show that $H$ satisfies the Lipschitz condition (4.7) with constants satisfying (4.8) in a sufficiently small neighborhood of $\mathcal{P}$ and with respect to the norm chosen. In fact, for $k_0$ sufficiently small, from (5.73), there is $0 < c_0 < 1$ so that

$$\left|\left|\left|\nabla_{\xi^0} H(\mathcal{P})\right|\right|\right|_{k_0} + \left|\left|\left|\nabla_{\xi^1} H(\mathcal{P})\right|\right|\right|_{k_0} \le c_0 < 1.$$

Since $H$ is $C^1$, there exists a norm-neighborhood $N_\varepsilon(\mathcal{P})$ of $\mathcal{P}$ as in Theorem 4.1, so that

$$\begin{aligned}\max_{(t;\zeta^0,\zeta^1;\xi^0,\xi^1)\in N_\varepsilon(\mathcal{P})} &\left|\left|\left|\nabla_{\xi^0} H\left(t;\zeta^0,\zeta^1;\xi^0,\xi^1\right)\right|\right|\right|_{k_0}\\ &+ \max_{(\hat{t};\hat{\zeta}^0,\hat{\zeta}^1;\hat{\xi}^0,\hat{\xi}^1)\in N_\varepsilon(\mathcal{P})} \left|\left|\left|\nabla_{\xi^1} H\left(\hat{t};\hat{\zeta}^0,\hat{\zeta}^1;\hat{\xi}^0,\hat{\xi}^1\right)\right|\right|\right|_{k_0} \le c_1,\end{aligned} \tag{5.74}$$

for some $c_0 < c_1 < 1$. Then by Proposition 4.5, the inequalities (4.7) and (4.8) hold with $C_0 = \max_{N_\varepsilon(\mathcal{P})} \left|\left|\left|\nabla_{\xi^0} H\right|\right|\right|$ and $C_1 = \max_{N_\varepsilon(\mathcal{P})} \left|\left|\left|\nabla_{\xi^1} H\right|\right|\right|$.

**Verification of** (4.9). If $k_0 < \dfrac{(\Delta_r - \Delta_b)^2}{4\Delta_r\Delta_b}$, then from Theorem 5.9 $|p_1| < 1$, and so $|h_1(\mathcal{P})| < 1$. Hence by continuity there exists a neighborhood of $\mathcal{P}$ so that $\left|h_1(t;\zeta^0,\zeta^1;\xi^0,\xi^1)\right| < 1$ for all $(t;\zeta^0,\zeta^1;\xi^0,\xi^1)$ in that neighborhood. Therefore, condition (4.9) in Theorem 4.1 is satisfied.

We conclude that for small values of $k_0$ Theorem 4.1 is applicable and therefore there exists a unique local solution to the system (5.4) with $Z'(0) = P$ and $P$ given in (5.56). Notice also that for such a solution $Z$, we have $z_1(0) = 0$, and $|z_1'(0)| = |h_1(\mathcal{P})| = |p_1| < 1$, then by reducing the neighborhood of zero, if necessary, we have that the solution satisfies $|z_1(t)| \le |t|$. Since we also showed in Theorem 5.9 that $p_1 \neq 0$, Theorem 5.7 is applicable and the proof of Theorem 5.13 is then complete. □

Summarizing the question of solvability of Problem B in the plane in a neighborhood of zero, we have:

(1) If $k_0 > \dfrac{(\Delta_r - \Delta_b)^2}{4\Delta_r\Delta_b}$ then Problem $B$ is not locally solvable.

(2) Let $A = (0, \rho_0)$, $B = (0, \rho_0 + d_0)$, and $0 < k_0 < C(r, b) < \dfrac{(\Delta_r - \Delta_b)^2}{4\Delta_r\Delta_b}$. Then there exists $\delta > 0$ and a unique lens $(L, S)$ with lower face $L = \{\rho(t)x(t)\}_{t\in[-\delta,\delta]}$, $x(t) = (\sin t, \cos t)$, and upper face $S = \{f_b(t)\}_{t\in[-\delta,\delta]}$, $f_b$ defined in Theorem 5.7, such that $L$ passes through $A$, $S$ passes through $B$, and so that $(L, S)$ refracts all rays emitted from $O$ with colors r and b and direction $x(t)$, $t \in [-\delta, \delta]$ into the vertical direction $e = (0, 1)$.

REMARK 5.14. In this final remark, we point out that Theorem 4.1 is not applicable to find solutions to the system (5.4) when $k_0 \le \dfrac{(\Delta_r - \Delta_b)^2}{4\Delta_r\Delta_b}$ and $k_0$ is away from 0. In this case we claim that there is no norm in $\mathbb{R}^5$ for which we can obtain (4.7) with $C_0$ and $C_1$ satisfying (4.8). In fact, from (5.55), (5.56), (5.70), and (5.72)

$$R_{\xi^0} + R_{\xi^1} = \sqrt{\frac{2\Delta_b}{\Delta_r + \Delta_b + \sqrt{\delta}}} + |p_1| \to \sqrt{\frac{2\Delta_b}{\Delta_r + \Delta_b}} + 1 > 1,$$

as $k_0 \to \dfrac{(\Delta_r - \Delta_b)^2}{4\Delta_r\Delta_b}$ from below. Hence for $k_0$ close to $\dfrac{(\Delta_r - \Delta_b)^2}{4\Delta_r\Delta_b}$, we have $R_{\xi^0} + R_{\xi^1} > 1$, and hence by Corollary 4.6 the claim follows.

## Acknowledgments

This project has received funding from the European Union's Horizon 2020 research and innovation program under the Marie Skłodowska-Curie grant agreement No. 665778.

## References

[BW59] Max Born and Emil Wolf, *Principles of optics: Electromagnetic theory of propagation, interference and diffraction of light*, Pergamon Press, New York-London-Paris-Los Angeles, 1959. MR0108202

[Can] Cannon. Lenses: Fluorite, aspherical and UD lenses, `http://cpn.canon-europe.com/content/education/infobank/lenses/fluorite_aspherical_and_ud_lenses.do`.

[Cau36] A. L. Cauchy. *Mémoire sur la dispersion de la lumière* `https://archive.org/details/mmoiresurladisp01caucgoog`. J. C. Calve, Prague, 1836.

[Dug78] James Dugundji, *Topology*, Allyn and Bacon, Inc., Boston, Mass.-London-Sydney, 1978. Reprinting of the 1966 original; Allyn and Bacon Series in Advanced Mathematics. MR0478089

[FM87] Avner Friedman and Bryce McLeod, *Optimal design of an optical lens*, Arch. Rational Mech. Anal. **99** (1987), no. 2, 147–164, DOI 10.1007/BF00275875. MR886934

[FMOT71] L. Fox, D. F. Mayers, J. R. Ockendon, and A. B. Tayler, *On a functional differential equation*, J. Inst. Math. Appl. **8** (1971), 271–307. MR301330

[GH09] Cristian E. Gutiérrez and Qingbo Huang, *The refractor problem in reshaping light beams*, Arch. Ration. Mech. Anal. **193** (2009), no. 2, 423–443, DOI 10.1007/s00205-008-0165-x. MR2525122

[GH14] Cristian E. Gutiérrez and Qingbo Huang, *The near field refractor*, Ann. Inst. H. Poincaré Anal. Non Linéaire **31** (2014), no. 4, 655–684, DOI 10.1016/j.anihpc.2013.07.001. MR3249808

[GS16] Cristian E. Gutiérrez and Ahmad Sabra, *Aspherical lens design and imaging*, SIAM J. Imaging Sci. **9** (2016), no. 1, 386–411, DOI 10.1137/15M1030807. MR3477314

[GS18] Cristian E. Gutiérrez and Ahmad Sabra, *Freeform lens design for scattering data with general radiant fields*, Arch. Ration. Mech. Anal. **228** (2018), no. 2, 341–399, DOI 10.1007/s00205-017-1196-y. MR3766979

[Gut13] C. E. Gutiérrez. Aspherical lens design. *Journal Optical Society of America A*, 30(9):1719–1726, 2013.

[Gut14] Cristian E. Gutiérrez, *Refraction problems in geometric optics*, Fully nonlinear PDEs in real and complex geometry and optics, Lecture Notes in Math., vol. 2087, Springer, Cham, 2014, pp. 95–150, DOI 10.1007/978-3-319-00942-1_3. MR3203560

[HJ85] Roger A. Horn and Charles R. Johnson, *Matrix analysis*, Cambridge University Press, Cambridge, 1985. MR832183

[ima] imatest. Chromatic aberration AKA color fringing `http://www.imatest.com/docs/sfr_chromatic/`.

[JW01] F. A. Jenkins and H. E. White. *Fundamental of Optics*. McGraw-Hill, 4th edition, 2001.

[Kar16] Aram L. Karakhanyan, *An inverse problem for the refractive surfaces with parallel lighting*, SIAM J. Math. Anal. **48** (2016), no. 1, 740–784, DOI 10.1137/140964941. MR3463050

[KJ10] R. Kingslake and R. B. Johnson. *Lens Design Fundamentals*. Academic Press and SPIE Press, 2nd edition, 2010.

[KM71] Tosio Kato and J. B. McLeod, *The functional-differential equation $y'(x) = ay(\lambda x) + by(x)$*, Bull. Amer. Math. Soc. **77** (1971), 891–937, DOI 10.1090/S0002-9904-1971-12805-7. MR283338

[LGM17] Roberto De Leo, Cristian E. Gutiérrez, and Henok Mawi, *On the numerical solution of the far field refractor problem*, Nonlinear Anal. **157** (2017), 123–145, DOI 10.1016/j.na.2017.03.009. MR3645803

[Lun64] R. K. Luneburg, *Mathematical theory of optics*, Foreword by Emil Wolf; supplementary notes by M. Herzberger, University of California Press, Berkeley, Calif., 1964. MR0172589

[Rog88] Joel C. W. Rogers, *Existence, uniqueness, and construction of the solution of a system of ordinary functional-differential equations, with application to the design of perfectly focusing symmetric lenses*, IMA J. Appl. Math. **41** (1988), no. 2, 105–134, DOI 10.1093/imamat/41.2.105. MR984002

[Sch05] K. Schwarzschild. Untersuchungen zur geometrischen Optik I. Einleitung in die Fehlertheorie optischer Instrumente auf Grund des Eikonalbegriffs. *Abhandlungen der Gesellschaft der Wissenschaften in Göttingen*, 4(1):1–31, 1905.

[Sch83] G. Schultz. Achromatic and sharp real imaging of a point by a single aspheric lens. *Applied Optics*, 22(20):3242–3248, 1983.

[vB94] B. van Brunt. Mathematical possibility of certain systems in geometrical optics. *J. Opt. Soc. Am. A*, 11(11):2905–2914, 1994.

[vBO92] B. van-Brunt and J. R. Ockendon, *A lens focusing light at two different wavelengths*, J. Math. Anal. Appl. **165** (1992), no. 1, 156–179, DOI 10.1016/0022-247X(92)90073-M. MR1151066

Department of Mathematics, Temple University, Philadelphia, Pennsylvania 19122
*Email address*: `gutierre@temple.edu`

Faculty of Mathematics, Informatics, and Mechanics, University of Warsaw, 13121L Poland; and American University of Beirut, P.O. Box 11-0236, Riad El-Solh, Beirut 1107 2020
*Email address*: `asabra@aub.edu.lb`

Contemporary Mathematics
Volume **748**, 2020
https://doi.org/10.1090/conm/748/15058

# Free boundary regularity near the fixed boundary for the fully nonlinear obstacle problem

Emanuel Indrei

Abstract. The interior free boundary theory for linear elliptic operators in higher dimensions was developed by Caffarelli [Acta Math. 139 (1977), pp. 155–184] in the low regularity context. In these notes, the up-to-the boundary free boundary regularity is discussed for nonlinear elliptic operators based on a different approach.

## 1. Introduction

Caffarelli proved that if $L$ is a linear uniformly elliptic operator and $u \geq 0$ solves

$$L(D^2u) = \chi_{\{u>0\}} \text{ in } B_1 \tag{1.1}$$

then for $x \in \Gamma = \partial\{u > 0\} \cap B_1$ with positive Lebesgue density for $\{u = 0\}$, i.e. satisfying

$$\liminf_{r \to 0^+} \frac{|B_r(x) \cap \{u = 0\}|}{|B_r(x)|} > 0,$$

there is a Lipschitz function $g$ such that $\Gamma \cap B_s(x)$ admits a representation with respect to $g$ in a coordinate system for some $s > 0$. The Lipschitz regularity can be improved to $C^1$ and higher regularity follows (up to analyticity) via a theorem of Kinderlehrer and Nirenberg [**KN77**]. Caffarelli's theorem is optimal in the sense that there exists a solution when $L = \Delta$ for which there is a free boundary point with zero Lebesgue density for $\{u = 0\}$ and in a neighborhood of the point the free boundary develops a cusp singularity and is not a graph in any system of coordinates [**Sch77**].

In a recent work [**Ind19a**], the author proved that for solutions of (1.1) with zero Dirichlet boundary data, with $L$ replaced by a convex fully nonlinear uniformly elliptic operator $F$, if $x \in \partial B_1 \cap \overline{\Gamma}$, then $\Gamma$ can be represented as the graph of a $C^1$ function in a neighborhood of $x$. There are two surprising differences between the interior and boundary result: first, there are no density assumptions in the boundary case (in particular, cusp-type singularities do not exist); second, there is an example which generates a free boundary which is $C^1$ with a specific Dini modulus of continuity for the free normal (see e.g. [**PSU12**, Remark 8.8]).

2010 *Mathematics Subject Classification.* Primary 35G20; Secondary 49N60.
*Key words and phrases.* Regularity of solutions; non-transversal intersection.

In his original approach, Caffarelli estimated pure second derivatives from below. The linear approach developed thereafter to handle regularity near the fixed boundary involves specific barrier constructions involving the operator and monotonicity formulas [**Ura96**, **SU03**, **And07**]. The nonlinear method is based on understanding a maximal mixed partial derivative along a preferred direction.

In what follows, $F$ satisfies

- $F(0) = 0$.
- $F$ is uniformly elliptic with ellipticity constants $\lambda_0$, $\lambda_1 > 0$ such that
$$\mathcal{P}^-(M-N) \leq F(M) - F(N) \leq \mathcal{P}^+(M-N),$$
where $M$ and $N$ are symmetric matrices and $\mathcal{P}^\pm$ are the Pucci operators
$$\mathcal{P}^-(M) := \inf_{\lambda_0 \leq N \leq \lambda_1} \operatorname{tr}(NM), \qquad \mathcal{P}^+(M) := \sup_{\lambda_0 \leq N \leq \lambda_1} \operatorname{tr}(NM).$$
- $F$ is convex and $C^1$.

Let $\Omega$ be an open set and $B_r^+ = \{x : |x| < r, x_n > 0\}$. A continuous function $u$ belongs to $P_r^+(0, M, \Omega)$ if $u$ satisfies in the viscosity sense

- $F(D^2u) = \chi_\Omega$ in $B_r^+$;
- $||u||_{L^\infty(B_r^+)} \leq M$;
- $u = 0$ on $\{x_n = 0\} \cap \overline{B_1^+} =: B_1'$.

In [**IM16a**] it was shown that $W^{2,p}$ solutions are $C^{1,1}$ (see also [**FS14**, **IM16b**] for the interior case). Furthermore, given $u \in P_r^+(0, M, \Omega)$, the free boundary is denoted by $\Gamma = \partial\Omega \cap B_r^+$.

A *blow-up limit of* $\{u_j\} \subset P_1^+(0, M, \Omega)$ is a limit of the form
$$\lim_{k\to\infty} \frac{u_{j_k}(s_k x)}{s_k^2},$$
where $\{j_k\}$ is a subsequence of $\{j\}$ and $s_k \to 0^+$.

In §2, non-transversal intersection is shown for $\Omega = (\{u \neq 0\} \cup \{\nabla u \neq 0\}) \cap \{x_n > 0\}$ and a problem in superconductivity is discussed in which $\Omega = \{\nabla u \neq 0\}) \cap \{x_n > 0\}$; in §3, $C^1$ regularity is proved when $u \geq 0$; last, some of the technical details are shown in the appendix §4.

## 2. Non-transversal intersection and classification of blow-up limits

One of the main results discussed in this section is the following.

THEOREM 2.1. *There exists $r_0 > 0$ and a modulus of continuity $\omega$ such that*
$$\Gamma(u) \cap B_{r_0}^+ \subset \{x : x_n \leq \omega(|x'|)|x'|\}$$
*for all $u \in P_1^+(0, M, \Omega)$ provided $0 \in \overline{\Gamma(u)}$ and $\Omega = (\{u \neq 0\} \cup \{\nabla u \neq 0\}) \cap \{x_n > 0\}$.*

If one varies the boundary data, then non-transversal intersection may not hold [**And07**, see Examples 3 & 4]. The difficulty in the fully nonlinear context is that monotonicity formulas are not available and a classification of blow-up limits requires a new approach: if blow-up limits are not half-space solutions, then a certain regularity property holds. More precisely:

PROPOSITION 2.2. *Suppose* $\Omega = (\{u \neq 0\} \cup \{\nabla u \neq 0\}) \cap \{x_n > 0\}$ *and* $\{u_j\} \subset P_1^+(0, M, \Omega)$. *If* $0 \in \overline{\{u_j \neq 0\}}$ *and* $\nabla u_j(0) = 0$, *then one of the following is true: (i) all blow-up limits of* $\{u_j\}$ *at the origin are of the form* $u_0(x) = bx_n^2$ *for* $b > 0$*; (ii) there exists* $\{u_{k_j}\} \subset \{u_j\}$ *such that for all* $R \geq 1$, *there exists* $j_R \in \mathbb{N}$ *such that for all* $j \geq j_R$,

$$u_{k_j} \in C^{2,\alpha}(B^+_{\frac{Rr_j}{4}}),$$

*where the sequence* $\{r_j\}$ *depends on* $\{u_j\}$.

The proof relies on the fact that if not all blow-up solutions are half-space solutions, then one can construct a specific sequence producing a limit of the form $ax_1x_n + bx_n^2$.

PROPOSITION 2.3. *Let* $\{u_j\} \subset P_1^+(0, M, \Omega)$ *and suppose* $0 \in \overline{\{u_j \neq 0\}}$, $\{\nabla u_j \neq 0\} \cap \{x_n > 0\} \subset \Omega$, $\nabla u_j(0) = 0$. *Then one of the following is true: (i) all blow-up limits of* $\{u_j\}$ *at the origin are of the form* $u_0(x) = bx_n^2$ *for some* $b > 0$*; (ii) there exists a blow-up limit of* $\{u_j\}$ *of the form* $ax_1x_n + bx_n^2$ *for* $a \neq 0$, $b \in \mathbb{R}$.

PROOF. Let

$$N := \limsup_{|x| \to 0, x_n > 0} \frac{1}{x_n} \sup_{u \in \{u_j\}} \sup_{e \in \mathbb{S}^{n-2} \cap e_n^\perp} \partial_e u(x)$$

and consider a sequence $\{x^k\}_{k \in \mathbb{N}}$ with $x_n^k > 0$, $u_{j_k} \in \{u_j\}$, and $e^k \in \mathbb{S}^{n-2} \cap e_n^\perp$ such that the previous limit is given by

$$\lim_{k \to \infty} \frac{1}{x_n^k} \partial_{e^k} u_{j_k}(x^k).$$

Note that $N < \infty$ by $C^{1,1}$ regularity for the class $P_1^+(0, M, \Omega)$ and the boundary condition (see [**IM16a**]). By compactness, $e^k \to e_1 \in \mathbb{S}^{n-2}$ (along a subsequence) so that up to a rotation,

$$N = \lim_{k \to \infty} \frac{1}{x_n^k} \partial_{x_1} u_{j_k}(x^k).$$

Next, if

$$\tilde{u}_j(x) := \frac{u_{k_j}(s_j x)}{s_j^2} \to u_0(x)$$

for some sequence $s_j \to 0^+$, where the convergence is in $C^{1,\alpha}_{loc}(\mathbb{R}^n_+)$ for any $\alpha \in [0, 1)$, $u_0 \in C^{1,1}(\mathbb{R}^n_+)$ satisfies the following PDE in the viscosity sense

$$\begin{cases} F(D^2 u_0) = 1 & \text{a.e. in } \mathbb{R}^n_+ \cap \Omega_0 \\ |\nabla u_0| = 0 & \text{in } \mathbb{R}^n_+ \backslash \Omega_0 \\ u = 0 & \text{on } \mathbb{R}^{n-1}_+, \end{cases} \tag{2.1}$$

where $\Omega_0 = \{\nabla u_0 \neq 0\} \cap \{x_n > 0\}$. Note that

$$N \geq \lim_j \left| \frac{\partial_{x_i} u_{k_j}(s_j x)}{s_j x_n} \right| = \lim_j \left| \frac{\partial_{x_i} \tilde{u}_j(x)}{x_n} \right| = \left| \frac{\partial_{x_i} u_0(x)}{x_n} \right| \tag{2.2}$$

for all $i \in \{1, \ldots, n-1\}$. If $N = 0$, then $\partial_{x_i} u_0 = 0$ for all $i \in \{1, \ldots, n-1\}$ so that $u_0(x) = u_0(x_n)$ and the conditions readily imply $u_0(x_n) = bx_n^2$. Since

$N$ does not depend on the sequence $\{s_j\}$ it follows that in this case all blow-up limits have the previously stated form. Suppose that $N > 0$, let $r_k = |x^k|$, and consider the re-scaling of $u_{j_k}$ with respect to $r_k$. Note that along a subsequence, $y^k := \frac{x^k}{r_k} \to y \in \mathbb{S}^{n-1}$. By the choice of $r_k$,

$$\lim_{k\to\infty} \frac{v(y^k)}{y_n^k} = \lim_{k\to\infty} \frac{\partial_{x_1}\tilde{u}_k(y^k)}{y_n^k} = \lim_{k\to\infty} \frac{\partial_{x_1}u_{j_k}(r_k y^k)}{r_k y_n^k} = N,$$

where $v = \partial_{x_1} u_0$. In particular,

$$v(y) = N y_n$$

and by an argument in [**IM16a**] (involving the boundary Harnack inequality), $u_0(x) = a x_1 x_n + b x_n^2$ with $a \neq 0$. □

PROOF OF PROPOSITION 2.2. Either all blow-up limits are of the form $u_0(x) = b x_n^2$ or there exists a subsequence

$$\tilde{u}_j(x) = \frac{u_{k_j}(r_j x)}{r_j^2}$$

producing a limit of the form $u_0(x) = a x_1 x_n + b x_n^2$ for $a > 0$ (up to a rotation). Let $c = c(a,b)$ be the constant from Lemma 4.4 and note that since $\tilde{u}_j \to u_0$ in $C^{1,\alpha}_{loc}$, there exists $j_0 = j_0(a,R) \in \mathbb{N}$ such that for every cylinder $S_{(\alpha,\beta)}(e_1)$ there exists $x \in S_{(\alpha,\beta)}(e_1) \cap B_R^+$ such that $|\nabla \tilde{u}_j(x)| \geq \frac{c}{2}$ for all $j \geq j_0$, where $R \geq 1$. Choose a constant $C_0 = C_0(a,b,R) > 0$ such that

$$C_0 \partial_{x_1} u_0 - u_0 \geq 0$$

in $B_R^+$ and $j_0' \geq j_0$ for which

$$C_0 \partial_{x_1} \tilde{u}_j - \tilde{u}_j \geq 0 \text{ in } B^+_{\frac{R}{2}} \tag{2.3}$$

whenever $j \geq j_0'$ by Lemma 4.1. Now fix $j \geq j_0'$ and suppose $z \in \Gamma_i(\tilde{u}_j) \cap B^+_{\frac{R}{2}}$. Then there exists a ball $B \subset \text{int}\{\tilde{u}_j = 0\} \cap B^+_{\frac{R}{2}}$ and a cylinder $S$ in the $e_1$- direction generated by $B$. Now select $x \in S \cap B_R^+$ for which $|\nabla \tilde{u}_j(x)| > 0$ and $-R < x_1 < -R/2$. In particular, there exists a small ball around $x$, say $\tilde{B}$ such that $F(D^2 \tilde{u}_j) = 1$ in $\tilde{B}$ and one may assume $\tilde{B} \subset \{\tilde{u}_j \neq 0\}$. Note that $\tilde{B}$ is contained in the cylinder $S$ and let $E_t = \tilde{B} + t e_1$ for $t \in \mathbb{R}$. If $t > 0$ is such that $\overline{E_t} \cap \{\tilde{u}_j = 0\} \neq \emptyset$, and for all $0 \leq s < t$, $E_s \cap \{\tilde{u}_j = 0\} = \emptyset$, choose $y \in \overline{E_t} \cap \{\tilde{u}_j = 0\}$. If $\tilde{u}_j > 0$ in $\tilde{B}$, then by (2.3) it follows that $\tilde{u}_j$ is strictly positive at a point in $\{\tilde{u}_j = 0\}$, a contradiction. Thus $\tilde{u}_j < 0$ in $\tilde{B}$. By convexity of $F$

$$a_{kl} \partial_{kl} \tilde{u}_j \geq 0 \text{ in } E_t.$$

Since $0 = \tilde{u}_j(y) > \tilde{u}_j(x)$ for $x \in E_t$ and $y$ satisfies an interior ball condition, then Hopf's lemma implies that $\frac{\partial}{\partial n}\tilde{u}_j(y) > 0$, where $n$ is the outer normal to the ball at $y$. If there exists $z \in B_\delta(y)$ such that $\tilde{u}_j(z) > 0$, then this contradicts the monotonicity, if $\delta > 0$ is sufficiently small: $\overline{E_\eta} \subset B \subset int\{\tilde{u}_j = 0\}$ for $\eta > 0$ large enough and since $\tilde{u}_j(z) > 0$, the monotonicity (2.3) implies that $\tilde{u}_j(z + e_1 s) > 0$, for some $s > 0$ such that $z + e_1 s \in \{\tilde{u}_j = 0\}$. Hence, $\tilde{u}_j \leq 0$ on $B_\delta(y)$ and thus $\nabla \tilde{u}_j(y) = 0$, a contradiction. The conclusion is that for $j \geq j_0'$,

$$\Gamma_i(\tilde{u}_j) \cap B^+_{\frac{R}{2}} = \emptyset.$$

In particular, $(B^+_{\frac{R}{2}} \setminus \Omega_j)^o = \emptyset$ and non-degeneracy implies that $|B^+_{\frac{R}{2}} \setminus \Omega_j| = 0$. Thus the $C^{1,1}$ function $\tilde{u}_j$ satisfies $F(D^2\tilde{u}_j) = 1$ in $B^+_{\frac{R}{2}}$ in the viscosity sense and the up to the boundary Evans-Krylov theorem (see e.g. [**Saf94**]) implies that $\tilde{u}_j \in C^{2,\alpha}(B^+_{\frac{R}{4}})$. In particular, $u_{k_j} \in C^{2,\alpha}(B^+_{\frac{Rr_j}{4}})$. □

THEOREM 2.4. *Suppose $u \in P_1^+(0, M, \Omega)$ and $\Omega = (\{u \neq 0\} \cup \{\nabla u \neq 0\}) \cap \{x_n > 0\}$. If $0 \in \overline{\{u \neq 0\}}$ and $\nabla u(0) = 0$, then the blow-up limit of $u$ at the origin has the form*

$$u_0(x) = a x_1 x_n + b x_n^2$$

*for $a, b \in \mathbb{R}$.*

PROOF. By Proposition 2.2, either $u_0(x) = bx_n^2$ or $D^2u(0)$ exists and the rescaling of $u$ is given by

$$u_j(x) = \frac{u(r_j x)}{r_j^2} = \langle x, D^2u(0)x\rangle + o(1).$$

Since $u_0(x', 0) = 0$ for $x' \in \mathbb{R}^{n-1}$, it follows that $u_0$ has the claimed form (up to a rotation). □

THEOREM 2.5. *Suppose $\Omega = (\{u \neq 0\} \cup \{\nabla u \neq 0\}) \cap \{x_n > 0\}$, $0 \in \overline{\Gamma}$, and $\{u_j\} \subset P_1^+(0, M, \Omega)$. Then the blow-up limit of $u_j$ at the origin has the form*

$$u_0(x) = bx_n^2$$

*for $b > 0$.*

PROOF. By Proposition 2.2, either $u_0(x) = bx_n^2$ or there exists a subsequence

$$u_{k_j}(x) \in C^{2,\alpha}(B^+_{\frac{Rr_j}{4}})$$

which contradicts that $F$ is continuous (consider a sequence of points approaching the free boundary from the set where the equation is satisfied with the right-hand-side being equal to one and from the complement). □

REMARK 2.1. There exist global solutions which are not blow-up solutions (at contact points).

PROOF OF THEOREM 2.1. It suffices to show that for any $\epsilon > 0$ there exists $\rho_\epsilon > 0$ such that $\Gamma(u) \cap B^+_{\rho_\epsilon} \subset B^+_{\rho_\epsilon} \setminus \mathcal{C}_\epsilon$, where $\mathcal{C}_\epsilon = \{x_n > \epsilon|x'|\}$. If not, then there exists $\epsilon > 0$ such that for all $k \in \mathbb{N}$ there exists $u_k \in P_1^+(0, M, \Omega)$ with

$$\Gamma(u_k) \cap B^+_{1/k} \cap \mathcal{C}_\epsilon \neq \emptyset, \tag{2.4}$$

where $0 \in \overline{\Gamma(u_k)}$. If all blow-ups of $\{u_k\}$ are half-space solutions. Let $x_k \in \Gamma(u_k) \cap B^+_{1/k} \cap \mathcal{C}_\epsilon$ and set $y_k = \frac{x_k}{r_k}$ with $r_k = |x_k|$. Consider $\tilde{u}_k(x) = \frac{u_k(r_k x)}{r_k^2}$ so that $y_k \in \Gamma(\tilde{u}_k)$, $\tilde{u}_k \to bx_n^2$, $y_k \to y \in \partial B_1 \cap \mathcal{C}_\epsilon$ (up to a subsequence), and $y \in \Gamma(u_0)$, a contradiction. Second, select a subsequence $\{u_{k_j}\}$ of $\{u_k\}$ such that for all $j \geq j_2$, $u_{k_j} \in C^{2,\alpha}(B^+_{\frac{r_j}{2}})$, where $j_2 \in \mathbb{N}$ and the sequence $\{r_j\}$ depends on $\{u_k\}$. Since $0 \in \overline{\Gamma(u_{k_j})}$, there exists

$$x_j \in \Gamma(u_{k_j}) \cap B^+_{\frac{r_j}{2}}$$

which contradicts the continuity of $F$ (consider a sequence of points approaching the free boundary from the set where the equation is satisfied with the right-hand-side being equal to one and from the complement). □

**2.1. An obstacle problem in superconductivity.** Equations of the type

$$F(D^2u, x) = g(x, u)\chi_{\{\nabla u \neq 0\}}$$

have been investigated in [**CS02**] and are based on physical models, e.g. the stationary equation for the mean-field theory of superconducting vortices when the scalar stream is a function of the scalar magnetic potential [**Cha95, CRS96, ESS98**]. It is shown that in certain configurations in two dimensions, the set $\{\nabla u = 0\}$ is convex. In a recent paper [**Ind19b**], the author proved non-transversal intersection for $\Omega = \{\nabla u \neq 0\} \cap \{x_2 > 0\}$. If $\{u < 0\}$ has sufficiently small density, non-transversal intersection follows from the techniques discussed above without a dimension restriction: suppose

$$\frac{|\{u < 0\} \cap B_r^+|}{|B_r^+|} \to 0$$

as $r \to 0^+$. A limit of the form

$$u_0(x) = \lim_{k\to\infty} \frac{u_{j_k}(s_k x)}{s_k^2}$$

satisfies $u_0 \geq 0$ and therefore cannot be $ax_1x_n + bx_n^2$ for $a \neq 0$. In particular, it must be a half-space solution by Proposition 2.3 and the non-transversal intersection follows as before. The assumption on the negativity set appeared in [**MM04**] where the authors considered the non-transversal intersection subject to additional assumptions on the operator and solution.

## 3. $C^1$ regularity

In the physical case when $u \geq 0$, the free boundary is $C^1$ without density assumptions.

THEOREM 3.1. *Let $u \in P_1^+(0, M, \Omega)$ be non-negative, $\Omega = (\{u \neq 0\} \cup \{\nabla u \neq 0\}) \cap \{x_n > 0\}$, and $0 \in \overline{\Gamma(u)}$. There exists $r_0 > 0$ such that $\Gamma$ is the graph of a $C^1$ function in $B_{r_0}^+$.*

PROOF. First, for any $\epsilon > 0$ there exists $r(\epsilon, M) > 0$ such that if $x^0 \in \Gamma(u) \cap B_{1/2}^+$ and $d = x_n^0 < r$, then

$$\sup_{B_{2d}^+(x^0)} |u - h| \leq \epsilon d^2, \quad \sup_{B_{2d}^+(x^0)} |\nabla u - \nabla h| \leq \epsilon d,$$

where

$$h(x) = b[(x_n - d)^+]^2,$$

and $b > 0$ depends on the ellipticity constants of $F$. If not, then there exists $\epsilon > 0$, non-negative $u_j \in P_1^+(0, M, \Omega)$, and $x^j \in \Gamma(u_j) \cap B_{1/2}^+$ with $d_j = x_n^j \to 0$, for which

$$\sup_{B_{2d_j}(x^j)^+} |u_j - b[(x_n - d_j)^+]^2| > \epsilon d_j^2,$$

or

$$\sup_{B_{2d_j}(x^j)^+} |\nabla u_j - 2b(x_n - d_j)^+| > \epsilon d_j.$$

Let $\tilde{u}_j(x) = \frac{u_j((x^j)'+d_j x)}{d_j^2}$ so that in particular

$$||\tilde{u}_j - h||_{C^1(B_2^+(e_n))} \geq \epsilon,$$

where $h(x) = b[(x_n-1)^+]^2$. Since $\tilde{u}_j(e_n) = |\nabla \tilde{u}_j(e_n)| = 0$, the $C^{1,1}$ regularity of $\tilde{u}_j$ implies that $|\tilde{u}_j(x)| \leq C|x-e_n|^2$. By passing to a subsequence, if necessary,

$$\tilde{u}_j \to u_0$$

where $u_0 \in C^{1,1}(\mathbb{R}^n_+)$ satisfies the following PDE in the viscosity sense

$$\begin{cases} F(D^2u_0) = 1 & \text{a.e. in } \mathbb{R}^n_+ \cap \Omega_0, \\ |\nabla u_0| = 0 = u_0 & \text{in } \mathbb{R}^n_+ \backslash \Omega_0, \\ u_0 = 0 & \text{on } \mathbb{R}^{n-1}_+. \end{cases} \tag{3.1}$$

Now let

$$N = \limsup_{|x|\to 0, x_n>0} \frac{1}{x_n} \sup_{u\in P_1^+\cap\{u\geq 0\}} \sup_{e\in\mathbb{S}^{n-2}\cap e_n^\perp} \sup_{y\in \overline{B^+_{1/2}}\cap\{x_n=0\}} \partial_e u(x+y)$$

and note that $N < \infty$ by $C^{1,1}$ regularity and the boundary condition: for any $e \in \mathbb{S}^{n-2}\cap e_n^\perp$ and $y \in \overline{B^+_{1/2}}\cap\{x_n=0\}$, it follows that $\partial_e u(x'+y) = 0$. Furthermore,

$$N \geq \lim_j \left|\frac{\partial_{x_i} u_j(d_j x + (x^j)')}{d_j x_n}\right| = \lim_j \left|\frac{\partial_{x_i}\tilde{u}_j(x)}{x_n}\right| = \left|\frac{\partial_{x_i} u_0(x)}{x_n}\right| \tag{3.2}$$

for all $i \in \{1,\dots,n-1\}$. In particular, let $v = \partial_{x_1} u_0$ so that in $\mathbb{R}^n_+$,

$$|v(x)| \leq N x_n. \tag{3.3}$$

If $N = 0$, then $\partial_{x_i} u_0 = 0$ for all $i \in \{1,\dots,n-1\}$ and therefore $u_0(x) = u_0(x_n)$. Since $e_n$ is a free boundary point, it follows that $u_0 = h$, a contradiction. Thus $N > 0$ and there is a sequence $\{x^k\}_{k\in\mathbb{N}}$ with $x_n^k > 0$, $u_k \in P_1^+(0,M,\Omega)$, $u_k \geq 0$, $y^k \in \overline{B^+_{1/2}} \cap \{x_n = 0\}$, and $e^k \in \mathbb{S}^{n-2}\cap e_n^\perp$ such that

$$N = \lim_{k\to\infty} \frac{1}{x_n^k}\partial_{e^k} u_k(x^k+y^k).$$

By compactness, $e^k \to e_1 \in \mathbb{S}^{n-2}$ (along a subsequence) so that up to a rotation,

$$N = \lim_{k\to\infty} \frac{1}{x_n^k}\partial_{x_1} u_k(x^k+y^k).$$

Let

$$\tilde{u}_k(x) = \frac{u_k(y^k + r_k x)}{r_k^2},$$

where $r_k = |x^k|$, $z^k = \frac{x^k}{r_k}$, and note that along a subsequence $z^k \to z \in \mathbb{S}^{n-1}$ and $\tilde{u}_k \to u_0$. It follows that $\partial_{x_1} u_0(z) = N z_n$ and proceeding as in [**IM16a**] one deduces that $u_0(x) = a x_1 x_n + c x_n + \tilde{b} x_n^2$ for $a \neq 0$ and $c, \tilde{b} \in \mathbb{R}$, contradicting that $u \geq 0$. This implies that in a neighborhood of the origin, there is a cone of fixed opening that can be placed below and above each free boundary point; therefore, the free boundary is Lipschitz continuous and thus $C^1$ by interior results [**FS14**, Theorem 1.3]. Since the intersection of $\Gamma$ and the origin occurs non-transversally, and

$$\sup_{B^+_{2d}(x^0)} |u-h| \leq \epsilon d^2, \quad \sup_{B^+_{2d}(x^0)} |\nabla u - \nabla h| \leq \epsilon d,$$

the aperture of the cones can be taken arbitrarily close to $\pi$. □

## 4. Appendix

LEMMA 4.1. *Let $u \in P_r^+(0, M, \Omega)$ where $\{u \neq 0\} \subset \Omega$, $e \in \mathbb{S}^{n-2} \cap e_n^\perp$, and suppose there exist non-negative constants $\epsilon_0, C_0$ such that $C_0\partial_e u - u \geq -\epsilon_0$ in $B_r^+$. Then there exists $c = c(n, \Lambda, r) > 0$ such that if $\epsilon_0 \leq c$, then $C_0\partial_e u - u \geq 0$ in $B_{\frac{r}{2}}^+$.*

PROOF. By convexity of $F$, there exist measurable uniformly elliptic coefficients $a_{ij}$ such that

$$F(D^2u(x+he)) - F(D^2u(x)) \geq a_{ij}(\partial_{ij}u(x+he) - \partial_{ij}u(x))$$

if $x \in \Omega$ provided $h$ is small enough. Therefore,

$$0 \geq a_{ij}\partial_{ij}\partial_e u \text{ in } \Omega.$$

Convexity also yields

$$a_{ij}\partial_{ij}u \geq F(D^2u(x)) - F(0) = 1 \text{ in } \Omega.$$

Suppose now that there exists $y \in B_{\frac{r}{2}}^+$ for which $C_0\partial_e u(y) - u(y) < 0$. Let $w(x) = C_0\partial_e u(x) - u(x) + \frac{|x-y|^2}{2n\Lambda}$. Since $\lambda Id \leq (a_{ij}) \leq \Lambda Id$, it follows by the above that $Lw \leq 0$ in $\Omega$ where $L = a_{ij}\partial_{ij}$. The maximum principle implies $\min_{\partial(\Omega\cap B_r^+)} w = \min_{\Omega\cap B_r^+} w < 0$. Note that $w \geq 0$ on $\partial\Omega$ and likewise on $\{x_n = 0\}$. Therefore, the minimum occurs on $\partial B_r$ and thus $0 > -\epsilon_0 + \frac{1}{8n\Lambda}r^2$, a contradiction if $\epsilon_0$ is small enough. □

REMARK 4.2. One may take $\epsilon_0 = cr^2$, where $c > 0$ depends only on the dimension and ellipticity constants of $F$.

REMARK 4.3. If $u \geq 0$, then $\partial_{e_n}u \geq 0$ on $\{x_n = 0\} \cap B_r$ and Lemma 4.1 holds therefore in this case for all $e \in \mathbb{S}^{n-1}$ such that $e \cdot e_n \geq 0$.

LEMMA 4.4. *Let $u_0(x) = ax_1x_n + bx_n^2$ with $a \neq 0$ and $R \geq 1$. Then there exists $c = c(a, b) > 0$ such that*

$$\inf_D |\nabla u_0(x)| \geq c,$$

*where $D = \{x = (x_1, x'', x_n) : R > |x| > R/2, |x''| \leq \delta(R)\}$ for some $\delta(R) > 0$.*

PROOF. Note $|\nabla u_0(x)|^2 = a^2x_n^2 + a^2x_1^2 + 2abx_1x_n + 4b^2x_n^2$ so that if $|x_n| > \frac{1}{3}$, then $|\nabla u_0(x)|^2 \geq \frac{a^2}{9}$. If $|x_n| \leq \frac{1}{3}$, then for points that satisfy $|x''| \leq \sqrt{\frac{5}{72}}R$, where $x'' = (x_2, x_3, \ldots, x_{n-1})$, it follows that

$$x_1^2 > \frac{5}{72}R^2.$$

If $b \neq 0$, let $\epsilon^2 \in (\frac{1}{a^2+4b^2}, \frac{1}{b^2})$. Then

$$\begin{aligned}|\nabla u_0(x)|^2 &\geq (a^2 + 4b^2 - \frac{1}{\epsilon^2})x_n^2 + (a^2 - \epsilon^2a^2b^2)x_1^2\\ &> (a^2 - \epsilon^2a^2b^2)(\frac{5}{72}R^2).\end{aligned}$$

□

LEMMA 4.5. *Let* $u_0(x) = ax_1x_n + bx_n^2$ *with* $a > 0$ *and* $R \geq 1$. *Then there exists* $C_0 = C_0(a, b, R) > 0$ *such that*

$$C_0\partial_{x_1}u_0(x) - u_0(x) \geq 0$$

*in* $B_R^+$.

PROOF. The condition is equivalent to $ax_n(C_0 - x_1) \geq bx_n^2$. Since $x_1 \leq R$ and $0 \leq x_n \leq R$, it follows that any $C_0 \geq \frac{b}{a}R + R$ satisfies the condition. □

## Acknowledgments

The author wishes to thank Donatella Danielli and Irina Mitrea for organizing the AMS Special Session "Harmonic Analysis and Partial Differential Equations" at Northeastern University.

## References

[And07] John Andersson, *On the regularity of a free boundary near contact points with a fixed boundary*, J. Differential Equations **232** (2007), no. 1, 285–302, DOI 10.1016/j.jde.2006.06.012. MR2281197

[Caf77] Luis A. Caffarelli, *The regularity of free boundaries in higher dimensions*, Acta Math. **139** (1977), no. 3-4, 155–184, DOI 10.1007/BF02392236. MR0454350

[Cha95] S. Jonathan Chapman, *A mean-field model of superconducting vortices in three dimensions*, SIAM J. Appl. Math. **55** (1995), no. 5, 1259–1274, DOI 10.1137/S0036139994263665. MR1349309

[CRS96] S. J. Chapman, J. Rubinstein, and M. Schatzman, *A mean-field model of superconducting vortices*, European J. Appl. Math. **7** (1996), no. 2, 97–111, DOI 10.1017/S0956792500002242. MR1388106

[CS02] L. Caffarelli and J. Salazar, *Solutions of fully nonlinear elliptic equations with patches of zero gradient: existence, regularity and convexity of level curves*, Trans. Amer. Math. Soc. **354** (2002), no. 8, 3095–3115, DOI 10.1090/S0002-9947-02-03009-X. MR1897393

[ESS98] Charles M. Elliott, Reiner Schätzle, and Barbara E. E. Stoth, *Viscosity solutions of a degenerate parabolic-elliptic system arising in the mean-field theory of superconductivity*, Arch. Ration. Mech. Anal. **145** (1998), no. 2, 99–127, DOI 10.1007/s002050050125. MR1664550

[FS14] Alessio Figalli and Henrik Shahgholian, *A general class of free boundary problems for fully nonlinear elliptic equations*, Arch. Ration. Mech. Anal. **213** (2014), no. 1, 269–286, DOI 10.1007/s00205-014-0734-0. MR3198649

[IM16a] Emanuel Indrei and Andreas Minne, *Nontransversal intersection of free and fixed boundaries for fully nonlinear elliptic operators in two dimensions*, Anal. PDE **9** (2016), no. 2, 487–502, DOI 10.2140/apde.2016.9.487. MR3513142

[IM16b] Emanuel Indrei and Andreas Minne, *Regularity of solutions to fully nonlinear elliptic and parabolic free boundary problems*, Ann. Inst. H. Poincaré Anal. Non Linéaire **33** (2016), no. 5, 1259–1277, DOI 10.1016/j.anihpc.2015.03.009. MR3542613

[Ind19a] Emanuel Indrei, *Boundary regularity and nontransversal intersection for the fully nonlinear obstacle problem*, Comm. Pure Appl. Math. **72** (2019), no. 7, 1459–1473, DOI 10.1002/cpa.21814. MR3957397

[Ind19b] Emanuel Indrei, *Non-transversal intersection of the free and fixed boundary in the mean-field theory of superconductivity*, Interfaces Free Bound. **21** (2019), no. 2, 267–272, DOI 10.4171/IFB/423. MR3986537

[KN77] D. Kinderlehrer and L. Nirenberg, *Regularity in free boundary problems*, Ann. Scuola Norm. Sup. Pisa Cl. Sci. (4) **4** (1977), no. 2, 373–391. MR0440187

[MM04] Norayr Matevosyan and Peter A. Markowich, *Behavior of the free boundary near contact points with the fixed boundary for nonlinear elliptic equations*, Monatsh. Math. **142** (2004), no. 1-2, 17–25, DOI 10.1007/s00605-004-0243-6. MR2065018

[PSU12] Arshak Petrosyan, Henrik Shahgholian, and Nina Uraltseva, *Regularity of free boundaries in obstacle-type problems*, Graduate Studies in Mathematics, vol. 136, American Mathematical Society, Providence, RI, 2012. MR2962060

[Saf94] M. V. Safonov, *On the boundary value problems for fully nonlinear elliptic equations of second order*, Mathematics Research Report No. MRR 049-94, Canberra: The Australian National University, 1994.

[Sch77] David G. Schaeffer, *Some examples of singularities in a free boundary*, Ann. Scuola Norm. Sup. Pisa Cl. Sci. (4) **4** (1977), no. 1, 133–144. MR0516201

[SU03] Henrik Shahgholian and Nina Uraltseva, *Regularity properties of a free boundary near contact points with the fixed boundary*, Duke Math. J. **116** (2003), no. 1, 1–34, DOI 10.1215/S0012-7094-03-11611-7. MR1950478

[Ura96] N. N. Ural′tseva, *$C^1$ regularity of the boundary of a noncoincident set in a problem with an obstacle* (Russian, with Russian summary), Algebra i Analiz **8** (1996), no. 2, 205–221; English transl., St. Petersburg Math. J. **8** (1997), no. 2, 341–353. MR1392033

Department of Mathematics, Purdue University, West Lafayette, Indiana 47907
*Email address*: eindrei@purdue.edu

Contemporary Mathematics
Volume **748**, 2020
https://doi.org/10.1090/conm/748/15059

# The Poisson integral formula for variable-coefficient elliptic systems in rough domains

Dorina Mitrea, Irina Mitrea, and Marius Mitrea

Abstract. Let $L$ be a second-order, homogeneous, divergence-form, elliptic system in $\mathbb{R}^n$ with complex-valued Lipschitz coefficients. The goal of this article is to prove a Poisson integral representation formula for solutions of the Dirichlet problem for $L$ in domains of a very general geometric nature, which involves the conormal derivative of the Green function for the transpose system $L^\top$ as integral kernel.

## 1. Introduction

Let $n \in \mathbb{N}$ denote the dimension of the Euclidean ambient space $\mathbb{R}^n$. Fix an integer $M \in \mathbb{N}$ along with an open set $\Omega$ and let $\mathrm{Lip}(\Omega)$ denote the class of Lipschitz functions in $\Omega$, that is

$$\mathrm{Lip}(\Omega) := \Big\{ f : \Omega \to \mathbb{C} : \sup_{x,y\in\Omega, x\neq y} \frac{|f(x)-f(y)|}{|x-y|} < \infty \Big\}. \tag{1.1}$$

Consider a second-order, homogeneous, divergence-form $M \times M$ system, with variable, complex-valued coefficients

$$\begin{gathered} a^{\alpha\beta}_{rs} \in \mathrm{Lip}(\Omega) \text{ indexed by} \\ \alpha, \beta \in \{1, \dots, M\} \text{ and } r, s \in \{1, \dots, n\}, \end{gathered} \tag{1.2}$$

written as

$$Lu := \Big( \partial_r \big( a^{\alpha\beta}_{rs} \partial_s u_\beta \big) \Big)_{1\leq\alpha\leq M}, \tag{1.3}$$

when acting on a sufficiently smooth vector-valued function $u = (u_\beta)_{1\leq\beta\leq M}$ in $\Omega$. Above and throughout the paper we will use the usual convention of summation over repeated indices. The system $L$ is also assumed to be elliptic, in the sense that

---

2010 *Mathematics Subject Classification.* Primary: 31A10, 35C15, 35J08, 35J47, 42B37. Secondary: 28A75, 31A20, 35J15, 35J57.

*Key words and phrases.* Poisson integral representation formula, nontangential maximal function, second-order elliptic system, Green function, harmonic measure, Ahlfors regularity.

The first author was supported in part by the Simons Foundation grant #426669.

The second author has been supported in part by Simons Foundation grant #616050 and NSF DMS grant 1900938.

The third author has been supported in part by the Simons Foundation grant #281566.

there exists a real number $c \in (0,\infty)$ such that the following Legendre-Hadamard condition is satisfied:

$$\begin{aligned} &\mathrm{Re}\left[a^{\alpha\beta}_{rs}(x)\xi_r\xi_s\overline{\eta_\alpha}\eta_\beta\right] \geq c|\xi|^2|\eta|^2 \quad \text{for every } \ x \in \Omega, \\ &\xi = (\xi_r)_{1\leq r\leq n} \in \mathbb{R}^n, \quad \text{and} \quad \eta = (\eta_\alpha)_{1\leq\alpha\leq M} \in \mathbb{C}^M, \end{aligned} \tag{1.4}$$

where 'bar' denotes complex conjugation. In particular, the (real) transpose of $L$, i.e., the system

$$L^\top u := \Big(\partial_r\big(a^{\beta\alpha}_{sr}\partial_s u_\beta\big)\Big)_{1\leq\alpha\leq M}, \tag{1.5}$$

is also Legendre-Hadamard elliptic.

The goal of this paper is to prove a Poisson integral representation formula for solutions of the Dirichlet problem for a variable-coefficient elliptic second-order system $L$, in a very general setting, under geometric and analytic hypotheses which are in the nature of best possible. The said integral representation formula involves the conormal derivative of the Green function for the transpose system $L^\top$ as integral kernel. For detailed definitions we refer the reader to Section 2.

THEOREM 1.1. *Let $\Omega \subset \mathbb{R}^n$, where $n \in \mathbb{N}$ with $n \geq 2$, be a bounded open set which is locally pathwise nontangentially accessible, has a lower Ahlfors regular boundary, and with the property that $\sigma := \mathcal{H}^{n-1}\lfloor\partial\Omega$ is a doubling measure on $\partial\Omega$. Suppose $L$ is a second-order, homogenous, divergence-form, elliptic $M \times M$ system in $\mathbb{R}^n$, written as in (1.3) for some coefficient tensor $A = \big(a^{\alpha\beta}_{rs}\big)_{\substack{1\leq r,s\leq n\\ 1\leq\alpha,\beta\leq M}}$ with individual entries in $\mathrm{Lip}(\Omega)$. Finally, fix some aperture parameter $\kappa \in (0,\infty)$, along with an arbitrary point $x_0 \in \Omega$, and suppose $0 < \rho < \frac{1}{4}\,\mathrm{dist}\,(x_0,\partial\Omega)$.*

*Then there exists some aperture parameter $\widetilde{\kappa} > 0$, which depends only on $\Omega$ and $\kappa$, with the following significance. Assume $G$ is a matrix-valued function satisfying*

$$\begin{cases} G \in \big[W^{1,1}_{\mathrm{loc}}(\Omega)\big]^{M\times M}, \\ L^\top G = -\delta_{x_0} I_{M\times M} \ \ \textit{in} \ \big[\mathcal{D}'(\Omega)\big]^{M\times M}, \\ (\nabla G)\Big|^{\widetilde{\kappa}-\mathrm{n.t.}}_{\partial\Omega} \ \ \textit{exists (in } \mathbb{C}^{n\cdot M^2}\textit{) at } \sigma\textit{-a.e. point on } \partial\Omega, \\ G\Big|^{\widetilde{\kappa}-\mathrm{n.t.}}_{\partial\Omega} = 0 \in \mathbb{C}^{M\times M} \ \ \textit{at } \sigma\textit{-a.e. point on } \partial\Omega, \end{cases} \tag{1.6}$$

*(where $L^\top$ acts on the columns of $G$ and $I_{M\times M}$ is the $M \times M$ identity matrix), and assume $u$ is a $\mathbb{C}^M$-valued function satisfying*

$$\begin{cases} u \in \big[\mathscr{C}^1(\Omega)\big]^M, \quad Lu = 0 \ \ \textit{in} \ \Omega, \\ u\big|^{\kappa-\mathrm{n.t.}}_{\partial\Omega} \ \ \textit{exists at } \sigma\textit{-a.e. point on } \partial\Omega, \\ \displaystyle\int_{\partial\Omega} \mathcal{N}^\rho_\kappa u \cdot \mathcal{N}^\rho_{\widetilde{\kappa}}(\nabla G)\,d\sigma < +\infty. \end{cases} \tag{1.7}$$

*Then one has the Poisson integral representation formula*

$$u(x_0) = -\int_{\partial_*\Omega} \big\langle u\big|^{\kappa-\mathrm{n.t.}}_{\partial\Omega}, \partial^{A^\top}_\nu G\big\rangle\,d\sigma, \tag{1.8}$$

*where $\nu$ denotes the geometric measure theoretic outward unit normal to $\Omega$ and $\partial_\nu^{A^\top}$ stands for the conormal derivative associated with the transpose coefficient tensor $A^\top$, acting on the columns of the matrix-valued function $G$.*

A glossary of terms employed in the statement of Theorem 1.1 is as follows: the class of locally pathwise nontangentially accessible sets is introduced in Definition 2.4, while lower Ahlfors regularity is defined in (2.4). Also, $\mathcal{H}^{n-1}$ is the $(n-1)$-dimensional Hausdorff measure in $\mathbb{R}^n$, and being a doubling measure is characterized in (2.5). As is customary, $\mathcal{D}'(\Omega)$ stands for the space of distributions in the open set $\Omega$, and $\delta_{x_0}$ is the Dirac distribution with mass at $x_0$. All boundary traces are taken in the pointwise nontangential sense, as discussed in Definition 2.2, whereas truncated nontangential maximal operators are formally introduced in (2.17). Next, the definition of the geometric measure theoretic boundary $\partial_*\Omega$ is recalled in (2.39). Moreover, the geometric measure theoretic outward unit normal appears in (2.43). The conormal derivative associated with some coefficient tensor is given in (2.48). The reader is also alerted to the fact that $L^\top$ denotes the (real) transpose of the system $L$ (cf. (1.5)), while $A^\top$ is the (real) transpose of the coefficient tensor $A$. Lastly, the pairing $\langle\cdot,\cdot\rangle$ appearing under the integral sign in (1.8) is the pointwise inner product between the vector $u\big|_{\partial\Omega}^{\kappa-\mathrm{n.t.}}$ and the columns of the matrix $\partial_\nu^{A^\top}G$.

A few comments about the very formulation of Theorem 1.1 are also in order. First, since we are assuming that $G\in\big[W^{1,1}_{\mathrm{loc}}(\Omega)\big]^{M\times M}$, the condition in the third line of (1.6) is meaningfully formulated. The fact that the integral condition formulated in the third line of (1.7) is meaningful is discussed in the early stages of the proof of Theorem 1.1 in §3. The said condition ensures that the integral in (1.8) is absolutely convergent, hence the conclusion in the theorem has a clear meaning. Remarkably, the integral condition formulated in the third line of (1.7) (which as just noted, is expected in view of the right-hand side in (1.8)) is the only size assumption we impose on the Green function and the null-solution $u$. If in place of the demand formulated in the first line of (1.7) we merely ask that $u\in\big[L^p_{\mathrm{loc}}(\Omega,\mathcal{L}^n)\big]^M$ for some $p\in(1,\infty)$ then elliptic regularity (cf. (2.51)) ensures that actually $u$ belongs to $\big[\mathscr{C}^1(\Omega)\big]^M$, as demanded in (1.7) to begin with.

The integral representation formula recorded in (1.8) readily implies the vanishing property $u(x_0)=0$ for any null-solution $u$ of the Dirichlet problem for the system $L$ in $\Omega$, in a formulation which uses the last condition in (1.7) as the demand imposed on the size of the solution.

In the most standard setting, namely when $\Omega:=B(0,1)\subset\mathbb{R}^n$ and $L:=\Delta$, the Laplacian in $\mathbb{R}^n$ (in which case $M:=1$ and $A:=I_{n\times n}$, the $n\times n$ identity matrix), Theorem 1.1 gives that whenever $\kappa>0$ and

$$u \text{ is a harmonic function in } B(0,1) \text{ satisfying } \int_{\partial B(0,1)}\mathcal{N}_\kappa u\,d\sigma<+\infty \tag{1.9}$$

with $\sigma:=\mathcal{H}^{n-1}\lfloor\partial B(0,1)$ denoting the surface measure on the unit sphere, we have the integral representation formula

$$u(x_0)=\int_{\partial B(0,1)}\big(u\big|_{\partial B(0,1)}^{\kappa-\mathrm{n.t.}}\big)(y)k_{x_0}(y)\,d\sigma(y)\quad\text{for each } x_0\in B(0,1), \tag{1.10}$$

where, with $\omega_{n-1} := \mathcal{H}^{n-1}(\partial B(0,1))$,

$$k_{x_0}(y) := \frac{1-|x_0|^2}{\omega_{n-1}|x_0-y|^n} \quad \text{with } y \in \partial B(0,1) \text{ and } x_0 \in B(0,1), \tag{1.11}$$

is the Poisson kernel for the Laplacian for the unit ball. Indeed, the assumptions in (1.9) guarantee the applicability of Fatou's theorem (cf., e.g., [**10**, Theorem 1.4.9, p. 16]) which ensures the existence of the nontangential boundary trace $u\big|^{\kappa-\mathrm{n.t.}}_{\partial B(0,1)}$ at $\sigma$-a.e. point on $\partial B(0,1)$. Also, since Green's function $G_{x_0}$ with pole at the point $x_0 \in B(0,1)$ for the Laplacian in the unit ball is given (assuming $n \geq 3$) at each point $y \in \overline{B(0,1)} \setminus \{x_0\}$ by

$$G_{x_0}(y) := \frac{1}{\omega_{n-1}(2-n)}\Big\{|x_0-y|^{2-n} - \big(|x_0||y-x_0/|x_0|^2|\big)^{2-n}\Big\}, \tag{1.12}$$

then all conditions stipulated in (1.6) are presently satisfied (for any aperture parameter $\widetilde{\kappa} > 0$). In addition, (1.12) implies that for each $\widetilde{\kappa} > 0$, each $x_0 \in B(0,1)$, and each $0 < \rho < \frac{1}{4}\operatorname{dist}\big(x_0, \partial B(0,1)\big)$ fixed, $\mathcal{N}^{\rho}_{\widetilde{\kappa}}(\nabla G_{x_0}) \approx 1$ on $\partial B(0,1)$. Consequently, the integral condition in the last line of (1.7) simply becomes the finiteness condition from (1.9). Finally, an elementary computation (cf., e.g., [**4**, p. 40]) shows that the conormal derivative of the Green function is

$$\partial_\nu^{A^\top} G_{x_0}(y) = y \cdot \nabla_y G_{x_0}(y) = -k_{x_0}(y) \quad \text{at each } y \in \partial B(0,1). \tag{1.13}$$

On account of these observations Theorem 1.1 applies, and formula (1.8) presently yields (1.10).

Several conclusions may be drawn from the above discussion. First, Theorem 1.1 may be regarded as a far-reaching generalization of the Poisson integral representation formula (1.9)-(1.10). Second, the fact that even when specialized to this basic setting there are no redundant hypotheses in our result (other than the second line in (1.7) which, as noted earlier now is a consequence of the availability of a Fatou theorem for harmonic functions in the unit ball), points to the optimality of Theorem 1.1. Third, the Poisson integral representation formula (1.10) may be proved directly, under the assumptions made in (1.9), by invoking the classical version of this result (valid for harmonic functions which are continuous up to, and including, the boundary) in $B(0,r)$ with $r \in (0,1)$, and then passing to limit in the corresponding version of (1.10) as $r \nearrow 1$. In such a scenario, the fact that $\mathcal{N}_\kappa u \in L^1\big(\partial B(0,1), \sigma\big)$ permits us to invoke Lebesgue's Dominated Convergence Theorem to conclude that (1.10) holds under the assumptions made in (1.9).

This being said, an approach similar in flavor, which involves approximating the boundary of our original domain $\Omega$ in Theorem 1.1 by "nice" subdomains on which the Poisson integral representation formula is known to hold is utterly unrealistic, given the degree of geometric generality considered in Theorem 1.1. Thus, we are forced to develop a new line of attack and bring in tools which can cope with the roughness directly. A case in point is the new, much more potent, version of the Divergence Theorem recently proved in [**16**], which we recall in Theorem 2.11 in which the boundary trace of the vector field is considered in the nontangential pointwise sense. In addition, we make essential use of elliptic regularity results and interior estimates for systems with variable coefficients exhibiting only a mild amount of smoothness from [**18**].

We wish to note that the geometric environment presently considered in Theorem 1.1 is vastly inclusive, allowing for the consideration of sets which are not NTA domains in the sense of D. Jerison and C. Kenig [**9**] (since both inner and outer cusps are allowed), as well as sets whose boundaries fail to be uniformly rectifiable in the sense of G. David and S. Semmes [**3**] (since any bounded open set with an Ahlfors regular boundary is allowed).

Let us consider in greater detail the scalar case, i.e., when $M = 1$, restricting ourselves to symmetric elliptic operators with bounded measurable real-valued coefficients. From the work of W. Littman, G. Stampacchia, and H. Weinberger [**13**] we know that under the assumption that $\Omega$ is regular for the Laplacian, one may solve the classical Dirichlet problem for $L$ with continuous boundary data $f$. Moreover, in such a setting the solution $u$ may be represented as

$$u(x_0) = \int_{\partial\Omega} f\, d\omega_L^{x_0}, \qquad x_0 \in \Omega, \tag{1.14}$$

where $\omega_L^{x_0}$ is the harmonic measure with pole at $x_0 \in \Omega$ associated with the operator $L$ and the domain $\Omega$. One of the basic issues in this regard is determining under what geometric/analytic hypotheses it is possible to express the said harmonic measure as

$$d\omega_L^{x_0} = -\partial_\nu^{A^\top} G_{x_0}\, d\sigma, \qquad x_0 \in \Omega. \tag{1.15}$$

For example, this issue is explicitly raised by J. Garnett and D. Marshall in [**6**, Question 2, p. 49]. In this regard, comparing (1.14) with (1.8) in Theorem 1.1 yields the following corollary which refines work in [**2**, Theorem 5.7, p. 79], [**11**, Proposition A.1.1, p. 382], [**11**, Proposition A.2.2, p. 385], and [**12**, Lemma 3.4, p. 16], where more restrictive classes of sets have been considered.

COROLLARY 1.2. *Let $\Omega \subset \mathbb{R}^n$, where $n \in \mathbb{N}$ with $n \geq 2$, be a bounded open set which is regular for the Laplacian, is locally pathwise nontangentially accessible, has a lower Ahlfors regular boundary, and for which $\sigma := \mathcal{H}^{n-1}\lfloor\partial\Omega$ is a doubling measure on $\partial\Omega$. Consider a second-order, homogeneous, symmetric, divergence-form operator with real-valued coefficients*

$$L := \operatorname{div} A\nabla, \quad \text{with} \quad A \in \big[\mathrm{Lip}(\Omega)\big]^{n\times n}, \tag{1.16}$$

*satisfying the ellipticity condition (for some $c \in (0,\infty)$)*

$$\langle A(x)\xi, \xi\rangle \geq c|\xi|^2 \quad \text{for every} \quad x \in \Omega \quad \text{and} \quad \xi \in \mathbb{R}^n. \tag{1.17}$$

*Also, fix a point $x_0 \in \Omega$ and assume that there exists a function $G_{x_0}$ which, for some aperture parameter $\kappa \in (0,\infty)$, satisfies*

$$\begin{cases} G_{x_0} \in W^{1,1}_{\mathrm{loc}}(\Omega) \\ L^\top G_{x_0} = -\delta_{x_0} \quad \text{in } \mathcal{D}'(\Omega), \\ (\nabla G_{x_0})\Big|^{\kappa-\mathrm{n.t.}}_{\partial\Omega} \quad \text{exists (in } \mathbb{C}^n\text{) at } \sigma\text{-a.e. point on } \partial\Omega, \\ G_{x_0}\Big|^{\kappa-\mathrm{n.t.}}_{\partial\Omega} = 0 \quad \text{at } \sigma\text{-a.e. point on } \partial\Omega, \\ \mathcal{N}_\kappa^\rho(\nabla G_{x_0}) \in L^1(\partial\Omega,\sigma) \quad \text{for some } \rho \in \big(0\,,\, \mathrm{dist}\,(x_0,\partial\Omega)\big). \end{cases} \tag{1.18}$$

*Then*

(1.19) *$\omega_L^{x_0}$, the harmonic measure for the operator $L$ in $\Omega$ with pole at $x_0$, is absolutely continuous with respect to the surface measure $\sigma = \mathcal{H}^{n-1}\lfloor\partial\Omega$.*

*Moreover, if $\nu$ denotes the geometric measure theoretic outward unit normal to $\Omega$, then the Radon-Nikodym derivative of the harmonic measure $\omega_L^{x_0}$ with respect to the surface measure $\sigma$ is given by*

$$
\begin{aligned}
\frac{d\omega_L^{x_0}}{d\sigma} &= -\mathbf{1}_{\partial_*\Omega} \cdot \partial_\nu^A G_{x_0} \quad \textit{at } \sigma\textit{-a.e. point on } \partial\Omega, \\
&\textit{where } \partial_\nu^A G_{x_0} := \big\langle \nu\,,\, A(\nabla G_{x_0})\big|_{\partial\Omega}^{\kappa-\mathrm{n.t.}}\big\rangle.
\end{aligned}
\tag{1.20}
$$

In this vein we wish to remark that Green functions associated with elliptic operators having bounded measurable real-valued coefficients in arbitrary open sets have been studied by M. Grüter and K.-O. Widman in [**7**]. In particular, in the case of domains of class $\mathscr{C}^{1,\alpha}$ with $\alpha \in (0,1)$ the existence of a Green function satisfying (1.6) is guaranteed by [**7**, Theorem 3.5(ii), p. 33]. See also [**8**, Step 3, p. 2832] for the construction of a Green function for the Laplacian satisfying the properties listed in (1.6) when $\Omega$ is a regular SKT domain in $\mathbb{R}^n$.

The proof of Theorem 1.1 is given in §3, after reviewing some background material and a number of preliminary results in §2.

We also indicate a further partial refinement. As mentioned earlier, a key ingredient in the proof of Theorem 1.1 is the local elliptic regularity result from [**18**, Proposition 3.1], which further implies interior estimates via a scaling argument (as in the proof of [**18**, Proposition 3.4]). For general systems such local elliptic regularity results typically require stronger regularity assumptions for its coefficients than in the case of scalar elliptic operators. Indeed, for scalar operators (i.e., in the case when $M = 1$) local elliptic regularity results have been established in [**17**, Proposition 2.3] assuming that the coefficients are only Hölder continuous of order $\alpha \in (0,1)$. As such, the scalar version of Theorem 1.1 (hence also Corollary 1.2) continues to hold when the regularity assumptions on the coefficients are relaxed from Lipschitz to Hölder.

## 2. Preliminary matters

We first describe some standard notation, and elaborate on conventions frequently employed. Throughout, we set $\mathbb{N}_0 := \mathbb{N} \cup \{0\}$, and we let $\mathbb{Q}_+$ denote the set of positive rational numbers. For each $n \in \mathbb{N}$ we let $\mathcal{L}^n$ stand for the Lebesgue measure in $\mathbb{R}^n$. Given an open set $\Omega \subseteq \mathbb{R}^n$ and an integrability exponent $p \in (0,\infty]$, we shall denote by $L^p(\Omega, \mathcal{L}^n)$ the Lebesgue space of $\mathcal{L}^n$-measurable $p$-th power integrable functions with respect to the measure $\mathcal{L}^n$ in $\Omega$. In the same setting, $L^p_{\mathrm{loc}}(\Omega, \mathcal{L}^n)$ is the space of functions $f$ defined in $\Omega$ with the property that $f|_{B(x,r)} \in L^p\big(B(x,r), \mathcal{L}^n\big)$ whenever $x \in \Omega$ and $r > 0$ is such that $\overline{B(x,r)} \subseteq \Omega$ (with $\overline{B(x,r)}$ denoting the closed ball centered at $x$ of radius $r$). As is customary, we shall denote by $\mathcal{D}'(\Omega)$ the space of distributions in $\Omega$. In particular, $L^1_{\mathrm{loc}}(\Omega, \mathcal{L}^n) \subseteq \mathcal{D}'(\Omega)$. For each $u \in \mathcal{D}'(\Omega)$ we then agree to denote by $\operatorname{regsupp} u$ (aka the regular support of $u$) the smallest relatively closed subset $S$ of $\Omega$ with the property that $u\big|_{\Omega\setminus S} \in L^1_{\mathrm{loc}}(\Omega \setminus S, \mathcal{L}^n)$. The notation $\mathscr{E}'(\Omega)$ is reserved for the subspace of $\mathcal{D}'(\Omega)$ consisting of distributions in $\Omega$ with compact support. Hence, if for each $x_o \in \Omega$ we denote by $\delta_{x_o}$ the Dirac distribution with mass at $x_o$, then $\delta_{x_o} \in \mathscr{E}'(\Omega)$. We also agree to let ${}_{(\mathscr{C}^\infty(\Omega))^*}\big(\cdot\,,\,\cdot\big)_{\mathscr{C}^\infty(\Omega)}$ stand for the pairing between compactly supported distributions and smooth functions in $\Omega$ (cf., e.g., [**15**]).

Given an arbitrary open set $\Omega \subseteq \mathbb{R}^n$ along with $p \in [1,\infty]$ and $k \in \mathbb{N}$, denote by $W^{k,p}(\Omega)$ the standard $L^p$-based Sobolev space of order $k$ in $\Omega$, consisting of locally integrable functions (with respect to the Lebesgue measure) in $\Omega$ whose partial derivatives of order $\leq k$ (considered in the sense of distributions in $\Omega$) are $p$-th power integrable functions in $\Omega$. Also, we denote by $W^{k,p}_{\rm loc}(\Omega)$ the local version of this space.

The symbol $\mathbf{1}_E$ is used to denote the characteristic function of a given set $E$. Finally, for any two nonempty closed subsets $E, F$ of the Euclidean space $\mathbb{R}^n$ define $\operatorname{dist}(E,F) := \inf\{|x-y| : x \in E,\ y \in F\}$. Given $s \geq 0$, by $\mathcal{H}^s$ we shall denote the $s$-dimensional Hausdorff measure in $\mathbb{R}^n$. As is well-known (cf., e.g., [**5**, Theorem 1, p. 61]),

$$\mathcal{H}^s \text{ is a Borel-regular measure in } \mathbb{R}^n, \text{ for each } s \in [0,\infty). \tag{2.1}$$

The following result, which pertains to the nature of the Hausdorff measure restricted to Hausdorff-measurable subsets of the Euclidean ambient, plays a role in this work (a proof may be found in [**16**]).

LEMMA 2.1. *Consider $s \in [0,\infty)$ and let $X \subseteq \mathbb{R}^n$ be a $\mathcal{H}^s$-measurable set having the property that $\mathcal{H}^s(X \cap K) < +\infty$ for every compact $K \subset \mathbb{R}^n$. If $\tau_{\mathbb{R}^n}\big|_X$ denotes the topology induced by the ambient Euclidean space $\mathbb{R}^n$ on the set $X$, then*

$$\mathcal{H}^s\lfloor X \text{ is a complete, locally finite, Borel-regular measure on } \big(X, \tau_{\mathbb{R}^n}\big|_X\big). \tag{2.2}$$

*Moreover,*

$$\begin{aligned}&\text{the measure } \mathcal{H}^s\lfloor X \text{ is separable, and for each } p \in (0,\infty) \text{ the}\\ &\text{Lebesgue space } L^p\big(X, \mathcal{H}^s\lfloor X\big) \text{ is separable.}\end{aligned} \tag{2.3}$$

A closed subset $\Sigma \subseteq \mathbb{R}^n$ is said to be lower Ahlfors regular provided there exists $c \in (0,\infty)$ such that

$$c\,r^{n-1} \leq \mathcal{H}^{n-1}\big(B(x,r) \cap \Sigma\big) \text{ for each } x \in \Sigma \text{ and } r \in (0, 2\operatorname{diam}\Sigma). \tag{2.4}$$

Also, a Borel measure $\mu$ on $\Sigma$ is said to be doubling if there exists some $C \in [1,\infty)$ such that

$$\begin{aligned}&0 < \mu\big(B(x,2r) \cap \Sigma\big) \leq C\mu\big(B(x,r) \cap \Sigma\big) < +\infty\\ &\qquad\qquad \text{for all } x \in \Sigma \text{ and all } r \in (0,\infty).\end{aligned} \tag{2.5}$$

To begin in earnest, fix an open, nonempty, proper subset $\Omega$ of $\mathbb{R}^n$ and denote by $\delta_{\partial\Omega}(x)$ the distance from an arbitrary point $x \in \mathbb{R}^n$ to $\partial\Omega$. Given $\kappa > 0$ arbitrary, we define the nontangential approach regions to $\partial\Omega$ of aperture parameter $\kappa$ by setting

$$\Gamma_\kappa(x) = \Gamma_{\Omega,\kappa}(x) := \big\{y \in \Omega : |x-y| < (1+\kappa)\delta_{\partial\Omega}(y)\big\}, \qquad \forall\, x \in \partial\Omega. \tag{2.6}$$

If $u : \Omega \to \mathbb{R}$ is an arbitrary Lebesgue measurable function define the nontangential maximal function of $u$ with aperture $\kappa$ as

$$\begin{aligned}&\mathcal{N}_\kappa u : \partial\Omega \longrightarrow [0,+\infty],\\ &(\mathcal{N}_\kappa u)(x) := \|u\|_{L^\infty(\Gamma_\kappa(x),\mathcal{L}^n)} \text{ for all } x \in \partial\Omega.\end{aligned} \tag{2.7}$$

In particular, one may check that

$$\sup_{x \in \partial\Omega} (\mathcal{N}_\kappa u)(x) = \|u\|_{L^\infty(\Omega,\mathcal{L}^n)}. \tag{2.8}$$

Of course,

$$\text{(2.9)}\qquad \begin{array}{c}\text{whenever } u\in\mathscr{C}^0(\Omega) \text{ one has}\\ (\mathcal{N}_\kappa u)(x)=\sup_{y\in\Gamma_\kappa(x)}|u(y)| \ \text{ for all } \ x\in\partial\Omega.\end{array}$$

More generally, if $u:\Omega\to\mathbb{R}$ is a Lebesgue measurable function and $E\subseteq\Omega$ is a Lebesgue measurable set, we denote by $\mathcal{N}_\kappa^E u$ the non-tangential maximal function of $u$ restricted to $E$

$$\text{(2.10)}\qquad \mathcal{N}_\kappa^E u:\partial\Omega\longrightarrow[0,+\infty]$$

defined at each $x\in\partial\Omega$ as

$$\text{(2.11)}\qquad (\mathcal{N}_\kappa^E u)(x):=\|u\|_{L^\infty(\Gamma_\kappa(x)\cap E,\mathcal{L}^n)}.$$

It turns out (cf. [**16**]) that

$$\text{(2.12)}\qquad \mathcal{N}_\kappa^E u:\partial\Omega\longrightarrow[0,+\infty] \ \text{ is a Borel-measurable function.}$$

In addition, an inspection of definitions shows that

(2.13) $\big(\mathcal{N}_\kappa^E u\big)(x)=\sup_{y\in\Gamma_\kappa(x)\cap E}|u(y)|$ at each point $x\in\partial\Omega$, whenever $E\subseteq\Omega$ is open and $u$ is a Lebesgue measurable function defined on $\Omega$ which happens to be actually continuous on the given set $E$,

and

(2.14) $\mathcal{N}_\kappa^E u\le\mathcal{N}_{\kappa'}^{E'}u$ everywhere on $\partial\Omega$, if $u:E\to\mathbb{C}$ is Lebesgue measurable, $\kappa'\ge\kappa>0$, and $E,E'\subseteq\Omega$ are any two Lebesgue measurable sets such that $E\subseteq E'$.

Also, for any Lebesgue measurable functions $u,w:\Omega\to\mathbb{R}$ and any Lebesgue measurable set $E\subseteq\Omega$, we have

$$\text{(2.15)}\qquad \begin{array}{c}\mathcal{N}_\kappa^E(uw)\le\big(\mathcal{N}_\kappa^E u\big)\cdot\big(\mathcal{N}_\kappa^E w\big) \ \text{ pointwise on } \ \partial\Omega,\\ \text{with the convention that } 0\cdot\infty=\infty\cdot 0=0 \text{ used in the right side,}\end{array}$$

and

$$\text{(2.16)}\qquad \mathcal{N}_\kappa^E u=\mathcal{N}_\kappa(u\mathbf{1}_E)\le\mathcal{N}_\kappa u \ \text{ on } \ \partial\Omega.$$

For ease of notation, for each $\varepsilon>0$ we agree to abbreviate

$$\text{(2.17)}\qquad \mathcal{N}_\kappa^\varepsilon u:=\mathcal{N}_\kappa^{\mathcal{O}_\varepsilon}u \ \text{ where } \ \mathcal{O}_\varepsilon:=\big\{x\in\Omega:\,\delta_{\partial\Omega}(x)<\varepsilon\big\}.$$

Hence, $\mathcal{N}_\kappa^\varepsilon$ is a truncated nontangential maximal operator at "height" $\varepsilon$, and this is the sense in which the last line in (1.7) should be interpreted.

Next, introduce

$$\text{(2.18)}\qquad A_\kappa(\partial\Omega):=\big\{x\in\partial\Omega:\,x\in\overline{\Gamma_\kappa(x)}\,\big\}.$$

Informally, $A_\kappa(\partial\Omega)$ consists of those boundary points which are "accessible" in a nontangential fashion (specifically, from within nontangential approach regions of aperture $\kappa$). Clearly,

$$\text{(2.19)}\qquad A_{\kappa_0}(\partial\Omega)\subseteq A_\kappa(\partial\Omega) \ \text{ whenever } \ 0<\kappa_0\le\kappa<\infty,$$

and it turns out (cf. [**16**]) that for each $\kappa>0$ we have

(2.20) $A_\kappa(\partial\Omega)$ is a $G_\delta$ set in $\partial\Omega$ (equipped with the relative topology, inherited from the Euclidean space $\mathbb{R}^n$); in particular, each set $A_\kappa(\partial\Omega)$ is Borelian, hence $\mathcal{H}^{n-1}$-measurable.

We next recall the definition of the nontangential boundary trace given in [**16**].

DEFINITION 2.2. Fix $\kappa > 0$ and let $u$ be a real-valued Lebesgue measurable function defined $\mathcal{L}^n$-a.e. in an open set $\Omega \subset \mathbb{R}^n$. Consider a point $x \in A_\kappa(\partial\Omega)$. Then one says that the nontangential limit of $u$ at $x$ exists, and its value is the number $a \in \mathbb{R}$, provided

(2.21) for every $\varepsilon > 0$ there exist $r_\varepsilon > 0$ and a Lebesgue measurable set $N_\varepsilon \subset \Gamma_\kappa(x)$ with $\mathcal{L}^n(N_\varepsilon) = 0$ such that $|u(y) - a| < \varepsilon$ for every point $y \in \big(\Gamma_\kappa(x) \cap B(x, r_\varepsilon)\big) \setminus N_\varepsilon$.

Whenever the nontangential limit of $u$ at $x$ exists, its value is denoted by $\big(u\big|_{\partial\Omega}^{\kappa-\text{n.t.}}\big)(x)$.

We make a few observations that are apparent from definitions. First, we emphasize that, in contrast with the "standard" definition of nontangential limit, no continuity assumption is a priori imposed on the function in question. Second,

(2.22) if $x \in A_\kappa(\partial\Omega)$ and $\big(u\big|_{\partial\Omega}^{\kappa-\text{n.t.}}\big)(x)$ exists, then for any $\kappa' \in (0, \kappa]$ we have $x \in A_{\kappa'}(\partial\Omega)$ and $\big(u\big|_{\partial\Omega}^{\kappa'-\text{n.t.}}\big)(x) = \big(u\big|_{\partial\Omega}^{\kappa-\text{n.t.}}\big)(x)$.

Third, we observe that the nontangential limit of $u$ at $x$ exists if and only if

(2.23) there exists a Lebesgue measurable set $N(x) \subset \Gamma_\kappa(x)$, with $\mathcal{L}^n(N(x)) = 0$, such that $\lim\limits_{(\Gamma_\kappa(x)\setminus N(x))\ni y\to x} u(y)$ exists.

Indeed, (2.23) implies (2.21) with $N_\varepsilon := N(x)$ and $a$ the value of the limit in (2.23). Conversely, (2.21) implies that $N(x) := \bigcup_{j\in\mathbb{N}} N_{1/j}$ does the job in (2.23).

In particular, the description of the nontangential limit in (2.23) implies that if (as is usually the case) one works with equivalence classes, obtained by identifying functions which coincide $\mathcal{L}^n$-a.e., the nontangential limit is independent of the specific choice of a representative in a given equivalence class.

Fourth, whenever the limit in (2.23) exists, it actually equals $\big(u\big|_{\partial\Omega}^{\kappa-\text{n.t.}}\big)(x)$. As a consequence, whenever $x \in A_\kappa(\partial\Omega)$ and the nontangential limit of $u$ at $x$ exists, for each $\varepsilon > 0$ we have

$$\Big|\big(u\big|_{\partial\Omega}^{\kappa-\text{n.t.}}\big)(x)\Big| \leq (\mathcal{N}_\kappa^\varepsilon u)(x) \leq (\mathcal{N}_\kappa u)(x). \tag{2.24}$$

Fifth, in the class of continuous functions the current definition of nontangential boundary limit takes a simpler form, namely if $x \in A_\kappa(\partial\Omega)$ then

$$\big(u\big|_{\partial\Omega}^{\kappa-\text{n.t.}}\big)(x) = \lim_{\Gamma_\kappa(x)\ni y\to x} u(y), \qquad \forall\, u \in \mathscr{C}^0(\Omega). \tag{2.25}$$

In particular,

$$u\big|_{\partial\Omega}^{\kappa-\text{n.t.}} = u\big|_{\partial\Omega} \quad \text{whenever} \quad u \in \mathscr{C}^0(\overline{\Omega}). \tag{2.26}$$

Sixth, whenever $x \in A_\kappa(\partial\Omega)$ and $u, w : \Omega \to \mathbb{R}$ are two Lebesgue measurable functions such that the nontangential limits of $u$ and $w$ at $x$ exist, it follows that $\big((uw)\big|_{\partial\Omega}^{\kappa-\text{n.t.}}\big)(x)$ exists as well, and

$$\Big((uw)\big|_{\partial\Omega}^{\kappa-\text{n.t.}}\Big)(x) = \big(u\big|_{\partial\Omega}^{\kappa-\text{n.t.}}\big)(x) \cdot \big(w\big|_{\partial\Omega}^{\kappa-\text{n.t.}}\big)(x). \tag{2.27}$$

Seventh,

(2.28) given a point $x \in A_\kappa(\partial\Omega)$ along with some $u \in L^\infty_{\text{loc}}(\Omega, \mathcal{L}^n)$ such that $\big(u\big|_{\partial\Omega}^{\kappa-\text{n.t.}}\big)(x)$ exists, it follows that $(\mathcal{N}_\kappa^\varepsilon u)(x) < +\infty$ for each truncation parameter $\varepsilon > 0$.

Following [**16**], we next define the nontangentially accessible boundary of an open set.

DEFINITION 2.3. Define the nontangentially accessible boundary of any given nonempty open proper subset $\Omega$ of $\mathbb{R}^n$ as

$$\partial_{_{\rm nta}}\Omega := \bigcap_{\kappa>0} A_\kappa(\partial\Omega) = \big\{x \in \partial\Omega : x \in \overline{\Gamma_\kappa(x)} \text{ for each } \kappa > 0\big\}. \tag{2.29}$$

Since (2.19) and (2.29) imply that

$$\text{for any open nonempty proper subset } \Omega \text{ of } \mathbb{R}^n \text{ we have } \partial_{_{\rm nta}}\Omega = \textstyle\bigcap_{\kappa\in\mathbb{Q}_+} A_\kappa(\partial\Omega), \tag{2.30}$$

we deduce from (2.30) and (2.20) that

(2.31) given any open nonempty proper subset $\Omega$ of $\mathbb{R}^n$, it follows that the set $\partial_{_{\rm nta}}\Omega$ is Borelian, hence $\mathcal{H}^{n-1}$-measurable.

We shall also need the notion of locally pathwise nontangentially accessible set, originally introduced in [**16**].

DEFINITION 2.4. Call an open nonempty proper subset $\Omega$ of $\mathbb{R}^n$ locally pathwise nontangentially accessible provided the following holds:

(2.32) given any $\kappa > 0$ there exist $\widetilde{\kappa} \geq \kappa$ along with $c \in [1,\infty)$ and $d > 0$ such that $\sigma$-a.e. point $x \in \partial\Omega$ has the property that any $y \in \Gamma_\kappa(x)$ with $\delta_{\partial\Omega}(y) < d$ may be joined by a rectifiable curve $\gamma_{x,y}$ satisfying $\gamma_{x,y} \setminus \{x\} \subset \Gamma_{\widetilde{\kappa}}(x)$ and whose length is $\leq c|x-y|$.

For example, with the class of one-sided NTA domains defined as the collection of all open sets satisfying an interior corkscrew condition and an interior Harnack chain condition (which may be thought of as some quantitative versions of openness and connectivity, respectively; cf. [**9**] for details), it follows that

(2.33) any one-sided NTA domain is a locally pathwise nontangentially accessible set.

Of course, being a locally pathwise nontangentially accessible set is a much weaker condition than being a one-sided NTA domain. For example, a partially slit disk is a locally pathwise nontangentially accessible set, but fails to satisfy the Harnack chain condition.

Our notion of local pathwise nontangential accessibility should be compared with the concept of semi-uniformity introduced by H. Aikawa and K. Hirata in [**1**]. The latter is a quantitative connectivity condition, which may be regarded as a more restrictive version of the uniform condition introduced by O. Martio and J. Sarvas in [**14**]. Lastly, we wish to remark that

(2.34) if $\Omega \subseteq \mathbb{R}^n$ is an open set which is locally pathwise nontangentially accessible then so is $\Omega \setminus \overline{B(x,r)}$ for each $x \in \Omega$ and each $r \in \big(0\,,\,\delta_{\partial\Omega}(x)\big)$.

It turns out that for any locally pathwise nontangentially accessible set with a doubling surface measure, the nontangentially accessible boundary has full (surface) measure in the topological boundary. Specifically, the following result is proved in [**16**].

PROPOSITION 2.5. *Let $\Omega \subset \mathbb{R}^n$ be an open set which is locally pathwise nontangentially accessible and has the property that $\sigma := \mathcal{H}^{n-1}\lfloor\partial\Omega$ is a doubling measure on $\partial\Omega$. Then*

$$\mathcal{H}^{n-1}\big(\partial\Omega \setminus \partial_{\rm nta}\Omega\big) = 0. \tag{2.35}$$

*As a consequence of (2.35), (2.29), and (2.20),*

$$\mathcal{H}^{n-1}\big(\partial\Omega \setminus A_\kappa(\partial\Omega)\big) = 0 \quad \textit{for each} \quad \kappa > 0. \tag{2.36}$$

We shall also need the following result which quantifies the rate at which the truncated nontangential maximal operator (2.17) vanishes as the truncation parameter goes to zero (for a proof see [**16**]).

PROPOSITION 2.6. *Suppose $\Omega \subset \mathbb{R}^n$ is a locally pathwise nontangentially accessible set, and abbreviate $\sigma := \mathcal{H}^{n-1}\lfloor\partial\Omega$. Also, fix $\kappa > 0$ arbitrary, then let $\widetilde{\kappa} \geq \kappa$ together with $d > 0$ and $c \in [1,\infty)$ be associated with $\kappa$ as in (2.32), and define $\theta := \big[c(1+\kappa)\big]^{-1} \in (0,1)$. Finally, consider a function $u \in \mathscr{C}^1(\Omega)$ with the property that*

$$u\Big|_{\partial\Omega}^{\widetilde{\kappa}-{\rm n.t.}} = 0 \quad \textit{at } \sigma\textit{-a.e. point on } \partial_{\rm nta}\Omega. \tag{2.37}$$

*Then for each $\varepsilon \in (0, d/\theta)$ one has*

$$\mathcal{N}_\kappa^{\theta\,\varepsilon} u \leq \varepsilon \cdot \mathcal{N}_{\widetilde{\kappa}}^{\varepsilon}(\nabla u) \quad \textit{at } \sigma\textit{-a.e. point on } \partial_{\rm nta}\Omega. \tag{2.38}$$

Going further, recall (cf., e.g., [**5**, Definition p. 208]) that the geometric measure theoretic boundary, denoted by $\partial_*\Omega$, of a Lebesgue measurable subset $\Omega$ of $\mathbb{R}^n$ is defined as

$$\begin{aligned}\partial_*\Omega := \Big\{x \in \mathbb{R}^n : \limsup_{r\to 0^+} \frac{\mathcal{L}^n(B(x,r)\cap\Omega)}{r^n} > 0 \ \text{ and} \\ \limsup_{r\to 0^+} \frac{\mathcal{L}^n(B(x,r)\setminus\Omega)}{r^n} > 0\Big\}.\end{aligned} \tag{2.39}$$

For example, tips of inner or outer cusps fail to belong to $\partial_*\Omega$.

In this vein, recall (cf. [**5**, Theorem 1, p. 222]) that

(2.40) given $\Omega \subseteq \mathbb{R}^n$ which is $\mathcal{L}^n$-measurable, the set $\Omega$ is of locally finite perimeter if and only if $\mathcal{H}^{n-1}(K \cap \partial_*\Omega) < \infty$ for each compact set $K \subset \mathbb{R}^n$.

In particular, as is visible from (2.5) and (2.40),

(2.41) if $\Omega \subseteq \mathbb{R}^n$ is $\mathcal{L}^n$-measurable and $\sigma := \mathcal{H}^{n-1}\lfloor\partial\Omega$ is a doubling measure on $\partial\Omega$ then the set $\Omega$ is of locally finite perimeter.

Also,

(2.42) if $\Omega \subseteq \mathbb{R}^n$ has locally finite perimeter then $\partial_*\Omega$ is a Borel set hence, in particular, $\mathcal{H}^{n-1}$-measurable.

As part of the classical theory of sets of locally finite perimeter (cf., e.g., [**5**]),

(2.43) given any set of locally finite perimeter $\Omega \subseteq \mathbb{R}^n$ one has $\nabla \mathbf{1}_\Omega = -\nu\,\mathcal{H}^{n-1}\lfloor\partial_*\Omega$ in $\big[\mathcal{D}'(\mathbb{R}^n)\big]^n$ where $\nu : \partial_*\Omega \to \mathbb{R}^n$ is a $\mathcal{H}^{n-1}$-measurable function with the property that $|\nu(x)| = 1$ at $\mathcal{H}^{n-1}$-a.e. point $x \in \partial_*\Omega$.

In such a scenario, we shall refer to $\nu$ as being the geometric measure theoretic outward unit normal to $\Omega$.

Our next proposition, which elaborates on the relationship between the geometric measure theoretic boundary and the nontangentially accessible boundary of open sets, has been proved in [**16**].

PROPOSITION 2.7. *Let $\Omega$ be an open subset of $\mathbb{R}^n$ with the property that $\partial\Omega$ is lower Ahlfors regular and the measure $\sigma := \mathcal{H}^{n-1}\lfloor\partial\Omega$ is doubling on $\partial\Omega$. Then*

$$\sigma\big(\partial_*\Omega \setminus \partial_{\text{nta}}\Omega\big) = 0. \tag{2.44}$$

Regarding the nature of (2.44), it is remarkable that intrinsic properties of the topological boundary $\partial\Omega$ (such as lower Ahlfors regularity and the surface measure being doubling) provide information about the "thickness" of the set $\Omega$ itself, by implying that almost all points in the geometric measure theoretic boundary $\partial_*\Omega$ may be approached nontangentially (with any fixed aperture parameter) from within the open set $\Omega$. Incidentally, it would be misleading to think of this as being merely a "soft" topological property, since there are quantitative estimates underpinning this implication.

Moving on, having fixed $n, M \in \mathbb{N}$, by a coefficient tensor we mean any block of the form

$$A = \big(a_{rs}^{\alpha\beta}\big)_{\substack{1\leq r,s\leq n\\ 1\leq\alpha,\beta\leq M}} \tag{2.45}$$

where each entry $a_{rs}^{\alpha\beta}$ is a complex-valued function. The (real) transpose of $A$ is then the coefficient tensor $A^\top$ given by

$$A^\top := \big(a_{sr}^{\beta\alpha}\big)_{\substack{1\leq r,s\leq n\\ 1\leq\alpha,\beta\leq M}}. \tag{2.46}$$

With a coefficient tensor $A$ as in (2.45), we shall associate a conormal derivative on the boundary of a given open set as follows.

DEFINITION 2.8. Suppose $\Omega \subseteq \mathbb{R}^n$ is an open set with the property that $\partial\Omega$ is lower Ahlfors regular and $\sigma := \mathcal{H}^{n-1}\lfloor\partial\Omega$ is a doubling measure on $\partial\Omega$. In particular, $\Omega$ is a set of locally finite perimeter (cf. (2.41)) and, as such, its geometric measure theoretic outward unit normal $\nu = (\nu_1, \dots, \nu_n)$ is defined $\sigma$-a.e. on $\partial_*\Omega$ (cf. (2.43)). Fix some some background aperture parameter $\kappa > 0$.

Consider a complex coefficient tensor $A$, expressed as in (2.45) with locally bounded entries, and suppose $u = (u_\beta)_{1\leq\beta\leq M} \in \big[W^{1,1}_{\text{loc}}(\Omega)\big]^M$ is a $\mathbb{C}^M$-valued function with the property that for each $\alpha \in \{1, \dots, M\}$ and $r \in \{1, \dots, n\}$ the nontangential limit

$$\big(a_{rs}^{\alpha\beta}\partial_s u_\beta\big)\Big|_{\partial\Omega}^{\kappa-\text{n.t.}} \text{ exists at } \sigma\text{-a.e. point on } \partial_*\Omega. \tag{2.47}$$

In such a setting, define the (pointwise) conormal derivative of $u$ (with respect to the coefficient tensor $A$ and the set $\Omega$) as the $\mathbb{C}^M$-valued function

$$\partial_\nu^A u := \Big(\nu_r\big(a_{rs}^{\alpha\beta}\partial_s u_\beta\big)\Big|_{\partial\Omega}^{\kappa-\text{n.t.}}\Big)_{1\leq\alpha\leq M} \quad \text{at } \sigma\text{-a.e. point on } \partial_*\Omega. \tag{2.48}$$

Two other important preliminary results are recorded in Theorems 2.9-2.10. The first theorem, dealing with the issue of local elliptic regularity for variable coefficient systems, has been proved in [**18**].

THEOREM 2.9. *Let $L$ be as in (1.2)-(1.4) and suppose $\Omega$ is an arbitrary open set in $\mathbb{R}^n$. Also, assume $1 < p \leq q < \infty$. Then*

$$u \in \big[L^p_{\text{loc}}(\Omega, \mathcal{L}^n)\big]^M \text{ and } Lu \in \big[L^q_{\text{loc}}(\Omega, \mathcal{L}^n)\big]^M \Longrightarrow u \in \big[W^{2,q}_{\text{loc}}(\Omega)\big]^M \tag{2.49}$$

*and a naturally accompanying estimate holds.*

The manner in which the membership $Lu \in \big[L^q_{\rm loc}(\Omega,\mathcal{L}^n)\big]^M$ in (2.49) should be understood is as follows: there exists $v = (v_\alpha)_{1\leq\alpha\leq M} \in \big[L^q_{\rm loc}(\Omega,\mathcal{L}^n)\big]^M$ such that for each $\varphi = (\varphi_\alpha)_{1\leq\alpha\leq M} \in \big[\mathscr{C}^\infty_0(\Omega)\big]^M$ we have

$$\int_\Omega u_\beta \partial_s\big(a^{\alpha\beta}_{rs}(\partial_r\varphi_\alpha)\big)\,d\mathcal{L}^n = \int_\Omega v_\alpha\varphi_\alpha\,d\mathcal{L}^n. \tag{2.50}$$

In relation to this, observe that $a^{\alpha\beta}_{rs}(\partial_r\varphi_\alpha)$ is a compactly supported function in $\mathrm{Lip}(\Omega)$ hence $\partial_s\big(a^{\alpha\beta}_{rs}(\partial_r\varphi_\alpha)\big)$ is a compactly supported essentially bounded function in $\Omega$. This makes the integral in the left side of (2.50) absolutely convergent.

As a consequence of (2.49) and classical embedding theorems,

$$\begin{array}{l}\text{null-solutions } u \in \big[L^p_{\rm loc}(\Omega,\mathcal{L}^n)\big]^M \text{ (with } 1<p<\infty\text{) of } L\\ \text{in } \Omega \text{ belong to } \big[\mathscr{C}^{1+\varepsilon}(\Omega,\mathcal{L}^n)\big]^M \text{ for each } \varepsilon\in[0,1).\end{array} \tag{2.51}$$

It has also been shown in [**18**] that if (1.2) is strengthened to $a^{\alpha\beta}_{rs} \in \mathscr{C}^{1+\gamma}(\mathbb{R}^n)$ for some $\gamma>0$ then we may allow $p=1$ in (2.49), hence also in (2.51).

The second important preliminary result alluded to above is described in the theorem below, dealing with interior estimates for null-solutions of variable coefficient elliptic systems.

THEOREM 2.10. *Let $L$ be as in* (1.2)-(1.4) *and suppose $\Omega$ is an arbitrary open set in $\mathbb{R}^n$. Also, fix some integrability exponent $p\in(1,\infty)$. Then there exists a constant $C = C(L,p)\in(0,\infty)$ with the property that if $u \in \big[L^p_{\rm loc}(\Omega,\mathcal{L}^n)\big]^M$ satisfies $Lu=0$ in $\Omega$ then for every $x\in\Omega$, every $c\in\mathbb{R}$, and every $r\in\big(0,\delta_{\partial\Omega}(x)/2\big)$ there holds*

$$\|\nabla u\|_{[L^\infty(B(x,r),\mathcal{L}^n)]^{n\cdot M}} \leq \frac{C}{r}\Big[\frac{1}{\mathcal{L}^n\big(B(x,2r)\big)}\int_{B(x,2r)}|u(z)-c|^p\,dz\Big]^{1/p}. \tag{2.52}$$

This is implied by [**18**, Proposition 3.4], where a more general result of this flavor is established (by allowing lower order terms in (1.3)).

We conclude this section by recalling a recent result from [**16**], where the following sharp version of the Divergence Theorem has been proved.

THEOREM 2.11. *Fix $n\in\mathbb{N}$ and let $\Omega$ be an open, bounded, nonempty subset of $\mathbb{R}^n$ with a lower Ahlfors regular boundary, and such that $\sigma := \mathcal{H}^{n-1}\lfloor\partial\Omega$ is a doubling measure on $\partial\Omega$. In particular, $\Omega$ is a set of locally finite perimeter, and its geometric measure theoretic outward unit normal $\nu$ is defined $\sigma$-a.e. on $\partial_*\Omega$ (which, up to a $\sigma$-nullset, is contained in $\partial_{\rm nta}\Omega$; cf.* (2.44)*). Fix an aperture parameter $\kappa\in(0,\infty)$ and assume that the vector field*

$$\vec{F} \in \big[\mathscr{E}'(\Omega) + L^1_{\rm loc}(\Omega,\mathcal{L}^n)\big]^n \subset \big[\mathcal{D}'(\Omega)\big]^n \tag{2.53}$$

*satisfies*

$$\begin{array}{c}\mathcal{N}^{\varepsilon}_{\kappa}\vec{F}\in L^1(\partial\Omega,\sigma) \text{ for some } 0<\varepsilon<\mathrm{dist}\,(\mathrm{regsupp}\,\vec{F},\,\partial\Omega),\\ \text{the nontangential trace } \vec{F}\big|^{\kappa-\mathrm{n.t.}}_{\partial\Omega} \text{ exists } \sigma\text{-a.e. on } \partial_{\rm nta}\Omega, \text{ and}\\ \mathrm{div}\,\vec{F} \text{ computed in the sense of distributions in } \Omega \text{ belongs to } \mathscr{E}'(\Omega).\end{array} \tag{2.54}$$

*Then for any $\kappa'>0$ the nontangential trace $\vec{F}\big|^{\kappa'-\mathrm{n.t.}}_{\partial\Omega}$ exists $\sigma$-a.e. on $\partial_{\rm nta}\Omega$ and is actually independent of $\kappa'$. Furthermore, with the dependence on the parameter*

*$\kappa'$ dropped, one has*

$$ {}_{(\mathscr{C}^\infty(\Omega))^*}\big(\operatorname{div}\vec{F}\,,\,1\big)_{\mathscr{C}^\infty(\Omega)}=\int_{\partial_*\Omega}\nu\cdot\big(\vec{F}\,\big|^{\rm n.t.}_{\partial\Omega}\big)\,d\sigma. \tag{2.55}$$

As discussed in [**16**], this result is in the nature of best possible. It is worth comparing Theorem 2.11 to the classical version of the De Giorgi-Federer Divergence Theorem which is valid in arbitrary sets of locally finite perimeter and for smooth compactly supported vector fields defined in the entire Euclidean space. By way of contrast, a key attribute of our result is the fact that in Theorem 2.11 the vector field $\vec{F}$ is only defined in $\Omega$. In particular, this renders the issue of making sense of its boundary trace delicate. We define the said boundary trace in the strong nontangential pointwise sense (cf. Definition 2.2), thus allowing for a much larger class of vector fields than the one accommodated by the De Giorgi-Federer Divergence Theorem.

## 3. Proof of main result

This section is devoted to presenting the proof of Theorem 1.1.

PROOF OF THEOREM 1.1. With the aperture parameter $\widetilde{\kappa}>0$ to be chosen later, suppose $G$ is a matrix value function as in (1.6). Throughout, write $\big(G_{\alpha\beta}\big)_{1\leq\alpha,\beta\leq M}$ for the entries of the $\mathbb{C}^{M\times M}$-valued function $G$. Then the second line in (1.6) implies

$$\big(L^\top G_{\bullet\beta}\big)_\alpha=-\delta_{x_0}\delta_{\alpha\beta}\ \text{ in }\ \mathcal{D}'(\Omega)\ \text{ for all }\ \alpha,\beta\in\{1,\dots,M\}, \tag{3.1}$$

where $\delta_{\alpha\beta}$ is the Kronecker symbol. In particular,

$$\big(L^\top G_{\bullet\beta}\big)_\alpha=0\in\mathbb{C}^M\ \text{ in }\ \Omega\setminus\{x_0\}\ \text{ for all }\ \alpha,\beta\in\{1,\dots,M\}. \tag{3.2}$$

Based on this and elliptic regularity (cf. Theorem 2.9) we see that

$$G\in\big[W^{2,q}_{\rm loc}(\Omega\setminus\{x_0\})\big]^{M\times M}\ \text{ for each }\ q\in[1,\infty). \tag{3.3}$$

In concert with standard Sobolev embedding results, this entails

$$G\in\big[\mathscr{C}^{1+\mu}_{\rm loc}(\Omega\setminus\{x_0\})\big]^{M\times M}\ \text{ for each }\ \mu\in[0,1), \tag{3.4}$$

in a quantitative fashion. Also, since we are assuming that $G\in\big[W^{1,1}_{\rm loc}(\Omega)\big]^{M\times M}$ and that the entries of the coefficient tensor $A$ are in $\mathscr{C}^0(\overline{\Omega})$ (thanks to (1.2) and the fact $\operatorname{Lip}(\Omega)=\{\psi|_\Omega:\ \psi\in\operatorname{Lip}(\mathbb{R}^n)\}$), the assumption made in the third line of (1.6) together with (2.26) and (2.27) allow us to define the conormal derivative $\partial^{A^\top}_\nu G_{\bullet\beta}$ as in (2.48) with $\kappa:=\widetilde{\kappa}$ in a meaningful fashion.

Next, from the first two lines in (1.7), the first and third line of (1.6), (2.28), and Proposition 2.5 we conclude that

$$\big(\mathcal{N}^\rho_\kappa u\big)(x)<+\infty\ \text{and}\ \big(\mathcal{N}^\rho_{\widetilde{\kappa}}(\nabla G)\big)(x)<+\infty\ \text{for }\sigma\text{-a.e. }x\in\partial\Omega. \tag{3.5}$$

In particular, (3.5) ensures that the product $\big(\mathcal{N}^\rho_\kappa u\big)(x)\cdot\big(\mathcal{N}^\rho_{\widetilde{\kappa}}(\nabla G)\big)(x)$ is a well-defined number in $[0,+\infty)$ for $\sigma$-a.e. $x\in\partial\Omega$. In concert with (2.12), this proves that $\mathcal{N}^\rho_\kappa u\cdot\mathcal{N}^\rho_{\widetilde{\kappa}}(\nabla G)$ is a well-defined, non-negative, $\sigma$-measurable function on $\partial\Omega$. Consequently, the integral condition formulated in the third line of (1.7) is meaningful. In turn, based on (2.24) and the hypothesis in the third line of (1.7) we see that the integral in (1.8) is absolutely convergent, so the conclusion in the theorem has a clear meaning.

To justify formula (1.8) we will apply the Divergence Formula in the version recorded in Theorem 2.11 to a suitably chosen vector field. Specifically, fix some $\beta \in \{1,\dots,M\}$ and define

$$\vec{F} := \Big( u_\alpha\, a^{\gamma\alpha}_{kj}\, \partial_k G_{\gamma\beta} - G_{\alpha\beta}\, a^{\alpha\gamma}_{jk}\, \partial_k u_\gamma \Big)_{1\leq j\leq n} \quad \text{at } \mathcal{L}^n\text{-a.e. point in } \Omega, \tag{3.6}$$

where $(u_\mu)_{1\leq\mu\leq M}$ are the scalar components of the $\mathbb{C}^M$-valued function $u$.

From (3.6), (3.3), and the fact that $u \in \big[\mathscr{C}^1(\Omega)\big]^M$ it follows that the vector field $\vec{F}$ is well-defined and

$$\vec{F} \in \big[L^1_{\rm loc}(\Omega, \mathcal{L}^n)\big]^n. \tag{3.7}$$

Moreover, in the sense of distributions in $\Omega$, we have

$$\begin{aligned} \operatorname{div}\vec{F} &= (\partial_j u_\alpha)\, a^{\gamma\alpha}_{kj}\,(\partial_k G_{\gamma\beta}) + u_\alpha \partial_j\big[a^{\gamma\alpha}_{kj}\,\partial_k G_{\gamma\beta}\big] \\ &\quad - (\partial_j G_{\alpha\beta})\, a^{\alpha\gamma}_{jk}\,(\partial_k u_\gamma) - G_{\alpha\beta}\partial_j\big[a^{\alpha\gamma}_{jk}\,\partial_k u_\gamma\big] \\ &=: I_1 + I_2 + I_3 + I_4, \end{aligned} \tag{3.8}$$

where the last equality defines the $I_i$'s. Changing variables $j' = k$, $k' = j$, $\alpha' = \gamma$, and $\gamma' = \alpha$ in $I_3$ yields

$$I_3 = -(\partial_{k'} G_{\gamma'\beta})\, a^{\gamma'\alpha'}_{k'j'}\,(\partial_{j'} u_{\alpha'}) = -I_1. \tag{3.9}$$

Recalling the fact that $u$ is a null-solution for $L$, we obtain

$$I_4 = -G_{\alpha\beta}\,(Lu)_\alpha = 0. \tag{3.10}$$

Also, the second line in (1.6) implies

$$I_2 = u_\alpha (L^\top G_{\cdot\beta})_\alpha = -u_\alpha \delta_{\alpha\beta}\delta_{x_0} = -u_\beta\,\delta_{x_0} = -u_\beta(x_0)\,\delta_{x_0}. \tag{3.11}$$

Collectively, (3.8)-(3.11) permit us to conclude that, in the sense of distributions in $\Omega$,

$$\operatorname{div}\vec{F} = -u_\beta(x_0)\,\delta_{x_0}. \tag{3.12}$$

In particular,

$$\operatorname{div}\vec{F} \in \mathscr{E}'(\Omega). \tag{3.13}$$

Moving on, recall from Proposition 2.5 that in the present context we have

$$\sigma\big(\partial\Omega \setminus \partial_{\rm nta}\Omega\big) = 0. \tag{3.14}$$

Starting with the aperture parameter $\kappa$ given in the statement of the theorem, choose $a \in (0,1)$ small enough so that $\kappa > 2a/(1-a)$, then define

$$\kappa' := \kappa(1-a) - 2a \in (0,\kappa). \tag{3.15}$$

Next, take $\widetilde{\kappa} \geq \kappa'$ along with $d > 0$ and $c \in [1,\infty)$ to be the parameters associated as in (2.32) with $\kappa := \kappa'$. If necessary, further increase $\widetilde{\kappa}$ as to ensure that $\widetilde{\kappa} \geq \kappa$. Also, set $\theta := [c(1+\kappa')]^{-1} \in (0,1)$, define

$$N_1 := \Big\{ x \in \partial\Omega : \mathcal{N}^{\rho}_{\widetilde{\kappa}}(\nabla G)(x) = +\infty \Big\} \tag{3.16}$$

and introduce

$$N_2 := \Big\{ x \in \partial_{\rm nta}\Omega : \big(u\big|^{\kappa-{\rm n.t.}}_{\partial\Omega}\big)(x) \ \text{ fails to exist} \Big\}. \tag{3.17}$$

Then (3.5) implies $\sigma(N_1)=0$. Also, from the second line in (1.7), (2.35), and (2.2) (presently used with $X:=\partial\Omega$ and $s:=n-1$) we see that $N_2$ is a $\sigma$-measurable set and $\sigma(N_2)=0$.

To proceed, fix $x\in\partial_{\rm nta}\Omega\setminus(N_1\cup N_2)$ and pick an arbitrary point

$$y\in\Gamma_{\kappa'}(x)\ \text{ with }\ \delta_{\partial\Omega}(y)<\min\big\{\theta\cdot\rho\,,\,d\big\}. \tag{3.18}$$

The property in the fourth line of (1.6) ensures that the condition formulated in (2.37) is satisfied by (the entries of) the matrix-valued function $G$. Granted this and (3.4), we may invoke (2.38) (for the function $(1-\psi)G$ where $\psi\in\mathscr{C}^\infty_0\big(B(x_0,\rho/100)\big)$ is a scalar-valued function satisfying $\psi\equiv 1$ near $x_0$) written with $\kappa'$ in place of $\kappa$ and the choice

$$\varepsilon:=\theta^{-1}\delta_{\partial\Omega}(y)\in\Big(0\,,\,\min\big\{\rho\,,\,d/\theta\big\}\Big), \tag{3.19}$$

and conclude the existence of a $\sigma$-measurable set $N_3\subseteq\partial\Omega$ satisfying $\sigma(N_3)=0$ along with some constant $C=C(\Omega,\kappa',\theta)\in(0,\infty)$ independent of $x$ and $y$, such that

$$\begin{aligned} |G(y)|&\leq\big(\mathcal{N}^{\theta\varepsilon}_{\kappa'}G\big)(x)\leq\varepsilon\cdot\mathcal{N}^{\varepsilon}_{\widetilde{\kappa}}(\nabla G)(x)\\ &\leq C\delta_{\partial\Omega}(y)\cdot\mathcal{N}^{\rho}_{\widetilde{\kappa}}(\nabla G)(x)\ \text{ provided }\ x\notin N_3, \end{aligned} \tag{3.20}$$

where the first inequality takes into account (2.13) as well as (3.4), and where the last inequality is a consequence of (3.19) and (2.14) (bearing in mind (2.17)). Next, the fact that $x$ belongs to the set $A_\kappa(\partial\Omega)\setminus N_2$ ensures that the nontangential boundary trace $\big(u\big|^{\kappa-{\rm n.t.}}_{\partial\Omega}\big)(x)$ exists in $\mathbb{C}^M$. Since $u\in\big[\mathscr{C}^1(\Omega)\big]^M$, we may apply Theorem 2.10 to $u$, with $x$ replaced by $y$, with $r:=a\cdot\delta_{\partial\Omega}(y)/2$, for the constant $c:=\big(u\big|^{\kappa-{\rm n.t.}}_{\partial\Omega}\big)(x)$, and with, say, $p:=2$. These choices guarantee the existence of some constant $C=C(L,n,a)\in(0,\infty)$ such that

$$|(\nabla u)(y)|\leq\frac{C}{\delta_{\partial\Omega}(y)}\left[\frac{1}{\mathcal{L}^n\big(B(y,a\cdot\delta_{\partial\Omega}(y))\big)}\int_{B(y,a\cdot\delta_{\partial\Omega}(y))}\Big|u-\big(u\big|^{\kappa-{\rm n.t.}}_{\partial\Omega}\big)(x)\Big|^2\,d\mathcal{L}^n\right]^{1/2}. \tag{3.21}$$

To proceed, by relying on the fact that the distance function $\delta_{\partial\Omega}$ is Lipschitz with constant $\leq 1$, for each point $z\in B\big(y\,,\,a\cdot\delta_{\partial\Omega}(y)\big)$ we may write

$$\delta_{\partial\Omega}(y)\leq\delta_{\partial\Omega}(z)+|z-y|<\delta_{\partial\Omega}(z)+a\cdot\delta_{\partial\Omega}(y)\Longrightarrow\delta_{\partial\Omega}(y)<\frac{1}{1-a}\delta_{\partial\Omega}(z). \tag{3.22}$$

Bearing in mind that $y$ is as in (3.18) and the formula for $\kappa'$ from (3.15), the final estimate in (3.22) permits us to estimate, for every $z\in B\big(y\,,\,a\cdot\delta_{\partial\Omega}(y)\big)$,

$$\begin{aligned} |z-x|&\leq|z-y|+|y-x|<a\cdot\delta_{\partial\Omega}(y)+(1+\kappa')\delta_{\partial\Omega}(y)\\ &=(1+\kappa'+a)\delta_{\partial\Omega}(y)<\frac{1+\kappa'+a}{1-a}\delta_{\partial\Omega}(z)=(1+\kappa)\,\delta_{\partial\Omega}(z). \end{aligned} \tag{3.23}$$

Hence,

$$\begin{aligned} &z\in\Gamma_\kappa(x)\ \text{ and }\ |z-x|<(1+\kappa'+a)\delta_{\partial\Omega}(y)\\ &\qquad\text{for each }\ z\in B\big(y\,,\,a\cdot\delta_{\partial\Omega}(y)\big). \end{aligned} \tag{3.24}$$

When used back in (3.21), this further implies

$$|(\nabla u)(y)| \leq C\delta_{\partial\Omega}(y)^{-1} \cdot \sup_{\substack{z\in\Gamma_\kappa(x)\\ |x-z|<(1+\kappa'+a)\delta_{\partial\Omega}(y)}} \Big|u(z) - \big(u\big|_{\partial\Omega}^{\kappa-\text{n.t.}}\big)(x)\Big|, \tag{3.25}$$

By combining (3.20) and (3.25) we arrive at

$$|G(y)||(\nabla u)(y)| \leq C\mathcal{N}_{\widetilde{\kappa}}^{\rho}(\nabla G)(x) \cdot \sup_{\substack{z\in\Gamma_\kappa(x)\\ |x-z|<(1+\kappa'+a)\delta_{\partial\Omega}(y)}} \Big|u(z) - \big(u\big|_{\partial\Omega}^{\kappa-\text{n.t.}}\big)(x)\Big|. \tag{3.26}$$

Consequently,

$$\lim_{\Gamma_{\kappa'}(x)\ni y\to x} |G(y)||(\nabla u)(y)| = 0 \quad \text{for each } \ x \in \partial_{\text{nta}}\Omega \setminus (N_1 \cup N_2 \cup N_3) \tag{3.27}$$

bearing in mind (3.16). In turn, from (3.6), (1.6), (1.7), (3.27), and (2.22) we conclude (upon recalling that $0 < \kappa' < \kappa \leq \widetilde{\kappa}$ and $\sigma(N_1 \cup N_2 \cup N_3) = 0$) that

$$\vec{F}\big|_{\partial\Omega}^{\kappa'-\text{n.t.}} \ \text{ exists at } \sigma\text{-a.e. point on } \ \partial_{\text{nta}}\Omega \tag{3.28}$$

and, in fact,

$$\begin{aligned}\vec{F}\Big|_{\partial\Omega}^{\kappa'-\text{n.t.}} &= \Big(\big(u_\alpha\big|_{\partial\Omega}^{\kappa'-\text{n.t.}}\big)a_{kj}^{\gamma\alpha}\big(\partial_k G_{\gamma\beta}\big)\big|_{\partial\Omega}^{\kappa'-\text{n.t.}}\Big)_{1\leq j\leq n} \\ &= \Big(\big(u_\alpha\big|_{\partial\Omega}^{\kappa-\text{n.t.}}\big)a_{kj}^{\gamma\alpha}\big(\partial_k G_{\gamma\beta}\big)\big|_{\partial\Omega}^{\widetilde{\kappa}-\text{n.t.}}\Big)_{1\leq j\leq n}.\end{aligned} \tag{3.29}$$

Next, fix an arbitrary point $x \in \partial_{\text{nta}}\Omega$ and, for each $y \in \Gamma_{\kappa'}(x)$ with

$$\delta_{\partial\Omega}(y) < \varepsilon_o := \min\big\{\theta\cdot\rho,\, d,\, \rho/(1+\kappa'+a)\big\}, \tag{3.30}$$

apply Theorem 2.10 to $u$ one more time, with $x$ replaced by $y$, with $r := a\cdot\delta_{\partial\Omega}(y)/2$, with $p := 2$, and now taking the constant $c := 0 \in \mathbb{C}^M$. Keeping in mind (3.24) we are therefore able to write

$$\begin{aligned}|(\nabla u)(y)| &\leq \frac{C}{\delta_{\partial\Omega}(y)}\left[\frac{1}{\mathcal{L}^n\big(B(y,a\cdot\delta_{\partial\Omega}(y))\big)}\int_{B(y,a\cdot\delta_{\partial\Omega}(y))}|u|^2\,d\mathcal{L}^n\right]^{1/2} \\ &\leq C\delta_{\partial\Omega}(y)^{-1}\cdot \sup_{\substack{z\in\Gamma_\kappa(x)\\ |x-z|<(1+\kappa'+a)\delta_{\partial\Omega}(y)}}|u(z)| \leq C\delta_{\partial\Omega}(y)^{-1}\cdot\big(\mathcal{N}_\kappa^\rho u\big)(x).\end{aligned} \tag{3.31}$$

Combining (3.20) with (3.31) then gives, on account of (3.14),

$$\mathcal{N}_{\kappa'}^{\varepsilon_o}\big(|G||\nabla u|\big) \leq C\big(\mathcal{N}_{\widetilde{\kappa}}^\rho(\nabla G)\big)\cdot\mathcal{N}_\kappa^\rho u \ \text{ at } \sigma\text{-a.e. point on } \ \partial\Omega. \tag{3.32}$$

Since thanks to (2.15), (2.14), $\rho > \varepsilon_o$, and $\widetilde{\kappa} > \kappa'$, we also have

$$\begin{aligned}\mathcal{N}_{\kappa'}^{\varepsilon_o}\big(|\nabla G||u|\big) &\leq \big(\mathcal{N}_{\kappa'}^{\varepsilon_o}(\nabla G)\big)\cdot\mathcal{N}_{\kappa'}^{\varepsilon_o}u \\ &\leq \big(\mathcal{N}_{\widetilde{\kappa}}^\rho(\nabla G)\big)\cdot\mathcal{N}_\kappa^\rho u \ \text{ at each point on } \ \partial\Omega,\end{aligned} \tag{3.33}$$

we conclude from (3.6), (3.32), (3.33), the fact that each $a_{rs}^{\alpha\beta}$ is a bounded function in $\Omega$, the third line in (1.7), and (2.12) that the truncated nontangential maximal function of $\vec{F}$ at height $\varepsilon_o$ and aperture $\kappa'$ satisfies

$$\mathcal{N}_{\kappa'}^{\varepsilon_o}\vec{F} \in L^1(\partial\Omega,\sigma). \tag{3.34}$$

Having established (3.7), (3.13), (3.34), and (3.28), Theorem 2.11 applies (with $\kappa$ replaced by $\kappa'$). Let $(\nu_j)_{1\leq j\leq n}$ denote the components of the geometric measure theoretic outward unit normal $\nu$ to $\Omega$. The Divergence Formula (2.55) combined with (3.12), the last line in (3.29), and (2.48) (written for $A^\top$ in place of $A$), then yields

$$
\begin{aligned}
-u_\beta(x_0) &= {}_{(\mathscr{C}^\infty(\Omega))^*}\big(\operatorname{div}\vec{F}\,,\,1\big)_{\mathscr{C}^\infty(\Omega)} = \int_{\partial_*\Omega} \nu\cdot\big(\vec{F}\big|_{\partial\Omega}^{\kappa'-\mathrm{n.t.}}\big)\,d\sigma \\
&= \int_{\partial_*\Omega} \big(u_\alpha\big|_{\partial\Omega}^{\kappa-\mathrm{n.t.}}\big)\nu_j a_{kj}^{\gamma\alpha}\big(\partial_k G_{\gamma\beta}\big)\big|_{\partial\Omega}^{\widetilde{\kappa}-\mathrm{n.t.}}\,d\sigma \\
&= \int_{\partial_*\Omega} \big\langle u\big|_{\partial\Omega}^{\kappa-\mathrm{n.t.}}\,,\,\partial_\nu^{A^\top} G_{\cdot\beta}\big\rangle\,d\sigma. 
\end{aligned}
\tag{3.35}
$$

Since $\beta\in\{1,\dots,M\}$ has been arbitrarily chosen, this finishes the proof of (1.8). □

## Acknowledgment

We close by thanking the referee for a careful reading of the manuscript and for a number of useful suggestions.

## References


[1] Hiroaki Aikawa and Kentaro Hirata, *Doubling conditions for harmonic measure in John domains* (English, with English and French summaries), Ann. Inst. Fourier (Grenoble) **58** (2008), no. 2, 429–445. MR2410379

[2] Luca Capogna, Carlos E. Kenig, and Loredana Lanzani, *Harmonic measure: Geometric and analytic points of view*, University Lecture Series, vol. 35, American Mathematical Society, Providence, RI, 2005. MR2139304

[3] G. David and S. Semmes, *Singular integrals and rectifiable sets in* $\mathbf{R}^n$*: Beyond Lipschitz graphs* (English, with French summary), Astérisque **193** (1991), 152. MR1113517

[4] Lawrence C. Evans, *Partial differential equations*, Graduate Studies in Mathematics, vol. 19, American Mathematical Society, Providence, RI, 1998. MR1625845

[5] Lawrence C. Evans and Ronald F. Gariepy, *Measure theory and fine properties of functions*, Studies in Advanced Mathematics, CRC Press, Boca Raton, FL, 1992. MR1158660

[6] John B. Garnett and Donald E. Marshall, *Harmonic measure*, New Mathematical Monographs, vol. 2, Cambridge University Press, Cambridge, 2005. MR2150803

[7] Michael Grüter and Kjell-Ove Widman, *The Green function for uniformly elliptic equations*, Manuscripta Math. **37** (1982), no. 3, 303–342, DOI 10.1007/BF01166225. MR657523

[8] Steve Hofmann, Marius Mitrea, and Michael Taylor, *Singular integrals and elliptic boundary problems on regular Semmes-Kenig-Toro domains*, Int. Math. Res. Not. IMRN **14** (2010), 2567–2865, DOI 10.1093/imrn/rnp214. MR2669659

[9] David S. Jerison and Carlos E. Kenig, *Boundary behavior of harmonic functions in non-tangentially accessible domains*, Adv. in Math. **46** (1982), no. 1, 80–147, DOI 10.1016/0001-8708(82)90055-X. MR676988

[10] Carlos E. Kenig, *Harmonic analysis techniques for second order elliptic boundary value problems*, CBMS Regional Conference Series in Mathematics, vol. 83, Published for the Conference Board of the Mathematical Sciences, Washington, DC; by the American Mathematical Society, Providence, RI, 1994. MR1282720

[11] Carlos E. Kenig and Tatiana Toro, *Poisson kernel characterization of Reifenberg flat chord arc domains* (English, with English and French summaries), Ann. Sci. École Norm. Sup. (4) **36** (2003), no. 3, 323–401, DOI 10.1016/S0012-9593(03)00012-0. MR1977823

[12] Carlos Kenig and Tatiana Toro, *Free boundary regularity below the continuous threshold: 2-phase problems*, J. Reine Angew. Math. **596** (2006), 1–44, DOI 10.1515/CRELLE.2006.050. MR2254803

[13] W. Littman, G. Stampacchia, and H. F. Weinberger, *Regular points for elliptic equations with discontinuous coefficients*, Ann. Scuola Norm. Sup. Pisa Cl. Sci. (3) **17** (1963), 43–77. MR161019

[14] O. Martio and J. Sarvas, *Injectivity theorems in plane and space*, Ann. Acad. Sci. Fenn. Ser. A I Math. **4** (1979), no. 2, 383–401, DOI 10.5186/aasfm.1978-79.0413. MR565886

[15] Dorina Mitrea, *Distributions, partial differential equations, and harmonic analysis*, Universitext, Springer, New York, 2013. MR3114783

[16] D. Mitrea, I. Mitrea and M. Mitrea, *A Sharp Divergence Theorem with Nontangential Pointwise Traces and Applications to Singular Integral Operators*, book manuscript, 2018.

[17] Marius Mitrea and Michael Taylor, *Potential theory on Lipschitz domains in Riemannian manifolds: Hölder continuous metric tensors*, Comm. Partial Differential Equations **25** (2000), no. 7-8, 1487–1536, DOI 10.1080/03605300008821557. MR1765156

[18] Marius Mitrea and Michael Taylor, *Potential theory on Lipschitz domains in Riemannian manifolds: Sobolev-Besov space results and the Poisson problem*, J. Funct. Anal. **176** (2000), no. 1, 1–79, DOI 10.1006/jfan.2000.3619. MR1781631

Department of Mathematics, Baylor University, Waco, Texas 76798
*Email address*: Dorina_Mitrea@baylor.edu

Department of Mathematics, Temple University, Philadelphia, Pennsylvania 19122
*Email address*: imitrea@temple.edu

Department of Mathematics, Baylor University, Waco, Texas 76798
*Email address*: Marius_Mitrea@baylor.edu

Contemporary Mathematics
Volume **748**, 2020
https://doi.org/10.1090/conm/748/15060

# Variations on quantum ergodic theorems, II

Michael Taylor

ABSTRACT. We establish quantum ergodic theorems, related to the microlocal behavior of eigenfunctions of a first-order positive, self-adjoint, elliptic pseudodifferential operator $\Lambda$ on a compact Riemannian manifold $M$, emphasizing results that hold without the hypothesis that the Hamiltonian flow generated by the symbol of $\Lambda$ be ergodic. This work consists of two parts. The first treats quantum ergodic theorems for observables that are pseudodifferential operators whose symbols might be discontinuous. The second treats quantum ergodic theorems in which we replace the Schrödinger picture by the Heisenberg picture.

CONTENTS



## 1. Introduction

Let $M$ be a compact Riemannian manifold, $\Lambda$ a positive, self-adjoint, first-order elliptic pseudodifferential operator ($\Lambda \in OPS^1(M)$, for example, $\Lambda = \sqrt{-\Delta}$, where $\Delta$ is the Laplace-Beltrami operator on $M$), and $\{\varphi_k\}$ an orthonormal basis of $L^2(M)$ consisting of eigenfunctions of $\Lambda$, with eigenvalues $\lambda_k \nearrow +\infty$, i.e.,

$$\Lambda\varphi_k = \lambda_k\varphi_k, \quad 0 \le \lambda_1 \le \lambda_2 \le \cdots \nearrow +\infty. \tag{1.0.1}$$

---

2010 *Mathematics Subject Classification.* 35P20, 35S05,81Q50.
*Key words and phrases.* quantum ergodic theorem, Weyl law, Egorov's theorem, eigenvalue, eigenfunction, Hamiltonian flow.
Work supported by NSF grant DMS-1500817.

If $X \subset T^*M$ denotes the set on which the principal symbol $\sigma_\Lambda$ of $\Lambda$ is equal to 1, then $X$ carries a natural Liouville measure $dS$, which we normalize to have total mass 1, and the Hamiltonian flow $\{G_t : t \in \mathbb{R}\}$ generated by $\sigma_\Lambda$ acts on $X$,

$$G_t : X \longrightarrow X, \tag{1.0.2}$$

preserving the Liouville measure $dS$. In [**1**] (following work of [**6**] and [**12**]) the following classical quantum ergodic theorem was established.

THEOREM 1.0.1. *Assume the flow* (1.0.2) *is ergodic. Then, for each* $A \in OPS^0(M)$, *with principal symbol* $a \in C^\infty(X)$,

$$\lim_{N\to\infty} \frac{1}{N} \sum_{j=1}^{N} \Bigl|(A\varphi_j, \varphi_j)_{L^2} - \overline{a}\Bigr|^2 = 0, \tag{1.0.3}$$

*where*

$$\overline{a} = \int_X a\, dS. \tag{1.0.4}$$

*Consequently, there is a subset* $\mathcal{N} \subset \mathbb{N}$, *of density* 0, *such that for each such* $A$,

$$\lim_{k\to\infty, k\notin\mathcal{N}} (A\varphi_k, \varphi_k)_{L^2} = \int_X a\, dS. \tag{1.0.5}$$

The result (1.0.5) provides a microlocalization of the local equidistribution phenomenon, which is the special case of (1.0.5) when $A$ is multiplication by a smooth function:

$$\lim_{k\to\infty, k\notin\mathcal{N}} \int_M b(x)|\varphi_k(x)|^2\, dV = \int_X b \circ \pi\, dS, \tag{1.0.6}$$

for $b \in C^\infty(M)$, $\pi : X \to M$ the natural projection, assuming the flow (1.0.2) is ergodic. In case $\Lambda = \sqrt{-\Delta}$, this takes the form

$$\lim_{k\to\infty, k\notin\mathcal{N}} \int_M b(x)|\varphi_k(x)|^2\, dV = \frac{1}{V(M)} \int_M b(x)\, dV. \tag{1.0.7}$$

Theorem 1.0.1 has been extended to compact Riemannian manifolds with piecewise smooth boundary in [**15**], in the setting where $\Lambda = \sqrt{-\Delta}$, with the Dirichlet boundary condition.

Theorem 1.0.1 has also been extended to cases where the flow (1.0.2) is not ergodic, in [**5**] and [**9**], and in [**4**] and [**3**]. Results of [**9**] treat cases where the ergodicity hypothesis on $\{G_t\}$ is replaced by the hypothesis that

$$Pa \in C(X), \tag{1.0.8}$$

where $P$ is the orthogonal projection of $L^2(X)$ onto the subspace of $G_t$-invariant elements. See Proposition 2.3.1 of this paper for a detailed statement. The paper [**9**] gives examples where $P$ maps $C(X)$ to $C(X)$ but it does not map $C^\infty(X)$ to $C^\infty(X)$, or even to the space of Hölder continuous functions. Thus one is motivated to quantize non-smooth functions in $C(X)$.

One of the goals of this paper is to take this further, quantizing discontinuous functions on $X$, and making use of such a quantization to produce further quantum ergodic theorems.

To give another motivation for dealing with discontinuous symbols, we mention the following simple extension of (1.0.6).

PROPOSITION 1.0.2. *In the setting of Theorem* 1.0.1, *the limiting result* (1.0.6) *holds whenever*

$$b \in \mathcal{R}(M). \tag{1.0.9}$$

In (1.0.9), $\mathcal{R}(M)$ denotes the space of bounded functions on $M$ that are *Riemann integrable.* To prove Proposition 1.0.2, it suffices to treat the case where $b$ is real valued. Then, given $\varepsilon > 0$, we can pick $b_1, b_2 \in C^\infty(M)$ such that

$$b_1 \le b \le b_2, \quad \text{and} \quad \int_M (b_2 - b_1)\, dV < \varepsilon, \tag{1.0.10}$$

hence

$$\int_X (b_2 - b_1) \circ \pi \, dS < C\varepsilon. \tag{1.0.11}$$

We know from Theorem 1.0.1 that (1.0.6) holds with $b$ replaced by $b_j$. Hence the lim inf and the lim sup of the left side of (1.0.6) are squeezed between

$$\int_X b_1 \circ \pi \, dS \quad \text{and} \quad \int_X b_2 \circ \pi \, dS, \tag{1.0.12}$$

and taking $\varepsilon \searrow 0$ yields (1.0.6) for $b \in \mathcal{R}(M)$.

In §2 we extend Theorem 1.0.1 to handle operators $A$ whose symbols are discontinuous on $X$. One ingredient is the extension of the notion of quantization of an element $a \in C^\infty(X)$ to quantization of more general functions. We show that Friedrichs quantization extends from

$$\mathrm{op}_F : C^\infty(X) \longrightarrow OPS^0_{1,0}(M) \tag{1.0.13}$$

to

$$\mathrm{op}_F : L^\infty(X) \longrightarrow \mathcal{L}(L^2(M)). \tag{1.0.14}$$

Among quantum ergodic theorems that we establish using such an extended quantization procedure, we obtain the following extension of Theorem 1.0.1.

THEOREM 1.0.3. *There is a subset $\mathcal{N} \subset \mathbb{N}$ of density* 0 *such that if*

$$a \in C(X), \quad Pa \in \mathcal{R}(X), \tag{1.0.15}$$

*then*

$$\lim_{k\to\infty, k\notin\mathcal{N}} (A\varphi_k, \varphi_k)_{L^2} - (A_p\varphi_k, \varphi_k)_{L^2} = 0, \tag{1.0.16}$$

*where*

$$A = \mathrm{op}_F(a), \quad A_p = \mathrm{op}_F(Pa). \tag{1.0.17}$$

See Proposition 2.3.2 for the proof of Theorem 1.0.3. In §2.3 we furthermore show that, if (1.0.2) is ergodic, then (1.0.5) holds for $A = \mathrm{op}_F(a)$ whenever $a \in \mathcal{R}(X)$. In addition, we have a result for cases where $\{G_t\}$ acts ergodically on an open subset $U \subset X$. Another result of §2, Theorem 2.4.1, extends Proposition 1.0.2 from $b \in \mathcal{R}(M)$ to $b \in L^\infty(M)$, under an hypothesis on the behavior of $L^p$-norms of the eigenfunctions $\varphi_j$.

In §3 we look at a different sort of quantum ergodic theorem, involving the long time averages of operator conjugates

$$e^{-it\Lambda}Ae^{it\Lambda}. \tag{1.0.18}$$

Such a study arose in [**7**], in cases where the flow $\{G_t\}$ is ergodic. Again, our goal is to treat cases where $\{G_t\}$ is not necessarily ergodic. Taking

$$A_T = \frac{1}{2T}\int_{-T}^{T} e^{-it\Lambda}Ae^{it\Lambda}\,dt, \tag{1.0.19}$$

we consider how, in a rather general setting, we have the convergence

$$A_T \longrightarrow \Pi(A), \tag{1.0.20}$$

in the strong operator topology on $\mathcal{L}(L^2(M))$, as $T \to \infty$, and identify $\Pi : \mathcal{L}(L^2(M)) \to \mathcal{L}(L^2(M))$ as a projection onto the space of operators that commute with the unitary group $\{e^{it\Lambda}\}$. Then we establish the following.

THEOREM 1.0.4. *If $a, Pa \in C(X)$, and $A = \mathrm{op}_F(a)$, $A_p = \mathrm{op}_F(Pa)$, then*

$$\lim_{N\to\infty} \frac{1}{d_N} \big\| [\Pi(A) - \Pi(A_p)]Q_N \big\|_{\mathrm{HS}}^2 = 0. \tag{1.0.21}$$

Here $Q_N$ is the spectral projection of $L^2(M)$ onto the linear span of the eigenspaces of $\Lambda$ associated with eigenvalues $\le N$, $d_N = \operatorname{Tr} Q_N$, and $\|\cdot\|_{\mathrm{HS}}$ denotes the Hilbert-Schmidt norm. In case the flow $\{G_t\}$ is ergodic on $X$, $Pa = \overline{a}$, $A_p = \overline{a}I$, and Theorem 1.0.4 then specializes to the operator ergodic theorem of [**7**]. See Proposition 3.2.5 for the proof of Theorem 1.0.4.

## 2. Quantum ergodic theorems with discontinuous symbols

As stated in the introduction, here we want to establish various quantum ergodic theorems, of the form (1.0.16), in cases that involve discontinuous symbols. The first order of business is to obtain a quantization of discontinuous symbols, and we take this up in §2.1. We use the Friedrichs quantization, $\mathrm{op}_F : C^\infty(X) \to OPS^0_{1,0}(M)$, which enjoys the positivity property

$$a \ge 0 \Longrightarrow \mathrm{op}_F(a) \ge 0. \tag{2.0.1}$$

This has a unique continuous extension to $\mathrm{op}_F : C(X) \to \mathcal{L}(L^2(M))$, and from there to

$$\mathrm{op}_F : L^\infty(X) \longrightarrow \mathcal{L}(L^2(M)), \tag{2.0.2}$$

still obeying (2.0.1). In this setting,

$$\begin{aligned} &a_\nu \to a \text{ weak}^* \text{ in } L^\infty(X) \\ &\quad\Longrightarrow \mathrm{op}_F(a_\nu) \to \mathrm{op}_F(a) \text{ in the weak operator topology.} \end{aligned} \tag{2.0.3}$$

In §2.2 we establish the Weyl law

$$\lim_{N\to\infty} \frac{1}{N}\sum_{k=1}^{N} (B\varphi_k, \varphi_k)_{L^2} = \int_X b\,dS, \tag{2.0.4}$$

given

$$b \in \mathcal{R}(X), \quad B = \mathrm{op}_F(b), \tag{2.0.5}$$

which is useful for the analysis in §2.3.

In §2.3 we establish quantum ergodic theorems. One result, Proposition 2.3.2 (which proves Theorem 1.0.3), is that there is a subset $\mathcal{N} \subset \mathbb{N}$, of density 0, such that if

$$a \in C(X), \quad Pa \in \mathcal{R}(X), \tag{2.0.6}$$

then

$$\lim_{k\to\infty, k\notin\mathcal{N}} (A\varphi_k, \varphi_k)_{L^2} - (A_p\varphi_k, \varphi_k)_{L^2} = 0, \tag{2.0.7}$$

where $A = \mathrm{op}_F(a)$, $A_p = \mathrm{op}_F(Pa)$. Another, Proposition 2.3.3. is the following extension of Proposition 1.0.2: under the hypothesis that (1.0.2) is ergodic, (1.0.5) holds for $A = \mathrm{op}_F(a)$ whenever $a \in \mathcal{R}(X)$. A further extension, Proposition 2.3.5, assumes $\{G_t\}$ acts ergodically on an open set $U \subset X$. In such a case, if

$$a, b \in \mathcal{R}(X) \ \text{ are supported on a compact subset of } \ U, \tag{2.0.8}$$

then

$$\int_X a\, dS = \int_X b\, dS \Longrightarrow \lim_{k\to\infty, k\notin\mathcal{N}} (A\varphi_k, \varphi_k)_{L^2} - (B\varphi_k, \varphi_k)_{L^2} = 0, \tag{2.0.9}$$

for $A = \mathrm{op}_F(a)$, $B = \mathrm{op}_F(b)$.

In §2.4, we show that, if the flow $\{G_t\}$ is ergodic on $X$, then the quantum ergodicity result (1.0.3) holds, with

$$Au(x) = a(x)u(x), \quad \forall\, a \in L^\infty(M), \tag{2.0.10}$$

under an hypothesis on the behavior of $L^p$-norms of the eigenfunctions $\varphi_j$.

**2.1. Quantization of discontinuous symbols.** Let $M$ be a compact Riemannian manifold, $X \subset T^*M$ as in §1. A *quantization* of $C^\infty(X)$ is a continuous linear map

$$\mathrm{op} : C^\infty(X) \longrightarrow OPS^0_{1,0}(M), \tag{2.1.1}$$

with the property that, given $a \in C^\infty(X)$, $A = \mathrm{op}(a)$ has principal symbol $a$ (mod $S^{-1}_{1,0}(M)$). We insist that $\mathrm{op}(1) = I$, the identity map. The existence of quantizations follows via local coordinate charts and partitions of unity from the calculus of pseudodifferential operators on Euclidean space. There are many different quantizations. Each one gives rise to a family of elements

$$\mu_{u,v} \in \mathcal{D}'(X), \quad \forall\, u, v \in L^2(M), \tag{2.1.2}$$

defined by

$$\langle a, \mu_{u,v}\rangle = (\mathrm{op}(a)u, v)_{L^2}. \tag{2.1.3}$$

Basic examples are "Kohn-Nirenberg" quantizations and "Weyl" quantizations:

$$\mathrm{op}_{KN},\ \mathrm{op}_W : C^\infty(X) \longrightarrow OPS^0(M) \subset OPS^0_{1,0}(M). \tag{2.1.4}$$

Another family, of particular interest to us here, is the family of "Friedrichs quantizations,"

$$\mathrm{op}_F : C^\infty(X) \longrightarrow OPS^0_{1,0}(M), \tag{2.1.5}$$

which has the property

$$a \ge 0 \Longrightarrow \mathrm{op}_F(a) \ge 0. \tag{2.1.6}$$

This is constructed on the Euclidean space level from $\mathrm{op}_{KN}$ via Friedrichs symmetrization. See [**8**], Chapter 7. It has the property that

$$a \in C^\infty(X) \Longrightarrow \mathrm{op}_F(a) - \mathrm{op}_{KN}(a) \in OPS^{-1}_{1,0}(M), \tag{2.1.7}$$

which plays an important role in the Friedrichs approach to the proof of the sharp Gårding inequality. From (2.1.6) it follows that

$$\|\mathrm{op}_F(a)\|_{\mathcal{L}(L^2)} \le \sup_X |a|, \tag{2.1.8}$$

and hence (2.1.5) has a unique continuous linear extension to

$$\mathrm{op}_F : C(X) \longrightarrow \mathcal{L}(L^2(M)). \tag{2.1.9}$$

The image of $C(X)$ in (2.1.9) is contained in the $C^*$-algebra of operators on $L^2(M)$ generated by $OPS^0(M)$, which we denote $\Psi(M)$. If we compose the map

$$\mathrm{op}_F : C(X) \longrightarrow \Psi(M) \tag{2.1.10}$$

with taking the quotient by $\mathcal{K}(L^2(M))$, the space of compact operators on $L^2(M)$, we get an isomorphism of $C^*$-algebras:

$$C(X) \xrightarrow{\approx} \Psi(M)/\mathcal{K}(L^2(M)). \tag{2.1.11}$$

We see from (2.1.6) that, for $u \in L^2(M)$, the distribution $\mu_u = \mu_{u,u} \in \mathcal{D}'(X)$ given by (2.1.3), with $u = v$, has the property that

$$a \ge 0 \Longrightarrow \langle a, \mu_u \rangle \ge 0. \tag{2.1.12}$$

Also $\langle 1, \mu_u \rangle = (u,u)_{L^2} = \|u\|^2_{L^2}$. Consequently, for each $u \in L^2(M)$,

$$\mu_u \text{ is a positive measure on } X, \text{ of mass } \|u\|^2_{L^2}. \tag{2.1.13}$$

Going further, we have from Cauchy's inequality that, for $u, v \in L^2(M)$,

$$\begin{aligned} |\langle a, \mu_{u,v} \rangle| &= |(\mathrm{op}_F(a)u, v)_{L^2}| \\ &\le \big(\sup_X |a|\big) \|u\|_{L^2} \|v\|_{L^2}, \end{aligned} \tag{2.1.14}$$

so

$$\begin{aligned} &\mu_{u,v} \text{ is a complex measure on } X, \\ &\qquad \text{of total mass } \le \|u\|_{L^2}\|v\|_{L^2}. \end{aligned} \tag{2.1.15}$$

Thus we can extend $\mathrm{op}_F$ from (2.1.9) to

$$\mathrm{op}_F : \mathcal{B}(X) \longrightarrow \mathcal{L}(L^2(M)), \tag{2.1.16}$$

where

$$\mathcal{B}(X) = \text{space of bounded Borel functions } a : X \to \mathbb{C}, \tag{2.1.17}$$

by

$$(\mathrm{op}_F(a)u, v) = \int_X a\, d\mu_{u,v}. \tag{2.1.18}$$

We continue to have (2.1.6) and (2.1.8), now for $a \in \mathcal{B}(X)$. We investigate the action of $\mathrm{op}_F$ on sequences $a_\nu$.

PROPOSITION 2.1.1. *Let $a_\nu \in \mathcal{B}(X)$, $\nu \in \mathbb{N}$. Assume $|a_\nu(z)| \le M < \infty$ for all $\nu, z$, and that*

$$a_\nu(z) \longrightarrow a(z), \quad \forall\, z \in X, \quad \text{as } \nu \to \infty. \tag{2.1.19}$$

*Then $\mathrm{op}_F(a_\nu) \to \mathrm{op}_F(a)$ in the weak operator topology, i.e., for all $u, v \in L^2(M)$,*

$$(\mathrm{op}_F(a_\nu)u, v)_{L^2} \longrightarrow (\mathrm{op}_F(a)u, v)_{L^2}, \quad \text{as } \nu \to \infty. \tag{2.1.20}$$

PROOF. The desired conclusion (2.1.20) is equivalent to

$$\int_X a_\nu \, d\mu_{u,v} \longrightarrow \int_X a \, d\mu_{u,v}, \quad \text{as } \nu \to \infty, \tag{2.1.21}$$

which follows from the Lebesgue dominated convergence theorem. □

We can improve the conclusion of Proposition 2.1.1, given the following result on regularity of the measures $\mu_{u,v}$. This result follows by applying integration by parts arguments to oscillatory integrals that yield the Euclidean space versions of $\mathrm{op}_F(a)$.

PROPOSITION 2.1.2. *If $u, v \in C^\infty(M)$, then $\mu_{u,v}$ is a smooth multiple of the Liouville measure on $X$, that is, for all $a \in C^\infty(X)$, hence for all $a \in \mathcal{B}(X)$,*

$$\int_X a \, d\mu_{u,v} = \int_X a\Phi(u,v)\, dS, \quad \text{given } u, v \in C^\infty(M), \tag{2.1.22}$$

*where*

$$\Phi : C^\infty(M) \times C^\infty(M) \longrightarrow C^\infty(X) \tag{2.1.23}$$

*is a continuous bilinear map.*

From here, we deduce that

$$a \in \mathcal{B}(X),\ a = 0 \ S\text{-a.e.} \implies (\mathrm{op}_F(a)u, v)_{L^2} = 0, \tag{2.1.24}$$

first for all $u, v \in C^\infty(M)$, and then for all $u, v \in L^2(M)$, by denseness of $C^\infty(M)$ in $L^2(M)$ and the estimate (2.1.8). Thus we can pass from (2.1.16) to

$$\mathrm{op}_F : L^\infty(X) \longrightarrow \mathcal{L}(L^2(M)), \tag{2.1.25}$$

satisfying

$$a \in L^\infty(X),\ a \ge 0 \ S\text{-a.e. on } X \implies \mathrm{op}_F(a) \ge 0, \tag{2.1.26}$$

and

$$a \in L^\infty(X) \implies \|\mathrm{op}_F(a)\|_{\mathcal{L}(L^2)} \le \|a\|_{L^\infty}. \tag{2.1.27}$$

We can then extend Proposition 2.1.1 as follows.

PROPOSITION 2.1.3. *Let $a_\nu \in L^\infty(X)$, $\nu \in \mathbb{N}$. Assume $\|a_\nu\|_{L^\infty} \le M < \infty$ for all $\nu$, and that*

$$a_\nu \longrightarrow a, \ S\text{-a.e. on } X, \text{ as } \nu \to \infty. \tag{2.1.28}$$

*Then $\mathrm{op}_F(a_\nu) \to \mathrm{op}_F(a)$ in the weak operator topology, i.e., for all $u, v \in L^2(M)$, (2.1.20) holds.*

PROOF. If $u, v \in C^\infty(M)$, then (2.1.20) follows from

$$\int_X a_\nu \, \Phi(u,v)\, dS \longrightarrow \int_X a\, \Phi(u,v)\, dS, \quad \text{as } \nu \to \infty, \tag{2.1.29}$$

which in turn follows from the Lebesgue dominated convergence theorem. Then (2.1.20) follows for general $u, v \in L^2(M)$ via denseness of $C^\infty(M)$ in $L^2(M)$ and the operator bounds $\| \operatorname{op}_F(a_\nu)\|_{\mathcal{L}(L^2)} \le M$. □

A similar argument yields the following.

PROPOSITION 2.1.4. *Let $a_\nu \in L^\infty(X)$ for $\nu \in \mathbb{N}$, and $a \in L^\infty(X)$. If $a_\nu \to a$ weak* in $L^\infty(X)$, then $\operatorname{op}_F(a_\nu) \to \operatorname{op}_F(a)$ in the weak operator topology.*

PROOF. Again, (2.1.29) holds for each $u, v \in C^\infty(M)$. From here, we follow the rest of the proof of Proposition 2.1.3. □

**2.2. Weyl law.** Recall that $M$ is a compact Riemannian manifold, the operator $\Lambda \in OPS^1(M)$ is a first-order, positive, self-adjoint elliptic operator, and $\{\varphi_k : k \in \mathbb{N}\}$ is an orthonormal basis of $L^2(M)$ consisting of eigenfunctions of $\Lambda$, with eigenvalues $\lambda_j \nearrow +\infty$. Also $X \subset T^*M$ is the level set on which the principal symbol of $\Lambda$ is equal to 1.

This section is devoted to the proof of the following.

PROPOSITION 2.2.1. *Given*

$$b \in \mathcal{R}(X), \quad B = \operatorname{op}_F(b), \tag{2.2.1}$$

*we have*

$$\lim_{N\to\infty} \frac{1}{N} \sum_{k=1}^N (B\varphi_k, \varphi_k)_{L^2} = \int_X b\, dS. \tag{2.2.2}$$

In (2.2.1), $\mathcal{R}(X)$ is the space of Riemann integrable functions on $X$. The result (2.2.2) is classical for $b \in C^\infty(X)$. In such a case, one can obtain an asymptotic expansion of

$$\operatorname{Tr} Be^{-t\Lambda}, \quad \text{as } t \searrow 0, \tag{2.2.3}$$

and deduce (2.2.2) via Karamata's Tauberian theorem (cf. [**8**], Chapter 12, §7).

The transition from $b \in C^\infty(X)$ to $b \in \mathcal{R}(X)$ is fairly straightforward, given the material of §2.1. It suffices to treat the case where $b \in \mathcal{R}(X)$ is real valued. Then, given $\varepsilon > 0$, we can pick $b_1, b_2 \in C^\infty(X)$ such that

$$b_1 \le b \le b_2, \quad \int_X (b_2 - b_1)\, dS < \varepsilon. \tag{2.2.4}$$

We have

$$\operatorname{op}_F(b_1) \le B \le \operatorname{op}_F(b_2), \tag{2.2.5}$$

and the classical result for $b_j \in C^\infty(X)$ gives

$$\lim_{N\to\infty} \frac{1}{N} \sum_{k=1}^N (\operatorname{op}_F(b_j)\varphi_k, \varphi_k)_{L^2} = \int_X b_j\, dS, \quad j = 1, 2. \tag{2.2.6}$$

Hence

$$
\begin{aligned}
\int_X b_1\,dS &\le \liminf_{N\to\infty} \frac{1}{N}\sum_{k=1}^{N}(B\varphi_k,\varphi_k)_{L^2} \\
&\le \limsup_{N\to\infty} \frac{1}{N}\sum_{k=1}^{N}(B\varphi_k,\varphi_k)_{L^2} \\
&\le \int_X b_2\,dS,
\end{aligned} \tag{2.2.7}
$$

and (2.2.2) follows, upon taking $\varepsilon \searrow 0$.

REMARK. It would be interesting to know if one can extend this result, replacing the hypothesis $b \in \mathcal{R}(X)$ by $b \in L^\infty(X)$.

**2.3. Quantum ergodic theorems.** Our goal here is to present various extensions of the following result, established in [**5**] and [**9**].

PROPOSITION 2.3.1. *Given $M$ and $\Lambda$, and $\varphi_k$ as described in §1, there is a subset $\mathcal{N} \subset \mathbb{N}$, of density 0, such that the following holds. Let $a \in C(X)$, $A = \mathrm{op}_F(a)$. Assume*

$$Pa = \overline{a} = \int_X a\,dS. \tag{2.3.1}$$

*Then*

$$\lim_{k\to\infty, k\notin\mathcal{N}} (A\varphi_k,\varphi_k)_{L^2} = \overline{a}. \tag{2.3.2}$$

*Going further, if we replace (2.3.1) by*

$$Pa \in C(X), \tag{2.3.3}$$

*then*

$$\lim_{k\to\infty, k\notin\mathcal{N}} (A\varphi_k,\varphi_k)_{L^2} - (\mathrm{op}_F(Pa)\varphi_k,\varphi_k)_{L^2} = 0. \tag{2.3.4}$$

That (2.3.1) $\Rightarrow$ (2.3.2), for $a \in C^\infty(X)$, was noted in [**5**], and the extension to $a \in C(X)$ was made in [**9**]. The result (2.3.3) $\Rightarrow$ (2.3.4) follows by applying the first part of the proposition to $B = \mathrm{op}_F(b)$, with $b = a - Pa$, for which we have $Pb = \overline{b} = 0$. Our first goal here is to establish the following more general result.

PROPOSITION 2.3.2. *There is a set $\mathcal{N} \subset \mathbb{N}$, of density 0, such that the following holds. Assume*

$$a \in C(X), \quad Pa \in \mathcal{R}(X), \tag{2.3.5}$$

*where $\mathcal{R}(X)$ is the space of Riemann integrable functions on $X$. Then (2.3.4) continues to hold.*

PROOF. The strategy will be to use results of §2.2 to extend the proof of Proposition 2.3.1 given in [**9**] to handle the more general situation described above.

We bring in the probability measures $\mu_k$ on $X$, given by $\mu_k = \mu_{\varphi_k} = \mu_{\varphi_k,\varphi_k}$. We will show that, given (2.3.5),

$$\lim_{N\to\infty} \frac{1}{N} \sum_{k=1}^{N} \Bigl| \int_X (a - Pa)\, d\mu_k \Bigr| = 0. \tag{2.3.6}$$

One ingredient is the following consequence of Egorov's theorem:

$$\int_X (a_T - a)\, d\mu_k \longrightarrow 0, \quad \text{as } k\to\infty, \quad \forall\, a\in C(X), \tag{2.3.7}$$

locally uniformly in $T\in[0,\infty)$, where

$$a_T = \frac{1}{T}\int_0^T a\circ G_t\, dt. \tag{2.3.8}$$

See (5.7) of [**9**]. From (2.3.7), we have

$$\int_X (a_T - Pa)\, d\mu_k - \int_X (a - Pa)\, d\mu_k \longrightarrow 0, \quad \text{as } k\to\infty, \tag{2.3.9}$$

locally uniformly in $T$. The next ingredient is the Weyl law:

$$\lim_{N\to\infty} \frac{1}{N} \sum_{k=1}^{N} \int_X b\, d\mu_k = \int_X b\, dS, \quad \forall\, b\in \mathcal{R}(X), \tag{2.3.10}$$

established in §2.2, as an extension of the classical version, for $b\in C^\infty(X)$. We apply this to

$$b = |a_T - Pa|, \tag{2.3.11}$$

which belongs to $\mathcal{R}(X)$ if (2.3.5) holds, and obtain

$$\lim_{N\to\infty} \frac{1}{N} \sum_{k=1}^{N} \int_X |a_T - Pa|\, d\mu_k = \int_X |a_T - Pa|\, dS. \tag{2.3.12}$$

Now the mean ergodic theorem implies that, for each $\varepsilon>0$, there exists $T_\varepsilon<\infty$ such that

$$\int_X |a_T - Pa|\, dS \le \varepsilon, \quad \forall\, T\ge T_\varepsilon. \tag{2.3.13}$$

Together, (2.3.9), (2.3.12), and (2.3.13) yield

$$\limsup_{N\to\infty} \frac{1}{N} \sum_{k=1}^{N} \Bigl| \int_X (a - Pa)\, d\mu_k \Bigr| \le \varepsilon, \quad \forall\, \varepsilon>0, \tag{2.3.14}$$

which implies (2.3.6).

We deduce (2.3.4) from (2.3.6) as follows. Note that

$$E(X) = \{a\in C(X) : Pa\in \mathcal{R}(X)\} \tag{2.3.15}$$

is a closed linear subspace of $C(X)$. This follows from the fact that

$$\sup_z |Pa_\nu(z) - Pa(z)| \le \sup_z |a(z) - a_\nu(z)|, \tag{2.3.16}$$

and that uniform limits of elements of $\mathcal{R}(X)$ also belong to $\mathcal{R}(X)$. Since $C(X)$ is separable, so is $E(X)$, and we can take a countable dense subset $\{a_\nu : \nu\in\mathbb{N}\}$ of

$E(X)$. It follows directly from (2.3.6), applied to $a_\nu$, that there exist $\mathcal{N}_\nu \subset \mathbb{N}$, of density 0, such that

$$\lim_{k\to\infty, k\notin\mathcal{N}_\nu} \int_X (a_\nu - Pa_\nu)\, d\mu_k = 0. \tag{2.3.17}$$

Now we can take a set $\mathcal{N} \subset \mathbb{N}$, of density 0, such that $\mathcal{N} \setminus \mathcal{N}_\nu$ is finite, for each $\nu$, so

$$\lim_{k\to\infty, k\notin\mathcal{N}} \int_X (a_\nu - Pa_\nu)\, d\mu_k = 0, \quad \forall \nu \in \mathbb{N}. \tag{2.3.18}$$

The denseness of $\{a_\nu : \nu \in \mathbb{N}\}$ in $E(X)$, together with (2.3.16), then implies

$$\lim_{k\to\infty, k\notin\mathcal{N}} \int_X (a - Pa)\, d\mu_k = 0, \quad \forall\, a \in E(X), \tag{2.3.19}$$

and this is equivalent to (2.3.4). □

We now relax the hypothesis that the symbol $a$ be continuous on $X$, first under an ergodicity hypothesis.

PROPOSITION 2.3.3. *Assume $\{G_t : t \in \mathbb{R}^+\}$ acts ergodically on $X$. Then there is a set $\mathcal{N} \subset \mathbb{N}$, of density 0, such that*

$$\lim_{k\to\infty, k\notin\mathcal{N}} (\mathrm{op}_F(a)\varphi_k, \varphi_k)_{L^2} = \int_X a\, dS, \tag{2.3.20}$$

*for all $a \in \mathcal{R}(X)$.*

PROOF. For $a \in C^\infty(X)$, this is the classical result given in Theorem 1.0.1. Note that it also follows from Proposition 2.3.1, since ergodicity implies $Pa = \overline{a}$. Now, given a real valued $a \in \mathcal{R}(X)$, and given $\varepsilon > 0$, we can pick

$$b_1, b_2 \in C^\infty(X) \text{ such that } b_1 \le a \le b_2 \text{ and } \int_X (b_2 - b_1)\, dS < \varepsilon. \tag{2.3.21}$$

We know that

$$\lim_{k\to\infty, k\notin\mathcal{N}} (\mathrm{op}_F(b_j)\varphi_k, \varphi_k)_{L^2} = \int_X b_j\, dS, \tag{2.3.22}$$

and that

$$\mathrm{op}_F(b_1) \le \mathrm{op}_F(a) \le \mathrm{op}_F(b_2), \tag{2.3.23}$$

so

$$\begin{aligned} \limsup_{k\to\infty, k\notin\mathcal{N}} (\mathrm{op}_F(a)\varphi_k, \varphi_k)_{L^2} &\le \int_X b_2\, dS, \quad \text{and} \\ \liminf_{k\to\infty, k\notin\mathcal{N}} (\mathrm{op}_F(a)\varphi_k, \varphi_k)_{L^2} &\ge \int_X b_1\, dS, \end{aligned} \tag{2.3.24}$$

and we have (2.3.20). □

The following is a local equidistribution result, associated with ergodicity on an open subset of $X$. Compare this with results of [**4**] and [**3**].

PROPOSITION 2.3.4. *Let $U \subset X$ be open and assume $G_t : U \to U$, and that the action on $U$ is ergodic. Let $a, b \in C(X)$ be supported on a compact subset of $U$, $A = \mathrm{op}_F(a)$, $B = \mathrm{op}_F(b)$. Take $\mathcal{N}$ as in Proposition 2.3.3. Then*

$$\begin{aligned}\int\limits_X a\,dS = \int\limits_X b\,dS \\ \Longrightarrow \lim_{k\to\infty, k\notin\mathcal{N}} (A\varphi_k, \varphi_k)_{L^2} - (B\varphi_k, \varphi_k)_{L^2} = 0.\end{aligned} \tag{2.3.25}$$

PROOF. The hypotheses yield

$$a - b \in C(X), \quad P(a-b) = 0, \tag{2.3.26}$$

so the conclusion (2.3.25) is a corollary of Proposition 2.3.1. □

We next extend the scope of Proposition 2.3.4, along the lines of Proposition 2.3.3.

PROPOSITION 2.3.5. *In the setting of Proposition 2.3.4, the implication (2.3.25) holds for all $a, b \in \mathcal{R}(X)$ that are supported on a compact subset of $U$.*

PROOF. It suffices to show that, given

$$\text{real valued } \ a \in \mathcal{R}(X), \ \ \mathrm{supp}\, a \subset K \subset\subset U, \tag{2.3.27}$$

we have

$$\int\limits_X a\,dS = 0 \Longrightarrow \lim_{k\to\infty, k\notin\mathcal{N}} (A\varphi_k, \varphi_k)_{L^2} = 0. \tag{2.3.28}$$

To get this, let $\varepsilon > 0$ and take $b_1, b_2 \in C(X)$ such that

$$\mathrm{supp}\, b_j \subset\subset U, \quad b_1 \le a \le b_2, \quad \int\limits_X (b_2 - b_1)\,dS < \varepsilon. \tag{2.3.29}$$

Next, take $g_j \in C(X)$, $\mathrm{supp}\, g_j \subset\subset U$, such that

$$\sup |g_j| \le C\varepsilon, \quad \text{and} \quad \int\limits_X (b_j - g_j)\,dS = 0. \tag{2.3.30}$$

Then, by Proposition 2.3.4,

$$((\mathrm{op}_F(b_j) - \mathrm{op}_F(g_j))\varphi_k, \varphi_k)_{L^2} \longrightarrow 0, \quad \text{as } \ k \to \infty, \ k \notin \mathcal{N}. \tag{2.3.31}$$

Now

$$\mathrm{op}_F(b_1) \le A \le \mathrm{op}_F(b_2), \tag{2.3.32}$$

so

$$\mathrm{op}_F(b_1 - g_1) - C\varepsilon I \le A \le \mathrm{op}_F(b_2 - g_2) + C\varepsilon I, \tag{2.3.33}$$

hence

$$\begin{aligned}\limsup_{k\to\infty, k\notin\mathcal{N}} (A\varphi_k, \varphi_k)_{L^2} \le C\varepsilon, \quad \text{and} \\ \liminf_{k\to\infty, k\notin\mathcal{N}} (A\varphi_k, \varphi_k)_{L^2} \ge -C\varepsilon.\end{aligned} \tag{2.3.34}$$

This gives (2.3.28). □

**2.4. $L^p$ eigenfunction estimates and quantum ergodic theorems.** Here we take a further look at multiplication operators,

$$Au(x) = a(x)u(x). \tag{2.4.1}$$

We have seen quantum ergodic results for $a \in \mathcal{R}(M)$ (cf. Proposition 1.0.2). Now we seek to extend such a result

$$\text{from } a \in \mathcal{R}(M) \text{ to } a \in L^\infty(M). \tag{2.4.2}$$

We will bring in an hypothesis involving $L^p$-estimates on the eigenfunctions $\varphi_j$, and establish the following.

THEOREM 2.4.1. *Assume the orthonormal basis $\{\varphi_j\}$ in (1.0.1) has the property that, for some $p > 2$, and each $K < \infty$,*

$$\begin{gathered}\{j \in \mathbb{N} : \|\varphi_j\|_{L^p} > K\} \text{ has upper density } \vartheta(K), \\ \text{and } \vartheta(K) \to 0 \text{ as } K \to \infty.\end{gathered} \tag{2.4.3}$$

*Then, for all $a \in L^\infty(M)$,*

$$\lim_{N\to\infty} \frac{1}{N} \sum_{j=1}^{N} \Bigl|(A\varphi_j, \varphi_j)_{L^2} - \overline{a}\Bigr|^2 = 0. \tag{2.4.4}$$

To start the proof of Theorem 2.4.1, fix $a \in L^\infty(M)$ and take a sequence of mollifications,

$$a_\nu = e^{(1/\nu)\Delta} a,$$

so

$$\begin{gathered}a_\nu \in C^\infty(M), \quad \overline{a}_\nu = \overline{a}, \quad \|a_\nu\|_{L^\infty} \le \|a\|_{L^\infty}, \\ \|a - a_\nu\|_{L^q} \to 0, \quad \text{with } \frac{1}{q} + \frac{2}{p} = 1.\end{gathered} \tag{2.4.5}$$

Let

$$\begin{aligned}\mathcal{G}_K &= \{j \in \mathbb{N} : \|\varphi_j\|_{L^p} \le K\}, \\ \mathcal{B}_K &= \{j \in \mathbb{N} : \|\varphi_j\|_{L^p} > K\}.\end{aligned} \tag{2.4.6}$$

Replace $A$ on the left side of (2.4.4) by $B_\nu$, given by

$$B_\nu u(x) = [a(x) - a_\nu(x)]u(x), \tag{2.4.7}$$

and split the resulting sum as follows:

$$\frac{1}{N} \sum_{j\le N, j\in\mathcal{G}_K} \Bigl|(B_\nu\varphi_j, \varphi_j)_{L^2}\Bigr|^2 + \frac{1}{N} \sum_{j\le N, j\in\mathcal{B}_K} \Bigl|(B_\nu\varphi_j, \varphi_j)_{L^2}\Bigr|^2. \tag{2.4.8}$$

Noting that

$$\Bigl|(B_\nu\varphi_j, \varphi_j)_{L^2}\Bigr| \le \|a - a_\nu\|_{L^q} \|\varphi_j\|_{L^p}^2, \tag{2.4.9}$$

we have

$$\begin{aligned}\frac{1}{N} \sum_{j\le N, j\in\mathcal{G}_K} \Bigl|(B_\nu\varphi_j, \varphi_j)_{L^2}\Bigr|^2 &\le \frac{1}{N} \sum_{j\le N, j\in\mathcal{G}_K} \|a - a_\nu\|_{L^q}^2 \|\varphi_j\|_{L^p}^4 \\ &\le K^4 \|a - a_\nu\|_{L^q}^2.\end{aligned} \tag{2.4.10}$$

Meanwhile,

$$\begin{aligned}\frac{1}{N}\sum_{j\le N, j\in\mathcal{B}_K}\Bigl|(B_\nu\varphi_j,\varphi_j)_{L^2}\Bigr|^2 &\le \frac{1}{N}\sum_{j\le N, j\in\mathcal{B}_K}\|a-a_\nu\|_{L^\infty}^2 \\ &\le 4\|a\|_{L^\infty}^2\cdot\frac{1}{N}\,\#\{j\le N: j\in\mathcal{B}_K\},\end{aligned}\tag{2.4.11}$$

so

$$\limsup_{N\to\infty}\frac{1}{N}\sum_{j\le N, j\in\mathcal{B}_K}\Bigl|(B_\nu\varphi_j,\varphi_j)_{L^2}\Bigr|^2 \le 4\|a\|_{L^\infty}^2\vartheta(K).\tag{2.4.12}$$

Combining (2.4.10) and (2.4.12), we have

$$\limsup_{N\to\infty}\frac{1}{N}\sum_{j=1}^{N}\Bigl|(B_\nu\varphi_j,\varphi_j)_{L^2}\Bigr|^2 \le K^4\|a-a_\nu\|_{L^q}^2+\|a\|_{L^\infty}^2\vartheta(K),\tag{2.4.13}$$

for all $K<\infty$. Using

$$\begin{aligned}\bigl|(A\varphi_j,\varphi_j)_{L^2}\bigr|^2 &= \bigl|(B_\nu\varphi_j,\varphi_j)_{L^2}+(a_\nu\varphi_j,\varphi_j)_{L^2}-\overline{a}_\nu\bigr|^2 \\ &\le 2\bigl|(B_\nu\varphi_j,\varphi_j)_{L^2}\bigr|^2+2\bigl|(a_\nu\varphi_j,\varphi_j)_{L^2}-\overline{a}_\nu\bigr|^2,\end{aligned}\tag{2.4.14}$$

plus the fact that (2.4.4) holds with $A$ replaced by $A_\nu u = a_\nu(x)u(x)$, we have

$$\begin{aligned}&\limsup_{N\to\infty}\frac{1}{N}\sum_{j=1}^{N}\Bigl|(A\varphi_j,\varphi_j)_{L^2}-\overline{a}\Bigr|^2 \\ &\le \inf_\nu\,\limsup_{N\to\infty}\frac{2}{N}\sum_{j=1}^{N}\Bigl|(B_\nu\varphi_j,\varphi_j)_{L^2}\Bigr|^2 \\ &\le 8\|a\|_{L^\infty}^2\vartheta(K),\end{aligned}\tag{2.4.15}$$

for all $K<\infty$. Letting $K\to\infty$, we obtain (2.4.4) from the hypothesis (2.4.3). This proves Theorem 2.4.1.

## 3. Quantum ergodic theorems for conjugates $e^{-it\Lambda}Ae^{it\Lambda}$

Take $M$ and $\Lambda$ as in §1. If $A$ is a bounded linear operator on $L^2(M)$ (we write $A\in\mathcal{L}(L^2(M))$), and $T\in(0,\infty)$, we set

$$A_T=\frac{1}{2T}\int_{-T}^{T}e^{-it\Lambda}Ae^{it\Lambda}\,dt.\tag{3.0.1}$$

Our goal is to study the behavior of $A_T$ as $T\to\infty$ and relate this study to classical ergodic theory. This approach to quantum ergodic theorems arose in [**7**] (see also [**13**]).

To get started, we recall the abstract mean ergodic theorem of von Neumann. Let $U^t$ be a strongly continuous unitary group on a Hilbert space $\mathcal{H}$, and set

$$\mathcal{A}_T f=\frac{1}{2T}\int_{-T}^{T}U^t f\,dt,\quad f\in\mathcal{H}.\tag{3.0.2}$$

Then $U^t = e^{itB}$, where $B$ is a self-adjoint operator on $\mathcal{H}$, and the spectral theorem yields

$$\begin{aligned}\mathcal{A}_T f &= \frac{1}{2T}\int_{-T}^{T} e^{itB} f\, dt \\ &= \frac{\sin TB}{TB} f,\end{aligned} \tag{3.0.3}$$

and hence, for each $f \in \mathcal{H}$,

$$\mathcal{A}_T f \longrightarrow P_0 f \quad \text{in } \mathcal{H}\text{-norm, as } \ T \to \infty, \tag{3.0.4}$$

where

$$\begin{aligned}P_0 &= \ \text{orthogonal projection of } \mathcal{H} \text{ onto } \ \operatorname{Ker} B, \\ \operatorname{Ker} B &= \{f \in \mathcal{H} : U^t f = f,\ \forall\, t \in \mathbb{R}\}.\end{aligned} \tag{3.0.5}$$

In other words, $\mathcal{A}_T$ converges to $P_0$ in the strong operator topology of $\mathcal{L}(\mathcal{H})$.

To relate the last paragraph to (3.0.1), we look at

$$W^t : \mathcal{L}(L^2(M)) \longrightarrow \mathcal{L}(L^2(M)), \quad W^t(A) = e^{-it\Lambda} A e^{it\Lambda}. \tag{3.0.6}$$

Clearly $\{W^t\}$ is a group of isometries of $\mathcal{L}(L^2(M))$, and, for each $A \in \mathcal{L}(L^2(M))$, $W^t(A)$ is continuous in $t$ with values in $\mathcal{L}(L^2(M))$, equipped with the strong operator topology. On the other hand, clearly $W^t(A)$ is continuous from $t \in \mathbb{R}$ to $\mathcal{L}(L^2(M))$, with the *norm* topology, if $A$ has finite rank, and hence if $A$ is compact. We also have

$$W^t(A)\langle\Lambda\rangle^{-\kappa} = W^t(A\langle\Lambda\rangle^{-\kappa}), \tag{3.0.7}$$

where $\langle\Lambda\rangle = \sqrt{1-\Delta}$. Now $\mathcal{L}(L^2(M))$ is not a Hilbert space, so it is convenient to focus on the Hilbert space

$$\mathcal{H} = \mathrm{HS}(L^2(M)), \tag{3.0.8}$$

of Hilbert-Schmidt operators on $L^2(M)$, a Hilbert space with inner product

$$(A, B)_{\mathrm{HS}} = \operatorname{Tr} B^* A. \tag{3.0.9}$$

The restriction $U^t = W^t|_{\mathrm{HS}(L^2(M))}$ is a strongly continuous group of unitary operators on $\mathrm{HS}(L^2(M))$:

$$U^t(A) = e^{it \operatorname{ad} \Lambda}(A). \tag{3.0.10}$$

Consequently, for $A \in \mathrm{HS}(L^2(M))$,

$$A_T = \frac{\sin T \operatorname{ad} \Lambda}{T \operatorname{ad} \Lambda} A. \tag{3.0.11}$$

We have the following result.

PROPOSITION 3.0.1. *If $A \in \mathrm{HS}(L^2(M))$, then, for $A_T$ as in* (3.0.1),

$$A_T \longrightarrow \Pi_0(A) \ \textit{in } \mathrm{HS}\textit{-norm}, \tag{3.0.12}$$

*where*

$$\begin{aligned}\Pi_0 &= \ \textit{orthogonal projection of } \mathrm{HS}(L^2(M)) \textit{ onto} \\ \mathcal{K}_0 &= \{A \in \mathrm{HS}(L^2(M)) : e^{-it\Lambda} A e^{it\Lambda} = A,\ \forall\, t \in \mathbb{R}\}.\end{aligned} \tag{3.0.13}$$

REMARK. Under the natural isomorphism $\mathrm{HS}(L^2(M)) \approx L^2(M \times M)$, we have

$$\operatorname{ad} \Lambda = \Lambda_x - \Lambda_y. \tag{3.0.14}$$

Using (3.0.7), we see that, whenever $\kappa > n/2$, with $n = \dim M$ (so $\langle\Lambda\rangle^{-\kappa}$ is Hilbert-Schmidt),

$$
\begin{aligned}
A \in \mathcal{L}(L^2(M)) \Longrightarrow A_T\langle\Lambda\rangle^{-\kappa} &= \big(A\langle\Lambda\rangle^{-\kappa}\big)_T \\
&\to \Pi_0(A\langle\Lambda\rangle^{-\kappa}),
\end{aligned} \tag{3.0.15}
$$

in HS-norm, and a fortiori in operator norm in $\mathcal{L}(L^2(M))$. Consequently, for all $A \in \mathcal{L}(L^2(M))$,

$$
\begin{aligned}
&A_T \longrightarrow \Pi(A) = \Pi_0(A\langle\Lambda\rangle^{-\kappa})\langle\Lambda\rangle^{\kappa} \\
&\text{in operator norm in } \mathcal{L}(H^\kappa(M), L^2(M)),
\end{aligned} \tag{3.0.16}
$$

for each $\kappa > n/2$, where $H^\kappa(M) = \mathcal{D}(\langle\Lambda\rangle^\kappa)$ is an $L^2$-Sobolev space. Hence the identity

$$
\Pi(A) = \Pi_0(A\langle\Lambda\rangle^{-\kappa})\langle\Lambda\rangle^{\kappa}, \quad \kappa > \frac{n}{2}, \tag{3.0.17}
$$

defines

$$
\Pi : \mathcal{L}(L^2(M)) \longrightarrow \mathcal{L}(H^\kappa(M), L^2(M)). \tag{3.0.18}
$$

The action of $\Pi$ is independent of $\kappa > n/2$. In addition, the uniform operator norm bounds $\|A_T\|_{\mathcal{L}(L^2)} \le \|A\|_{\mathcal{L}(L^2)}$, plus denseness of $H^\kappa(M)$ in $L^2(M)$, yield

$$
\Pi : \mathcal{L}(L^2(M)) \longrightarrow \mathcal{L}(L^2(M)), \quad \|\Pi(A)\|_{\mathcal{L}(L^2(M))} \le \|A\|_{\mathcal{L}(L^2(M))}. \tag{3.0.19}
$$

Going further, using this denseness and uniform bounds on $A_T$, we have:

PROPOSITION 3.0.2. *For $A \in \mathcal{L}(L^2(M))$, $A_T$ as in* (3.0.2),

$$
A_T \longrightarrow \Pi(A) \ \ \textit{in the strong operator topology of } \ \mathcal{L}(L^2(M)). \tag{3.0.20}
$$

There is another formula for $\Pi(A)$, which will prove useful. To state it let

$$
\begin{aligned}
P_\lambda &= \ \text{orthogonal projection of } L^2(M) \text{ onto } \mathrm{Eigen}(\Lambda, \lambda), \\
\Sigma_N &= \{\lambda \in \operatorname{Spec}\Lambda : \lambda \le N\},
\end{aligned} \tag{3.0.21}
$$

and set

$$
S_N(A) = \sum_{\lambda \in \Sigma_N} P_\lambda A P_\lambda. \tag{3.0.22}
$$

We see that $\|S_N(A)\|_{\mathcal{L}(L^2)} \le \|A\|_{\mathcal{L}(L^2)}$, that $S_N(A\langle\Lambda\rangle^{-\kappa}) = S_N(A)\langle\Lambda\rangle^{-\kappa}$, and that

$$
S_N(A) \longrightarrow \Pi(A) \text{ in HS-norm, if } A \in \mathrm{HS}(L^2(M)). \tag{3.0.23}
$$

It follows that

$$
\begin{aligned}
S_N(A) \longrightarrow \Pi(A), &\text{ in the strong operator topology,} \\
&\text{for all } A \in \mathcal{L}(L^2(M)).
\end{aligned} \tag{3.0.24}
$$

In other words,

$$
\Pi(A) = \sum_{\lambda \in \operatorname{Spec}\Lambda} P_\lambda A P_\lambda. \tag{3.0.25}
$$

Compare this with (2.24) of [**13**], and also the material in [**7**].

Our main goal here will be to prove the following (cf. Proposition 3.2.5).

THEOREM 3.0.3. *If* $a, Pa \in C(X)$, $A = \mathrm{op}_F(a)$, *and* $A_p = \mathrm{op}_F(Pa)$, *then*

$$\lim_{N\to\infty} \frac{1}{d_N} \big\| [\Pi(A) - \Pi(A_p)]Q_N \big\|_{\mathrm{HS}}^2 = 0. \tag{3.0.26}$$

Here, $X \subset T^*M \setminus 0$ is as in (1.0.2), $P : L^2(X) \to L^2(X)$ is the orthogonal projection onto the space of functions on $X$ invariant under the flow in (1.0.2),

$$Q_N = \sum_{\lambda\le N} P_\lambda, \quad d_N = \mathrm{Tr}\, Q_N, \tag{3.0.27}$$

and $\mathrm{op}_F : C(X) \to \mathcal{L}(L^2(M))$ is the Friedrichs quantization operator discussed in §2.1. A particular case of (3.0.26) is

$$a \in C(X),\ Pa = \overline{a} \Rightarrow \lim_{N\to\infty} \frac{1}{d_N} \big\| [\Pi(A) - \overline{a}I]Q_N \big\|_{\mathrm{HS}}^2 = 0, \tag{3.0.28}$$

where $\overline{a}$ is the mean value of $a$ over $X$. The hypothesis $Pa = \overline{a}$ for all $a \in C(X)$ holds provided the flow $G_t$ on $X$ is ergodic. In the ergodic case, (3.0.28) is due to [**7**] (see also [**13**]). Our extension beyond the case of an ergodic flow is done in the spirit of [**5**] and [**8**]. In this connection, we mention the following result (cf. Proposition 3.2.6).

COROLLARY 3.0.4. *Let* $U \subset X$ *be open and assume the flow* $G_t : U \to U$ *and that the action on* $U$ *is ergodic. Let* $a, b \in C(X)$ *be supported in a compact subset of* $U$, $A = \mathrm{op}_F(a)$, $B = \mathrm{op}_F(b)$. *Then*

$$\begin{aligned} &\int_X a\, dS = \int_X b\, dS \\ &\qquad \Longrightarrow \lim_{N\to\infty} \frac{1}{d_N} \big\| [\Pi(A) - \Pi(B)]Q_N \big\|_{\mathrm{HS}}^2 = 0. \end{aligned} \tag{3.0.29}$$

As indicated above, these results are established in §3.2. In §3.1 we set up tools needed to accomplish this, including complements on the Weyl law and Egorov's theorem. A key ingredient in the proof of Theorem 3.0.3 is that, given $a, Pa \in C(X)$,

$$\lim_{N\to\infty} \frac{1}{N} \sum_{k=1}^N \Big| (A\varphi_k, \varphi_k) - (A_p\varphi_k, \varphi_k) \Big|^2 = 0, \tag{3.0.30}$$

where $\{\varphi_k\}$ is an orthonormal basis of $L^2(M)$ satisfying $\Lambda\varphi_k = \lambda_k\varphi_k$, $\lambda_k \nearrow \infty$. Cf. (3.2.16). If the flow $\{G_t\}$ on $X$ is ergodic, then $Pa = \overline{a}$, and (3.0.30) is a standard version of quantum ergodicity (cf. [**1**]).

REMARK. While the conjugate $e^{-it\Lambda}Ae^{it\Lambda}$ is natural to work with due to its connection to Egorov's theorem, in case $\Lambda = \sqrt{-\Delta}$, it is also quite natural to consider

$$e^{it\Delta}Ae^{-it\Delta}, \tag{3.0.31}$$

in view of its quantum mechanical significance, and to replace (3.0.1) by

$$A_T^\# = \frac{1}{2T} \int_{-T}^T e^{it\Delta}Ae^{-it\Delta}\, dt. \tag{3.0.32}$$

Arguments parallel to those leading to Proposition 3.0.2 also yield

$$A_T^\# \longrightarrow \Pi(A), \tag{3.0.33}$$

in the strong operator topology, for each $A \in \mathcal{L}(L^2(M))$. The limit here is the same as in (3.0.20), as one verifies that it satisfies (3.0.25). Having (3.0.33), we can interpret Theorem 3.0.3 as a quantum ergodic theorem in the Heisenberg picture, while Theorems 1.0.1 and 1.0.3 are set in the Schrödinger picture.

**3.1. Complements to Weyl laws and Egorov's theorem.** In preparation for §3.2, we discuss some complements to results on the Weyl law and Egorov's theorem given in §2. First, we recall Proposition 2.2.1:

PROPOSITION 3.1.1. *We have*

$$\lim_{N\to\infty} \frac{1}{N} \sum_{k=1}^{N} (B\varphi_k, \varphi_k)_{L^2} = \int_X b\, dS, \tag{3.1.1}$$

*for* $B = \mathrm{op}_F(b)$, $b \in C^\infty(X)$, *where* $dS$ *is the Liouville measure on* $X$, *normalized so that* $\int_X dS = 1$. *More generally,* (3.1.1) *holds for all* $b \in \mathcal{R}(X)$.

We can rewrite (3.1.1) as follows. Let

$$\begin{aligned} Q_N = &\text{ orthogonal projection of } L^2(M) \\ &\text{ onto } \mathrm{Span}\{\varphi_k : 1 \le k \le N\}. \end{aligned} \tag{3.1.2}$$

Then (3.1.1) says

$$\lim_{N\to\infty} \frac{1}{N} \operatorname{Tr} BQ_N = \int_X b\, dS. \tag{3.1.3}$$

The next result is a Weyl/Szegö type result.

PROPOSITION 3.1.2. *We have*

$$\lim_{N\to\infty} \frac{1}{N} \sum_{k=1}^{N} \|B\varphi_k\|_{L^2}^2 = \int_X |b|^2\, dS, \tag{3.1.4}$$

*for* $B = \mathrm{op}_F(b)$, $b \in C^\infty(X)$. *More generally,* (3.1.4) *holds for* $b \in C(X)$.

PROOF. The left side of (3.1.4) is equal to

$$\begin{aligned} \lim_{N\to\infty} \frac{1}{N} \|BQ_N\|_{\mathrm{HS}}^2 &= \lim_{N\to\infty} \frac{1}{N} \operatorname{Tr} Q_N B^* B Q_N \\ &= \lim_{N\to\infty} \frac{1}{N} \operatorname{Tr} B^* B Q_N. \end{aligned} \tag{3.1.5}$$

If $b \in C^\infty(X)$, then

$$B^*B = \mathrm{op}_F(|b|^2), \quad \mathrm{mod}\ OPS^{-1}_{1,0}(M), \tag{3.1.6}$$

and the result follows from Proposition 3.1.1, with $b$ replaced by $|b|^2 \in C^\infty(X)$. The extension of (3.1.4) to $b \in C(X)$ follows from the denseness of $C^\infty(X)$ in $C(X)$ and the estimate (2.1.27). □

REMARK. Unlike Proposition 3.1.1, we have not extended Proposition 3.1.2 to work for $b \in \mathcal{R}(X)$.

Another important tool is Egorov's theorem, which implies

$$e^{-it\Lambda}\,\mathrm{op}_F(a)e^{it\Lambda} - \mathrm{op}_F(a \circ G_t) \text{ is compact on } L^2(M), \quad \forall\, t \in \mathbb{R}, \tag{3.1.7}$$

for all $a \in C(X)$, where $G_t : X \to X$ is the flow in (1.0.2), generated by the Hamiltonian vector field associated to the principal symbol of $\Lambda$, a smooth flow on $X$ that preserves the Liouville measure. For $a \in C^\infty(X)$, this difference belongs to $OPS^{-1}_{1,0}(M)$ for all $t$. Extension of (3.1.7) to $a \in C(X)$ follows readily from the denseness of $C^\infty(X)$ in $C(X)$ and (2.1.27). As a consequence, we have the following.

PROPOSITION 3.1.3. *Let $a \in C(X)$, and, for $T \in (0,\infty)$, set*

$$a_T = \frac{1}{2T}\int_{-T}^{T} a \circ G_t \, dt, \tag{3.1.8}$$

*and*

$$A_T = \frac{1}{2T}\int_{-T}^{T} e^{-it\Lambda} A e^{it\Lambda}\, dt, \quad A = \mathrm{op}_F(a). \tag{3.1.9}$$

*Then, for each such $T$,*

$$A_T - \mathrm{op}_F(a_T) \ \text{ is compact on } L^2(M). \tag{3.1.10}$$

COROLLARY 3.1.4. *In the setting of Proposition 3.1.3,*

$$\lim_{N\to\infty} \frac{1}{N}\sum_{k=1}^{N} \big\|[A_T - \mathrm{op}_F(a_T)]\varphi_k\big\|^2_{L^2} = 0, \quad \forall\, T < \infty. \tag{3.1.11}$$

PROOF. Given $\{\varphi_k\}$ is an orthonormal set in $L^2(M)$, then $\varphi_k \to 0$ weakly, as $k \to \infty$, so $\|K\varphi_k\|_{L^2} \to 0$ as $k \to \infty$ for each compact operator on $L^2(M)$. Hence, for each $T < \infty$,

$$\big\|[A_T - \mathrm{op}_F(a_T)]\varphi_k\big\|_{L^2} \longrightarrow 0 \ \text{ as } \ k \to \infty, \tag{3.1.12}$$

and (3.1.11) follows. □

**3.2. Proof of Theorem 3.0.3.** As further preparation for the proof of Theorem 3.0.3, we recall some classical ergodic theorems, as applied to the group $G_t : X \to X$ of measure preserving homeomorphisms of $X$. Von Neumann's mean ergodic theorem yields for $a_T$ in (3.1.8)

$$a_T \longrightarrow Pa, \quad \text{in } L^2(X)\text{-norm}, \tag{3.2.1}$$

as $T \to \infty$, for all $a \in L^2(X)$, where

$$\begin{aligned} P = {}& \text{orthogonal projection of } L^2(X) \text{ onto} \\ & \{a \in L^2(X) : a \circ G_t = a,\ \forall\, t \in \mathbb{R}\}. \end{aligned} \tag{3.2.2}$$

Meanwhile, Birkhoff's ergodic theorem yields

$$a_T \longrightarrow Pa, \quad \text{a.e. on } X, \tag{3.2.3}$$

for all $a \in L^1(X)$. We also have

$$P : L^p(X) \longrightarrow L^p(X), \quad \forall\, p \in [1,\infty], \tag{3.2.4}$$

and

$$a_T \longrightarrow Pa \ \text{ in } L^p\text{-norm, for } a \in L^p(X), \quad 1 \le p < \infty, \tag{3.2.5}$$

while

$$\begin{aligned} a \in L^\infty(X) &\Longrightarrow a_T \to Pa \text{ pointwise a.e. and boundedly} \\ &\Longrightarrow a_T \to Pa \text{ weak}^* \text{ in } L^\infty(X), \end{aligned} \tag{3.2.6}$$

as $T \to \infty$. In light of Proposition 2.1.4, we deduce from (3.2.6) that

$$\operatorname{op}_F(a_T) \longrightarrow \operatorname{op}_F(Pa) \ \text{ in the weak operator topology of } \mathcal{L}(L^2(M)), \tag{3.2.7}$$

as $T \to \infty$, given $a \in L^\infty(X)$.

The relevance of (3.2.1) in particular arises from applying Proposition 3.1.2 to

$$b = a_T - Pa. \tag{3.2.8}$$

We have

PROPOSITION 3.2.1. *If $a \in C(X)$, then*

$$\lim_{N\to\infty} \frac{1}{N} \sum_{k=1}^{N} \bigl\|\operatorname{op}_F(a_T - Pa)\varphi_k\bigr\|^2_{L^2(M)} = \int_X |a_T - Pa|^2 \, dS, \tag{3.2.9}$$

*provided that also*

$$Pa \in C(X). \tag{3.2.10}$$

We can use this proposition, in combination with Corollary 3.1.4, to establish the following.

PROPOSITION 3.2.2. *If $a, Pa \in C(X)$, $A = \operatorname{op}_F(a)$, and $A_T$ is as in* (3.1.9), *then*

$$\limsup_{N\to\infty} \frac{1}{N} \sum_{k=1}^{N} \bigl\|[A_T - \operatorname{op}_F(Pa)]\varphi_k\bigr\|^2_{L^2(M)} \le 2 \int_X |a_T - Pa|^2 \, dS, \tag{3.2.11}$$

*for each $T < \infty$.*

PROOF. Write

$$A_T - \operatorname{op}_F(Pa) = [A_T - \operatorname{op}_F(a_T)] + [\operatorname{op}_F(a_T) - \operatorname{op}_F(Pa)], \tag{3.2.12}$$

and use $(\alpha + \beta)^2 \le 2\alpha^2 + 2\beta^2$ for $\alpha, \beta \ge 0$, to dominate the left side of (3.2.11) by

$$\begin{aligned} &\limsup_{N\to\infty} \frac{2}{N} \sum_{k=1}^{N} \bigl\|[A_T - \operatorname{op}_F(a_T)]\varphi_k\bigr\|^2_{L^2} \\ &+ \limsup_{N\to\infty} \frac{2}{N} \sum_{k=1}^{N} \bigl\|\operatorname{op}_F(a_T - Pa)\varphi_k\bigr\|^2_{L^2}. \end{aligned} \tag{3.2.13}$$

Apply (3.1.11) to the first limsup in (3.2.13). Then apply Proposition 3.1.2 with $b = a_T - Pa$ to get

$$\lim_{N\to\infty} \frac{1}{N} \sum_{k=1}^{N} \bigl\|\operatorname{op}_F(a_T - Pa)\varphi_k\bigr\|^2_{L^2} = \int_X |a_T - Pa|^2 \, dS. \tag{3.2.14}$$

Then we have (3.2.11). □

We pursue consequences of Proposition 3.2.2. Since $\|B\varphi_k\|^2_{L^2} \ge |(B\varphi_k, \varphi_k)|^2$ and $(A_T\varphi_k, \varphi_k) = (A\varphi_k, \varphi_k)$, for $\varphi_k$ as in (1.0.1), we deduce from (3.2.11) that

$$\limsup_{N\to\infty} \frac{1}{N} \sum_{k=1}^{N} \bigl|([A - \operatorname{op}_F(Pa)]\varphi_k, \varphi_k)\bigr|^2 \le 2 \int_X |a_T - Pa|^2 \, dS, \tag{3.2.15}$$

for each $T < \infty$. Taking $T \to \infty$, we have the following result, closely related to Proposition 2.3.1.

COROLLARY 3.2.3. *In the setting of Proposition* 3.2.2,

$$\lim_{N\to\infty}\frac{1}{N}\sum_{k=1}^{N}\Big|(A\varphi_k,\varphi_k)-(\mathrm{op}_F(Pa)\varphi_k,\varphi_k)\Big|^2=0. \tag{3.2.16}$$

To proceed with consequences of (3.2.11), let us rewrite it as

$$\limsup_{N\to\infty}\frac{1}{d_N}\sum_{\lambda\in\Sigma_N}\big\|[A_T-\mathrm{op}_F(Pa)]P_\lambda\big\|_{\mathrm{HS}}^2\le 2\int_X|a_T-Pa|^2\,dS, \tag{3.2.17}$$

where, as in (3.0.21), $P_\lambda$ is the orthogonal projection of $L^2(M)$ onto $\mathrm{Eigen}(\Lambda,\lambda)$, and

$$d_N=\sum_{\lambda\le N}\dim\mathrm{Eigen}(\Lambda,\lambda)=\mathrm{Tr}\sum_{\lambda\le N}P_\lambda. \tag{3.2.18}$$

Now

$$P_\lambda e^{-it\Lambda}Ae^{it\Lambda}P_\lambda=P_\lambda AP_\lambda,\quad\forall\, t\in\mathbb{R}, \tag{3.2.19}$$

hence

$$P_\lambda A_TP_\lambda=P_\lambda AP_\lambda. \tag{3.2.20}$$

Hence, for all $T<\infty$,

$$\begin{aligned}\|P_\lambda AP_\lambda\|_{\mathrm{HS}}^2&=\|P_\lambda A_TP_\lambda\|_{\mathrm{HS}}^2\\&\le\|A_TP_\lambda\|_{\mathrm{HS}}^2.\end{aligned} \tag{3.2.21}$$

Consequently, by (3.2.17),

$$\begin{aligned}Pa=0\Longrightarrow\limsup_{N\to\infty}\frac{1}{d_N}\sum_{\lambda\in\Sigma_N}&\|P_\lambda AP_\lambda\|_{\mathrm{HS}}^2\\&\le\inf_{T<\infty}\int_X|a_T-Pa|^2\,dS\\&=0.\end{aligned} \tag{3.2.22}$$

We hence have the following.

PROPOSITION 3.2.4. *In the setting of Proposition* 3.2.2, *if also* $Pa=0$, *then*

$$\lim_{N\to\infty}\frac{1}{d_N}\sum_{\lambda\in\Sigma_N}\|P_\lambda AP_\lambda\|_{\mathrm{HS}}^2=0. \tag{3.2.23}$$

If we apply this result with $a$ replaced by $a-Pa$, we have Theorem 3.0.3:

PROPOSITION 3.2.5. *If* $a,Pa\in C(X)$, $A=\mathrm{op}_F(a)$, *then*

$$\lim_{N\to\infty}\frac{1}{d_N}\sum_{\lambda\in\Sigma_N}\big\|P_\lambda[A-\mathrm{op}_F(Pa)]P_\lambda\big\|_{\mathrm{HS}}^2=0. \tag{3.2.24}$$

*That is,*

$$\lim_{N\to\infty}\frac{1}{d_N}\big\|[\Pi(A)-\Pi(\mathrm{op}_F(Pa))]Q_N\big\|_{\mathrm{HS}}^2=0, \tag{3.2.25}$$

*where*

$$Q_N=\sum_{\lambda\le N}P_\lambda. \tag{3.2.26}$$

Note that

$$Pa = \overline{a} \Rightarrow \mathrm{op}_F(Pa) = \overline{a}I \Rightarrow \Pi(\mathrm{op}_F(Pa)) = \overline{a}I. \tag{3.2.27}$$

This always holds when $\{G_t\}$ is ergodic. In such a case, the conclusion of Proposition 3.2.5, i.e.,

$$\lim_{N\to\infty} \frac{1}{d_N} \big\| [\Pi(A) - \overline{a}I]Q_N \big\|_{\mathrm{HS}}^2 = 0, \tag{3.2.28}$$

is contained in Theorem 2 of [**13**].

The following result (potentially) addresses cases where $\{G_t\}$ is not ergodic. It refines Proposition 2.3.4.

PROPOSITION 3.2.6. *Let $U \subset X$ be open and assume $G_t : U \to U$, and that the action on $U$ is ergodic. Let $a, b \in C(X)$ be supported on a compact subset of $U$, $A = \mathrm{op}_F(a)$, $B = \mathrm{op}_F(b)$. Then*

$$\begin{aligned} &\int_X a\, dS = \int_X b\, dS \\ &\Longrightarrow \lim_{N\to\infty} \frac{1}{d_N} \big\| [\Pi(A) - \Pi(B)]Q_N \big\|_{\mathrm{HS}}^2 = 0. \end{aligned} \tag{3.2.29}$$

PROOF. Under these hypotheses, $P(a-b) = 0$, so (3.2.25) applies to $A-B$. □

REMARK. It would be interesting to seek a corresponding refinement of Proposition 2.3.5, which allows $a, b \in \mathcal{R}(X)$.

**3.3. Comments and examples.** We make some further comments relating Theorem 3.0.3 to Theorems 1.0.1 and 1.0.3.

First, if all the eigenvalues of $\Lambda$ are simple, then, for each $A \in \mathcal{L}(L^2(M))$, $\Pi(A)$ is the diagonal part of $A$ with respect to the orthonormal basis $\{\varphi_j\}$. Then, in the ergodic case, (3.2.28) is equivalent to (1.0.3), and more generally the conclusion (3.2.25) is equivalent to (3.2.16).

On the other hand, there are many examples of compact Riemannian manifolds with ergodic geodesic flow that have infinitely many eigenspaces of dimension $> 1$. For example, take a compact Riemann surface $M$, of genus $g \geq 2$, with a noncommutative group $\Gamma$ of conformal automorphisms, and endow it with the Poincaré metric, so $\Gamma$ acts as a group of isometries. Then $\Gamma$ acts on each eigenspace of $\sqrt{-\Delta}$, and one can show that each irreducible representation of $\Gamma$ is contained in infinitely many of them. (See [**10**].)

We turn to some cases where the geodesic flow is not ergodic. Consider the flat tori $\mathbb{T}^n = \mathbb{R}^n/(2\pi\mathbb{Z}^n)$, for which the geodesic flow is integrable. For these manifolds, one readily sees that (3.2.27) applies to multiplication operators, since, for $M = \mathbb{T}^n$,

$$a(x,\xi) = a(x) \Longrightarrow Pa = \overline{a}. \tag{3.3.1}$$

(See §7 of [**9**].) One can take the orthonormal basis of eigenfunctions of $\Delta$,

$$e_\alpha(x) = e^{i\alpha\cdot x}, \quad \alpha \in \mathbb{Z}^n, \tag{3.3.2}$$

for $L^2(\mathbb{T}^n)$, with inner product $(u,v)_{L^2} = (2\pi)^{-n} \int_{\mathbb{T}^n} u(x)\overline{v(x)}\, dx$. In this basis we obviously have $(Ae_\alpha, e_\alpha)_{L^2} = \overline{a}$, for $Au(x) = a(x)u(x)$, which trivially agrees with (1.0.5). On the other hand, as is well known, for each $n \geq 2$ the eigenspaces of $\Delta$ have arbitrarily large dimension. (Counting these dimensions is equivalent to the notorious lattice point problem. See [**11**].) Hence it is of interest to see how Proposition 3.2.5 applies in this situation.

The summand in (3.2.24), for $Au(x) = a(x)u(x)$, $Pa = \overline{a}$, is

$$\|P_\lambda(A - \overline{a}I)P_\lambda\|_{\mathrm{HS}}^2 = \sum_{|\alpha|^2=|\beta|^2=\lambda^2,\alpha\neq\beta} |\hat{a}(\alpha-\beta)|^2. \tag{3.3.3}$$

Hence what appears on the left side of (3.2.25) is

$$\begin{aligned}
&\frac{1}{d_N}\big\|\,[\Pi(A) - \overline{a}I]Q_N\big\|_{\mathrm{HS}}^2 \\
&\qquad = \frac{1}{d_N}\sum_{|\alpha|\le N}\ \sum_{\{\beta:|\beta|=|\alpha|,\beta\neq\alpha\}} |\hat{a}(\alpha-\beta)|^2 \\
&\qquad = \frac{1}{d_N}\sum_{|\alpha|\le N}\ \sum_{\{\gamma:\gamma\neq 0,|\alpha+\gamma|=|\alpha|\}} |\hat{a}(\gamma)|^2 \\
&\qquad = \sum_{\gamma\in\mathbb{Z}^n\setminus 0} \frac{\nu_N(\gamma)}{d_N}\,|\hat{a}(\gamma)|^2,
\end{aligned} \tag{3.3.4}$$

where

$$\nu_N(\gamma) = \#\{\alpha\in\mathbb{Z}^n : |\alpha|\le N \text{ and } |\alpha| = |\alpha+\gamma|\}. \tag{3.3.5}$$

Note that $d_N = \#\{\alpha\in\mathbb{Z}^n : |\alpha|\le N\}$, so

$$\frac{\nu_N(\gamma)}{d_N} \le 1, \quad \forall\, N\in\mathbb{N},\ \gamma\in\mathbb{Z}^n. \tag{3.3.6}$$

Furthermore,

$$\lim_{N\to\infty} \frac{\nu_N(\gamma)}{d_N} = 0, \quad \forall\,\gamma\in\mathbb{Z}^n\setminus 0. \tag{3.3.7}$$

This gives a direct verification of (3.2.25) in this case. That is, the quantity (3.3.4) tends to 0 as $N\to\infty$, for each $a\in C(\mathbb{T}^n)$. In fact, this quantity tends to 0 for each $a\in L^2(\mathbb{T}^n)$.

Next we take a look at $S^n$, the unit sphere in $\mathbb{R}^{n+1}$, with its standard metric. Then the geodesic flow $\{G_t\}$ is periodic of period $2\pi$. It is convenient to take

$$\Lambda = \Big(-\Delta + \Big(\frac{n-1}{2}\Big)^2\Big)^{1/2} - \frac{n-2}{2}, \tag{3.3.8}$$

so

$$\operatorname{Spec}\Lambda = \{k\in\mathbb{Z} : k\ge 0\}, \tag{3.3.9}$$

and $e^{it\Lambda}$ is also periodic of period $2\pi$. Then, given $A\in\mathcal{L}(L^2(S^n))$,

$$\Pi(A) = \frac{1}{2\pi}\int_0^{2\pi} e^{-it\Lambda}Ae^{it\Lambda}\,dt. \tag{3.3.10}$$

In case $a\in C^\infty(S^*S^n)$, we have

$$Pa(x,\xi) = \frac{1}{2\pi}\int_0^{2\pi} a(G_t(x,\xi))\,dt, \tag{3.3.11}$$

and it is a straightforward consequence of Egorov's theorem that, if $A = \mathrm{op}_F(a)$,

$$\Pi(A) - \mathrm{op}_F(Pa) \in OPS^{-1}_{1,0}(S^n). \tag{3.3.12}$$

This is more precise than (3.2.25). In fact, we can apply (3.3.12) with $a$ replaced by $Pa$, to get

$$\Pi(\mathrm{op}_F(a)) - \mathrm{op}_F(Pa) \in OPS^{-1}_{1,0}(S^n), \tag{3.3.13}$$

and then subtract, to get

$$\Pi(A - \mathrm{op}_F(Pa)) \in OPS^{-1}_{1,0}(S^n), \tag{3.3.14}$$

which in turn implies (3.2.25). Indeed, it is elementary that, if $B \in \mathcal{L}(L^2(M))$,

$$\Pi(B) \text{ compact on } L^2(M) \Rightarrow \lim_{N\to\infty} \frac{1}{d_N} \|Q_N \Pi(B) Q_N\|^2_{\mathrm{HS}} = 0. \tag{3.3.15}$$

## References


[1] Y. Colin de Verdière, *Ergodicité et fonctions propres du laplacien* (French, with English summary), Comm. Math. Phys. **102** (1985), no. 3, 497–502. MR818831

[2] Victor J. Donnay, *Geodesic flow on the two-sphere. I. Positive measure entropy*, Ergodic Theory Dynam. Systems **8** (1988), no. 4, 531–553, DOI 10.1017/S0143385700004685. MR980796

[3] Jeffrey Galkowski, *Quantum ergodicity for a class of mixed systems*, J. Spectr. Theory **4** (2014), no. 1, 65–85, DOI 10.4171/JST/62. MR3181386

[4] Gabriel Rivière, *Remarks on quantum ergodicity*, J. Mod. Dyn. **7** (2013), no. 1, 119–133, DOI 10.3934/jmd.2013.7.119. MR3071468

[5] Robert Schrader and Michael E. Taylor, *Semiclassical asymptotics, gauge fields, and quantum chaos*, J. Funct. Anal. **83** (1989), no. 2, 258–316, DOI 10.1016/0022-1236(89)90021-9. MR995750

[6] A. I. Šnirel′man, *Ergodic properties of eigenfunctions* (Russian), Uspehi Mat. Nauk **29** (1974), no. 6(180), 181–182. MR0402834

[7] Toshikazu Sunada, *Quantum ergodicity*, Progress in inverse spectral geometry, Trends Math., Birkhäuser, Basel, 1997, pp. 175–196. MR1731156

[8] Michael E. Taylor, *Pseudodifferential operators*, Princeton Mathematical Series, vol. 34, Princeton University Press, Princeton, N.J., 1981. MR618463

[9] Michael Taylor, *Variations on quantum ergodic theorems*, Potential Anal. **43** (2015), no. 4, 625–651, DOI 10.1007/s11118-015-9489-y. MR3432452

[10] M. Taylor, Multiple eigenvalues of operators with noncommutative symmetry groups. Lecture notes, available at http://mtaylor.web.unc.edu/notes (item 4, spectral theory).

[11] M. Taylor, Flat 2D tori with sparse spectra. Lecture notes, available at the website http://mtaylor.web.unc.edu/notes (item 4, spectral theory).

[12] Steven Zelditch, *Uniform distribution of eigenfunctions on compact hyperbolic surfaces*, Duke Math. J. **55** (1987), no. 4, 919–941, DOI 10.1215/S0012-7094-87-05546-3. MR916129

[13] Steven Zelditch, *Quantum ergodicity of $C^*$ dynamical systems*, Comm. Math. Phys. **177** (1996), no. 2, 507–528. MR1384146

[14] Steven Zelditch, *Quantum mixing*, J. Funct. Anal. **140** (1996), no. 1, 68–86, DOI 10.1006/jfan.1996.0098. MR1404574

[15] Steven Zelditch and Maciej Zworski, *Ergodicity of eigenfunctions for ergodic billiards*, Comm. Math. Phys. **175** (1996), no. 3, 673–682. MR1372814



Department of Mathematics, University of North Carolina, Chapel Hill, North Carolina, 27599

*Email address*: met@math.unc.edu

## Selected Published Titles in This Series

748 **Donatella Danielli and Irina Mitrea, Editors,** Advances in Harmonic Analysis and Partial Differential Equations, 2020

746 **Erica Flapan and Helen Wong, Editors,** Topology and Geometry of Biopolymers, 2020

745 **Federico Binda, Marc Levine, Manh Toan Nguyen, and Oliver Röndigs, Editors,** Motivic Homotopy Theory and Refined Enumerative Geometry, 2020

744 **Pieter Moree, Anke Pohl, L'ubomír Snoha, and Tom Ward, Editors,** Dynamics: Topology and Numbers, 2020

743 **H. Garth Dales, Dmitry Khavinson, and Javad Mashreghi, Editors,** Complex Analysis and Spectral Theory, 2020

742 **Francisco-Jesús Castro-Jiménez, David Bradley Massey, Bernard Teissier, and Meral Tosun, Editors,** A Panorama of Singularities, 2020

741 **Houssam Abdul-Rahman, Robert Sims, and Amanda Young, Editors,** Analytic Trends in Mathematical Physics, 2020

740 **Alina Bucur and David Zureick-Brown, Editors,** Analytic Methods in Arithmetic Geometry, 2019

739 **Yaiza Canzani, Linan Chen, and Dmitry Jakobson, Editors,** Probabilistic Methods in Geometry, Topology and Spectral Theory, 2019

738 **Shrikrishna G. Dani, Surender K. Jain, Jugal K. Verma, and Meenakshi P. Wasadikar, Editors,** Contributions in Algebra and Algebraic Geometry, 2019

737 **Fernanda Botelho, Editor,** Recent Trends in Operator Theory and Applications, 2019

736 **Jane Hawkins, Rachel L. Rossetti, and Jim Wiseman, Editors,** Dynamical Systems and Random Processes, 2019

735 **Yanir A. Rubinstein and Bernard Shiffman, Editors,** Advances in Complex Geometry, 2019

734 **Peter Kuchment and Evgeny Semenov, Editors,** Differential Equations, Mathematical Physics, and Applications, 2019

733 **Peter Kuchment and Evgeny Semenov, Editors,** Functional Analysis and Geometry, 2019

732 **Samuele Anni, Jay Jorgenson, Lejla Smajlović, and Lynne Walling, Editors,** Automorphic Forms and Related Topics, 2019

731 **Robert G. Niemeyer, Erin P. J. Pearse, John A. Rock, and Tony Samuel, Editors,** Horizons of Fractal Geometry and Complex Dimensions, 2019

730 **Alberto Facchini, Lorna Gregory, Sonia L'Innocente, and Marcus Tressl, Editors,** Model Theory of Modules, Algebras and Categories, 2019

729 **Daniel G. Davis, Hans-Werner Henn, J. F. Jardine, Mark W. Johnson, and Charles Rezk, Editors,** Homotopy Theory: Tools and Applications, 2019

728 **Nicolás Andruskiewitsch and Dmitri Nikshych, Editors,** Tensor Categories and Hopf Algebras, 2019

727 **André Leroy, Christian Lomp, Sergio López-Permouth, and Frédérique Oggier, Editors,** Rings, Modules and Codes, 2019

726 **Eugene Plotkin, Editor,** Groups, Algebras and Identities, 2019

725 **Shijun Zheng, Marius Beceanu, Jerry Bona, Geng Chen, Tuoc Van Phan, and Avy Soffer, Editors,** Nonlinear Dispersive Waves and Fluids, 2019

724 **Lubjana Beshaj and Tony Shaska, Editors,** Algebraic Curves and Their Applications, 2019

723 **Donatella Danielli, Arshak Petrosyan, and Camelia A. Pop, Editors,** New Developments in the Analysis of Nonlocal Operators, 2019

For a complete list of titles in this series, visit the AMS Bookstore at **www.ams.org/bookstore/conmseries/**.